T0231482

Self-Assessment Color Review
Equine Internal Medicine
Second Edition

Self-Assessment Color Review

Equine Internal Medicine

Second Edition

Tim S. Mair
BVSc, PhD, DipECEIM, MRCVS
Bell Equine Veterinary Clinic
Maidstone, UK

Thomas J. Divers
DVM, DACVIM, DACVECC
College of Veterinary Medicine
Cornell University, Ithaca, USA

CRC Press
Taylor & Francis Group
Boca Raton London New York

CRC Press is an imprint of the
Taylor & Francis Group, an **informa** business

CRC Press
Taylor & Francis Group
6000 Broken Sound Parkway NW, Suite 300
Boca Raton, FL 33487-2742

© 2016 by Taylor & Francis Group, LLC
CRC Press is an imprint of Taylor & Francis Group, an Informa business

No claim to original U.S. Government works

Printed on acid-free paper
Version Date: 20150519

International Standard Book Number-13: 978-1-4822-2535-8 (Paperback)

Visit the Taylor & Francis Web site at
http://www.taylorandfrancis.com

and the CRC Press Web site at
http://www.crcpress.com

Preface

This book presents to its readers two hundred and one interesting and challenging clinical cases from our files. Most cases are accompanied by photographs of the horses, imaging or endoscopy findings, blood or fluid smears, other ancillary tests and, in some cases, pathologic findings. We ask pertinent questions with each case to help guide the reader through the case workup and treatments. In-depth answers for each case are provided in the second half of the book. All the cases are new to this second edition. We have included for most cases a brief discussion of each patient's disease and one or more pertinent references. The majority of our cases are disorders that may be found in both North America and Europe. However, a small number of cases unique to only one of these two continents are included because they represent important or common diseases on that particular continent.

We take this opportunity to thank Jill Northcott, editor at CRC Press, and her staff for their excellent work in preparing our reports and photographs for publication. Special thanks to Drs. Nora Grenager and Eva Chase Conant for reviewing and editing the scientific content of each case; their help was invaluable. Additional thanks go to internal medicine colleagues at Cornell University (Drs. Ainsworth, Perkins, and Felippe) and to veterinarians who completed their internal medicine residencies at Cornell University, provided several of the photographs and valuable information for use in the book, and assisted with the management of the cases: Drs. Monica Figueiredo, Rachel Gardner, Jen Gold, Amy Johnson, Alexa Burton, Katharyn Mitchell, Terri Ollivett, Toby Pinn, Dominic Dawson, Katie Mullen, Sally Ness, and Emily Barrell. We also wish to acknowledge and thank Drs. Laura Stokes-Green, Joy Tomlinson, Ashley Watts, Emily Barrell, and Ed Jedrzejewski for the information and photographs on cases 46, 78, 2 and 70, 105, and 86, respectively. We also thank colleagues and previous interns and residents at Bell Equine Veterinary Clinic, especially Ceri Sherlock and Edd Knowles.

Special thanks are due to Dr. Keith Montgomery, former ophthalmology resident at Cornell, who supplied the "what's your diagnosis" ophthalmology cases that are placed throughout the text.

We hope you will find these case presentations interesting, challenging, and of educational value.

Tim S. Mair
Thomas J. Divers

Preface

Dedication

This edition of *Equine Internal Medicine: Self-Assessment Color Review* is dedicated to T. Douglas Byars, DVM, DACVECC, DACVIM, who passed away in July 2014 after a prolonged illness. Dr. Byars was a legend in equine internal medicine and known around the world as the "go to person" for consultation on serious or complicated medical illnesses in the horse. Dr. Byars, although based in Lexington, Kentucky, spent a considerable amount of time traveling to Europe and elsewhere in the world to consult on difficult medical cases. He also spent innumerable hours on the phone with veterinarians worldwide offering advice on cases. Many cases in this current *Equine Internal Medicine* text would be similar to cases Dr. Byars was frequently asked to treat or consult on.

Dr. Byars was 'The Internist' in the United States most responsible for leading equine internal medicine diplomates into private practice. He was the first board-certified veterinary internist to leave a university practice and develop the highly successful private practice equine internal medicine hospital, now known as the Hagyard Medical Institute. Doug worked at Hagyard's for nearly 25 years and his success in that internal medicine specialty practice opened the doors for other internists in the Lexington area and throughout the world. For this and many other services to veterinary internal medicine, Doug was in 2007 awarded the Robert W. Kirk Award by the American College of Veterinary Internal Medicine (ACVIM), the highest award given to any veterinary internist (large or small animal). Dr. Byars should also be credited for developing the equine specialty area in the American College of Veterinary Emergency and Critical Care (ACVECC); he was one of the initial three, exam-boarded diplomates in that College. His pursuit of critical care specialty training developed because of his realization that his practice was as much about critical care as it was internal medicine.

Dr. Byars was a staunch believer in being a voice for the horse and always wanted to do what was best for every horse. During his last year of life, his work was mostly limited to caring for retired Thoroughbreds. When asked to take that last job, he was told he would not be called at night; Doug's response was "I will only take the job if you do call me at night when a horse needs my help." Doug was a strong promoter of the entire horse industry, a great and trusted friend of many horse owners and trainers and a mentor to many young veterinarians finding their way in equine practice. Although his work was mostly in hospital-based practice, Doug always made an effort to visit farms to follow up on his cases and to spend time with the farm managers and the stable hands that cared for the horses. He was a good and dedicated friend to all, caring for people but caring as much for horses. His professional ethics and morals were unquestionably strong; anyone, regardless of their wealth or status, who suggested a less than ethical approach to a case would immediately experience an angered Dr. Byars.

Doug was a trusted and true friend to both authors of this book and to hundreds of veterinarians worldwide. We learned so much from him and we laughed a lot together over many years. We and many other veterinarians worldwide would like to say "thank you Doug, you are greatly missed!"

Abbreviations

AKI acute kidney injury (formerly acute renal failure)
AST aspartate aminotransferase
bpm beats/breaths per minute
BCS body condition score
BUN blood urea nitrogen
CBC complete blood count
CK creatine kinase
CN cranial nerve
CNS central nervous system
Cr creatinine
CRI constant rate infusion
CRT capillary refill time
CSM cervical stenotic myelopathy
CT computed tomography
DIC disseminated intravascular coagulation
ECG electrocardiogram
epg eggs per gram
EPM equine protozoal myeloencephalitis
GA general anesthesia
GI gastrointestinal
HPA hypothalamus/pituitary/ adrenal (axis)
HR heart rate
IM intramuscular/intramuscularly
IV intravenous/intravenously
MIC mean inhibitory concentration
MPV mean platelet volume
MRSA methicillin-resistant *Staphylococcus aureus*
NSAID non-steroidal anti- inflammatory drug
OD oculus dexter (right eye)
OS oculus sinister (left eye)
OU oculus uterque (both eyes)
$PaCO_2$ partial pressure of arterial carbon dioxide
PaO_2 partial pressure of arterial oxygen
PCV packed cell volume
PMN polymorphonuclear neutrophil
PO per os, orally
PvO_2 partial pressure of venous oxygen
RBC red blood cell
RR respiratory rate
SC subcutaneous/subcutaneously
SG specific gravity
SvO_2 venous oxygen saturation
TP total protein
TPN total parenteral nutrition
TS total solids

Broad Classification of Cases

Abdomen
2, 5, 50, 59, 78, 182, 178

Cardiovascular system
9, 12, 26, 27, 32, 41, 53, 55, 81, 91, 93, 103, 106, 111, 147

Dental
71, 105

Dermatology
15, 21, 42, 44, 64, 69, 76, 77, 101, 106, 126, 132, 141, 146, 148, 153, 164, 187, 190, 193, 194

Endocrine system
10

Gastrointestinal tract
8, 11, 15, 30, 36, 43, 52, 58, 61, 65, 66, 70, 73, 77, 97, 109, 112, 113, 121, 127, 130, 131, 138, 149, 171, 201

Hematopoietic system
90

Immunology
4, 22, 47

Infectious diseases
38, 119, 159, 196

Liver
34, 47, 169, 183

Mammary glands
42, 125

Metabolic
34, 82, 95, 129, 135, 151

Musculoskeletal
6, 10, 16, 18, 19, 24, 25, 28, 31, 33, 34, 51, 54, 55, 63, 70, 71, 72, 75, 83, 87, 92, 108, 125, 189

Neonatology
13, 17, 20, 22, 25, 30, 45, 59, 74

Neoplasia
3, 5, 7, 35, 43, 49, 53, 56, 57, 78, 84, 86, 89, 94, 110, 117, 120, 125, 128, 137, 143, 156, 160, 165, 167, 168, 172, 173, 175, 184

Nephrology
16

Nervous system
1, 6, 13, 14, 19, 23, 37, 49, 62, 67, 72, 87, 96, 102, 123, 136, 152, 199

Ophthalmology
38, 46, 48, 79, 85, 86, 88, 99, 107, 114, 116, 122, 134, 144, 150, 155, 157, 162, 166, 174, 179, 198

Parasites
68, 100, 124, 181

Reproductive system
9, 39, 60

Respiratory tract
4, 20, 23, 24, 32, 35, 36, 37, 40, 74, 86, 98, 104, 115, 118, 132, 133, 139, 140, 145, 154, 158, 161, 163, 170, 176, 177, 180, 185, 186, 188, 191, 192, 195, 197, 200

Urinary tract
7, 17, 28, 29, 33, 80

CASE 1 A 6-year-old Appaloosa gelding would not voluntarily come into the barn at feeding time one day. The horse had appeared normal prior to that time, and there was no recent history of injury, illness, or medication. The next day the owner noticed a head tilt to the left and ataxia (**1.1**).

Hospital examination revealed a severe left-sided head tilt, base-wide stance, and severe balance loss. Horizontal nystagmus with a quick phase to the right was noted. There was ventral strabismus in the left eye and dorsal strabismus in the right eye. There was no evidence of depression, limb weakness, or facial nerve abnormality.

1 Based on the age and acute onset of clinical signs, what would be the top differential diagnoses?
2 How can it be determined if the vestibular dysfunction is central or peripheral?
3 What diagnostics should be performed?

CASE 2 An 8-year-old Thoroughbred gelding developed acute anorexia and a fever of 103.2°F (39.6°C). It was winter and there was snow on the pasture. There had been no recent shipping, exposure to other horses, or any other typical risk factors for infectious disease. The deworming history was good. This horse had a similar complaint in the spring 5 years earlier (**2.1**), at which time *Actinobacillus* spp. peritonitis was

confirmed and treated successfully. He was reported to have been in good health since that time. On presentation, physical examination was unremarkable other than anorexia, fever, mild tachycardia, and absence of gut sounds.

1 Based on the history in this case, and without availability of ultrasound in the field, what diagnostic procedures should be performed?

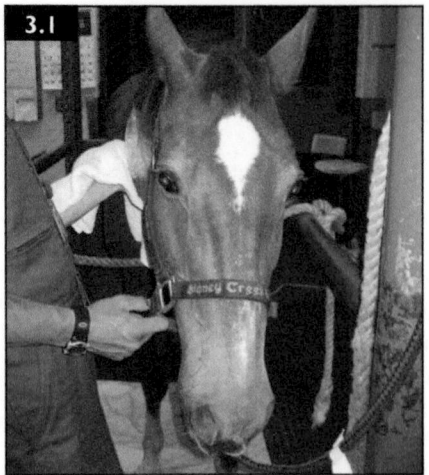

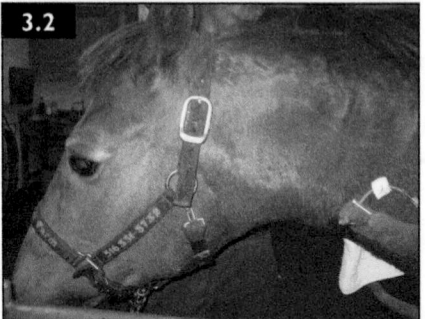

CASE 3 A 15-year-old Appendix mare presented for intermittent bilateral epistaxis of 5 days' duration (**3.1**). Abnormalities detected on clinical examination included tachycardia of 60 bpm and marked swelling of both the left cranial neck and the left hindlimb (**3.2, 3.3**). The mare was only 1/5 lame on the left hindlimb in spite of the marked swelling. The jugular veins were normal on visual examination and palpation. There was normal airflow through the nostrils, and no petechiations or nasal discharge were seen.

Endoscopy was performed (**3.4**) and the platelet count was later found to be 13,000 cells/µl (13 × 10^9/l). The prothrombin and partial thromboplastin times were normal. PCV was 22%, TP was 5.5 mg/dl (55 g/l). All other CBC and serum biochemistry values were normal except for a mild increase in muscle enzymes (CK = 888 U/l; AST = 720 U/l). A Coggin's test for equine infectious anemia was negative and there was no history of

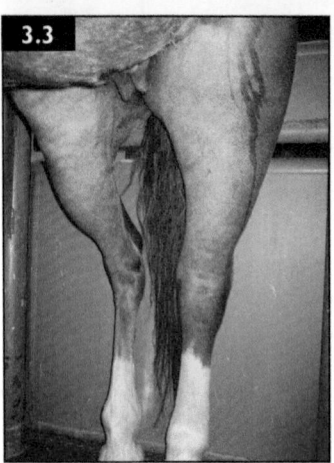

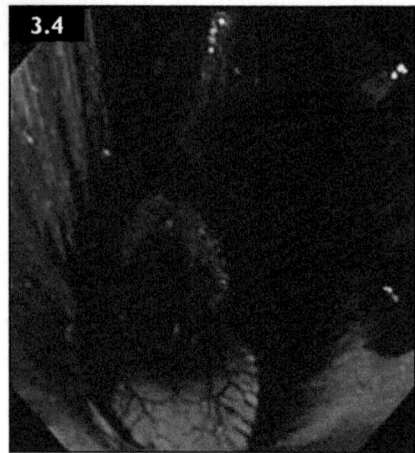

prior drug administration within the past 3 months.

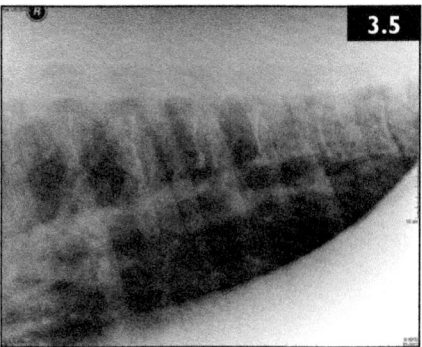

1 What question should be asked regarding the swelling of the left hindlimb and neck?
2 Are the swellings most likely septic or non-septic?
3 List some differential diagnoses for intermittent bilateral epistaxis.
4 What diagnostic test should be chosen next?
5 Given the severe thrombocytopenia, yet no evidence of infectious disease or recent drug administration, what general type of disease process would be the top differential?
6 What diagnostic test should be chosen next?
7 What further diagnostics would help evaluate the thrombocytopenia more specifically?
8 Radiography of the lungs was performed to look for any metastases (**3.5**). What is the interpretation?
9 What is the most common neoplasm to invade muscle and also result in thrombocytopenia?

CASE 4 A 2-month-old Standardbred foal presented because of a 4-day history of labored breathing, coughing, nasal discharge, and mild depression. The foal had been febrile, with temperatures as high as 103.6°F (39.8°C). The foal was treated for 3 days with ceftiofur, gentamycin, and ketofen without noticeable improvement.

On examination the foal was bright and alert with harsh lung sounds bilaterally throughout the thorax and crackles and wheezes in the cranioventral thorax (using a re-breathing bag). Thoracic ultrasound examination revealed bilateral, very small areas of cranioventral lung consolidation.

1 What is the primary differential diagnosis?
2 What are the most common etiologic causes for these signs?
3 What diagnostic procedure should be performed next to gain further information?
4 What treatment should be recommended based on these results?

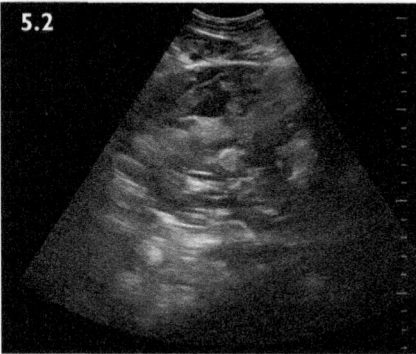

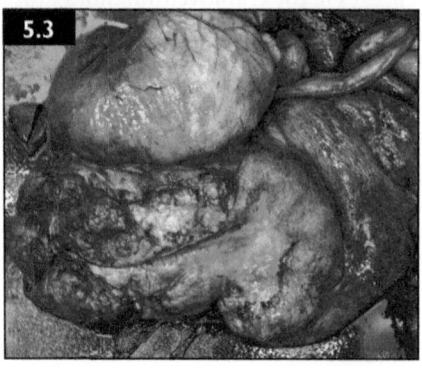

CASE 5 A 9-year-old Thoroughbred was examined for a 2-month history of intermittent inappetence and weight loss. Previous treatments had included 2 weeks of omeprazole and deworming with ivermectin/praziquantel, without a clinical response. Routine initial examination, including abdominal palpation per rectum, was unremarkable except for a low BCS (3/9) and a 3/6 systolic murmur over the left heart base with a HR of 48 bpm. Gastroscopic examination was normal and blood was submitted for CBC, blood chemistries, and urinalysis. Initial laboratory abnormalities included: PCV = 27%; color of plasma = white (**5.1** shows the spun down EDTA whole blood and separated plasma); Cl⁻ = 88 mEq/l (88 mmol/); plasma protein = 8.1 g/dl (81 g/l); Cr = 2.0 mg/dl (152.5 µmol/l); anion gap = 22 mEq/l (22 mmol/l). Ultrasound of the abdomen revealed the following appearance of the spleen and left cranial abdomen (**5.2**). Peritoneal fluid examination revealed: color = reddish; TP = 5.3 g/dl (53 g/l); RBCs = 230,500 cells/µl (0.2 × 10¹²/l); nucleated cell count = 4,300/µl (4.3 × 10⁹/l): 7% large round cells; 35% non-degenerate neutrophils; 32% small lymphocytes; 26% macrophages.

1 What is the most unusual blood laboratory observation in this mature horse, and how might it be associated with the clinical complaint?
2 What additional laboratory test(s) should be performed based on these findings?

3 What are the two major physiologic pathway abnormalities that might have caused the marked hyperlipemia?

4 With regard to the ultrasound findings, what is the most likely diagnosis?

5 In light of the postmortem findings (5.3), what is the most likely mechanism for this horse's hyperlipemia?

CASE 6 A 7-year-old Quarter Horse mare presented for a chronic, foul-smelling, unilateral nasal discharge. Surgery was performed under GA in right lateral recumbency. On recovery, the mare was reluctant to bear weight on the left forelimb but could extend the leg when she walked, and the left triceps muscle felt very hard on palpation. Recovery of function in the left forelimb was prolonged and painful; it was 7 days before the mare would bear weight on the leg (6.1). As she began to bear weight, an unusual sweating pattern occurred on the left side of the body, beginning at the withers and extending caudally (6.2).

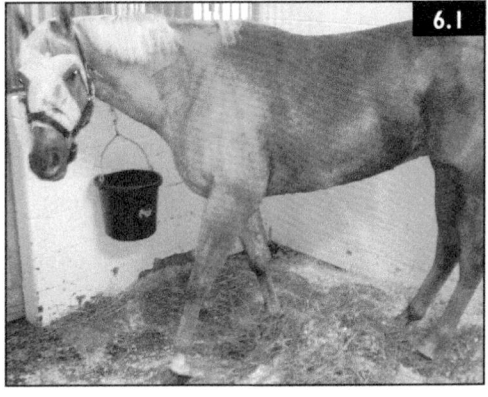

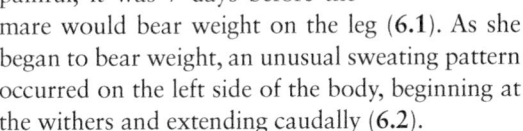

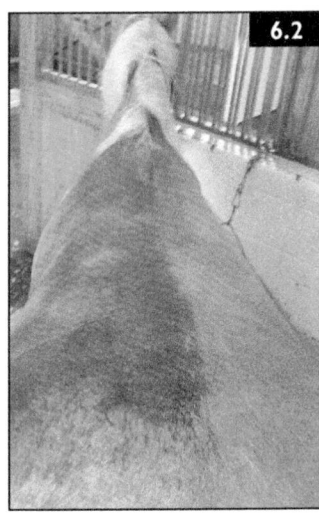

1 What is the most likely diagnosis for the chronic, fetid-smelling, unilateral nasal discharge?

2 How could this diagnosis be confirmed?

3 What treatment was probably provided?

4 What was the likely cause of the complication in the left forelimb?

5 What serum chemistry tests would help confirm this diagnosis?

6 What could cause the unusual sweating pattern?

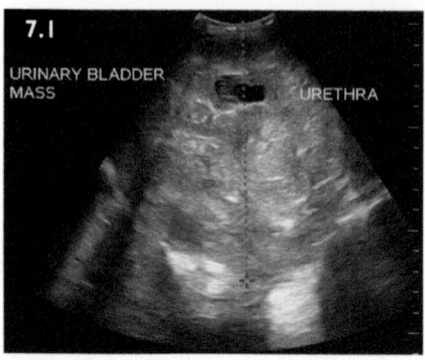

CASE 7 A 21-year-old Quarter Horse gelding was examined for pollakiuria and dysuria, including urine dribbling. The owners had treated the gelding with trimethoprim sulfa for 1 week without any change in clinical signs. His vital parameters (temperature, pulse, respiration) were all normal, although a loud diastolic murmur was heard at the base of the heart as the stethoscope was passed cranially under the left shoulder muscle. The horse was observed to urinate a small amount and the urine looked grossly normal; a free-catch sample was collected in a container for analysis.

1 What would be a reasonable clinical workup for this case?
2 An ultrasound image is shown (**7.1**). What is the interpretation?
3 What procedure could be performed to better visualize the urinary bladder?
4 What is the most likely diagnosis based on all the information?
5 What are the two main bladder neoplasms in the horse?
6 Which one is most likely in this horse, and why?
7 How could the diagnosis be confirmed?

CASE 8 A 10-month-old Arabian colt was examined because of rectal prolapse and poor body condition (**8.1**). The colt had a BCS of 2/9 (Henneke System), had not shed his winter coat, and chewing lice (*Damalinia* sp.) were found during the examination. The remainder of the clinical examination, a CBC, and blood biochemistries were unremarkable. The colt had been dewormed with ivermectin and praziquantel after the prolapse was noted by the referring veterinarian 2 days earlier, and fecal examination for parasites on admission was negative. A 'clump' of 1 × 2 cm flat parasites could be seen attached to the mucosa adjacent to the prolapse.

1 How should the rectal prolapse be treated?
2 What is the most likely cause of the rectal prolapse in this colt?

CASE 9 A 12-year-old Thoroughbred mare with an 8-hour-old foal by her side presented for colic, trembling, and abnormal sweating (**9.1**). The foal appeared to be normal and, although the foaling was assisted, the foal had presented in a normal anterior presentation and it was not a dystocia. On examination the mare was restless, sweating, and wanting to lie down but not violently colicky. She was mildly hyperthermic (T = 101.9°F [38.8°C]), tachycardic (HR = 80 bpm) and tachypneic (RR = 56 bpm). Mucous membranes were pale and slightly tacky with a prolonged CRT of 3 seconds. Abdominal palpation per rectum revealed no intestinal distension but a swelling was palpated in the right lateral abdomen, just over the brim of the pelvis. Transabdominal ultrasound revealed increased free abdominal fluid of mixed echogenicity with some swirling movement to a maximum depth of 30 cm (**9.2**). A rapid assessment test revealed a normal PCV and TS of 42% and 6.0 g/dl (60 g/l), respectively.

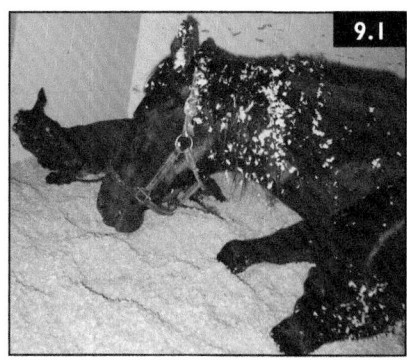

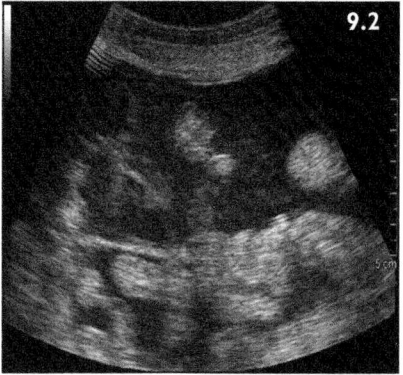

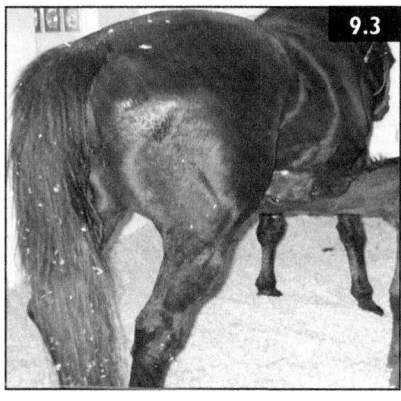

1 What is the most likely diagnosis in this mare?
2 What factors have been associated with an increased occurrence of this condition?
3 How can the blood volume of a horse be calculated, and what percentage of this can be lost before a drop in blood pressure occurs?
4 What treatments should be employed in a case like this and what should be avoided (**9.3**)? What other clinical signs and bloodwork could help determine the mare's perfusion status?
5 Why are the PCV and TS normal in this case?

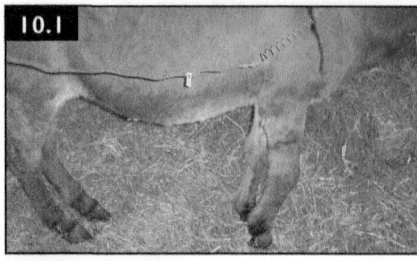

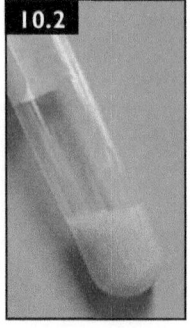

CASE 10 A 26-year-old donkey jenny presented for further diagnosis and management of shifting weight in the hindlimbs, reluctance to move, placement of the hindlimbs abnormally far under the body (**10.1**), and progressive anorexia of 2 weeks' duration. The donkey had been treated with ceftiofur and flunixin meglumine for the previous 10 days with no response. She had been on the same farm for the past 12 years and was kept with three other donkeys and eight horses, who were all healthy. No significant health problems had been noted previously and the donkey had excellent routine vaccination, deworming, and dental care. On presentation the donkey was in good to fat body condition and was depressed. She was mildly hypothermic (T = 97.6°F [36.4°C]) and tachycardic (HR = 96 bpm) with a normal respiratory rate (RR = 24 bpm). Mucous membranes were dark pink and tacky with a prolonged CRT of 3.5 seconds. GI sounds were absent and abdominal palpation per rectum revealed dry manure and a pelvic flexure impaction. A rapid assessment test revealed an elevated PCV of 53% and elevated TS of 8.6 g/dl (86 g/l). After centrifugation, the serum in the microhematocrit tube was opaque with a whitish discoloration (**10.2**).

1 What does the abnormal serum color mean, and what biochemical test would provide more information concerning this abnormality? Is this condition important, and what therapy should be used to treat it?
2 The donkey is dehydrated and may have been so for a prolonged period, since we know from the history that the illness has been progressing over 2 weeks. What organ system is of immediate concern with respect to secondary effects, and how should this be investigated further?
3 The donkey stopped shifting weight in the hindlimbs after abaxial sesamoid local nerve blocks were performed in both hindlimbs. What condition is most likely responsible for this clinical sign of weight-shifting, and what could be done to further assess this and make the donkey more comfortable?

After 5 days of hospitalization and aggressive, intensive medical therapy, the donkey improved clinically: her vital signs normalized, the pelvic flexure impaction resolved, hematologic parameters returned to within normal limits, and the laminitis pain was controlled. However, the donkey was still anorexic, quiet, and depressed.

4 What could be the cause of this, and how could it be remedied?

CASE 11 A 2-year-old Arabian gelding from upstate New York presented in the summer after 4 days of intermittent fever (up to 104.5°F [40.3°C]). The horse's temperature was taken twice daily and a single dose of phenylbutazone (2.2 mg/kg PO) had been given five times (whenever the temperature was >103°F

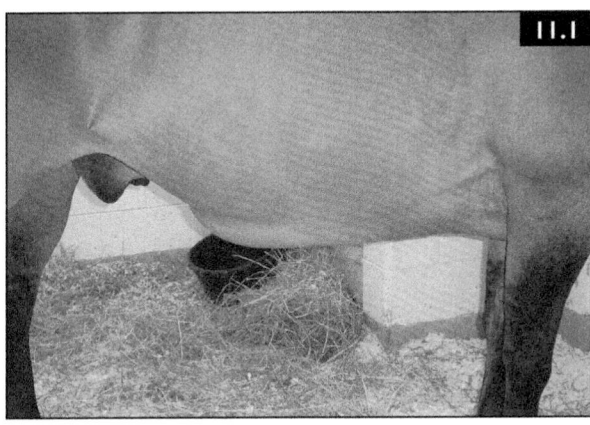

[39.4°C]). Twenty-five other horses were housed at the training facility and none were reported to be sick. The horse was in fit body condition and had been performing well in training for 4 months. The last time the horse had worked was 24 hours prior to development of the fever. On presentation the horse was very quiet with hyperemic mucous membranes and a prolonged CRT of 2.5 seconds. The GI sounds were hypermotile in all quadrants. Abdominal palpation per rectum revealed normal manure and no abnormalities. The horse was normothermic (T = 100.6°F [38.1°C]), tachycardic (HR = 60 bpm), and tachypneic (RR = 40 bpm). There was a large plaque of non-painful pitting edema on the ventral abdomen and thorax (**11.1**). CBC abnormalities included an elevated PCV (49%), leukopenia (4,100 cells/µl [4.1 × 10⁹/l]) with a profound neutropenia (100 cells/µl [0.1 × 10⁹/l]), immature band neutrophils (500 cells/µl [0.5 × 10⁹/l]) with moderate toxic changes, and thrombocytopenia (64,000 cells/µl [64 × 10⁹/l]). Biochemistry abnormalities included hyponatremia (124 mEq/l), hypochloremia (92 mEq/l), hypoproteinemia (5.2 g/dl [52 g/l]) due to a hypoalbuminemia (1.9 g/dl [19 g/l]), decreased colloid oncotic pressure (9.8 mmHg), and a decreased bicarbonate (20 mEq/l). Urinalysis, BUN, and creatinine were all normal.

1 The horse presented to an equine referral hospital. Where should this patient be housed, and why?
2 What is the primary differential diagnosis in this horse based on the history, location, time of year, clinical signs, and hematology?
3 What is the treatment of choice for this disease?
4 What important secondary complications can occur in these cases?

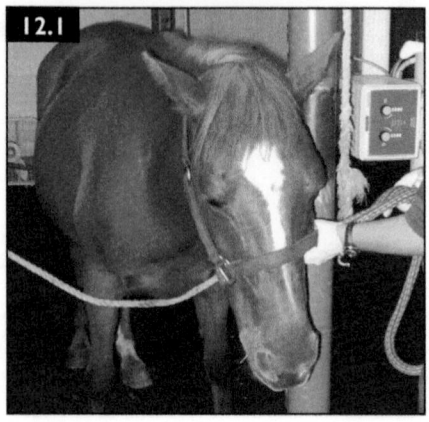

CASE 12 A 9-year-old Arabian mare (**12.1**) used for barrel racing was examined for a 12-month history of reduced exercise tolerance. Four months prior to presentation the mare had developed fever, lameness, ventral edema, and a heart murmur. Most recently, exercise intolerance was severe with obvious nostril flare and a RR of 70 bpm when walking. The mare was severely depressed and ataxic in all four limbs, with bilateral jugular vein distension and pulsation, HR of 70 bpm (>90 after walking), and a 4/6 systolic heart murmur best heard over the left pulmonic valve area. The lungs auscultated within normal limits. Echocardiography was performed to determine the size, function of the chambers, and anatomic or functional abnormalities of the valves (**12.2**, a long-axis view from the right chest, shows that the right ventricle and atrium appeared enlarged and the left ventricle and atrium appeared smaller than normal; **12.3** shows the velocity flow over the tricuspid valve; **12.4** and **12.5** illustrate the size comparison between the aorta and the pulmonary artery). A CBC was performed to look for evidence of an infectious or inflammatory process because the most severe clinical signs in the horse correlated with the onset of fever. Pertinent laboratory findings included: leukocytosis (11,200 cells/µl [11.2×10^9/l]), PMNs = 7,800/µl (7.8×10^9/l), fibrinogen = 300 mg/dl (3.0 g/l). Arterial blood gas: pH 7.419, PCO_2 = 39.7 mmHg, PaO_2 = 100 mmHg, HCO_3 = 25.6 mEq/l, O_2 saturation = 98%, lactate = 1.88 mmol/l. cTnI = 0.00, AST = 675 IU/l, GGT = 276 IU/l, SDH = 14 IU/l.

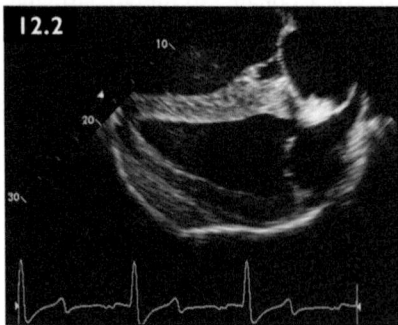

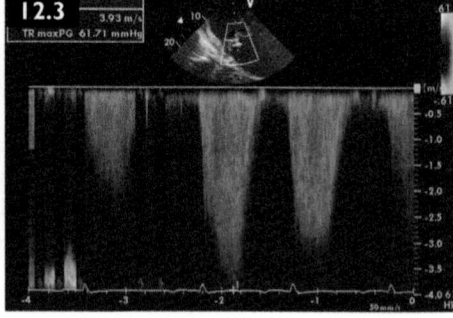

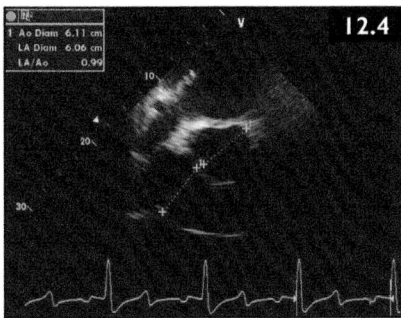

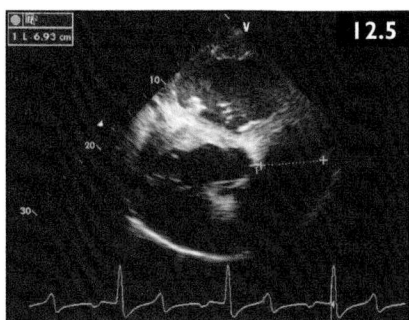

1 Interpret the echocardiographic findings.
2 What do the CBC abnormalities suggest?
3 Which of the following is the most likely diagnosis: cor pulmonale due to pulmonic valvular stenosis; cor pulmonale due to diffuse pulmonary parenchymal disease; cor pulmonale due to obstruction of the pulmonary artery?

CASE 13 A 2-day-old Thoroughbred colt presented because of milk observed at both nostrils after nursing (**13.1**). The foal was delivered without complications, the umbilicus was dipped with chlorhexidine at 15 and 30 minutes after birth, and an enema was given within 1 hour after birth. The foal stood normally and nursed within 1 hour of being delivered. An IM injection of vitamin E and selenium was given on the first day of life. Clinical examination on day 2 was normal other than the milk observed at both nostrils.

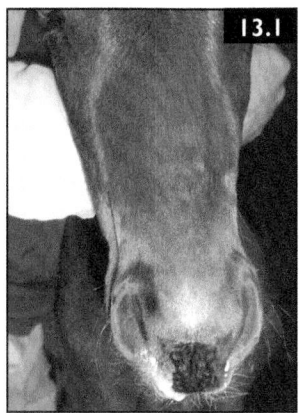

1 What are the rule outs for milk refluxing into the nasopharynx and nostrils in a newborn foal?
2 What diagnostic procedures should be performed next to investigate the observation of milk refluxing from the nostrils?
3 What abnormality is seen in this endoscopic image of the larynx (**13.2**), and how might it explain the clinical signs and future athletic use of the foal?
4 What routine test should be performed to make sure the foal had adequate absorption of maternal antibodies? How quickly after nursing can the test be accurately performed?

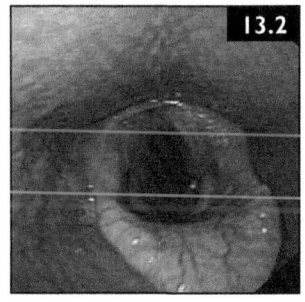

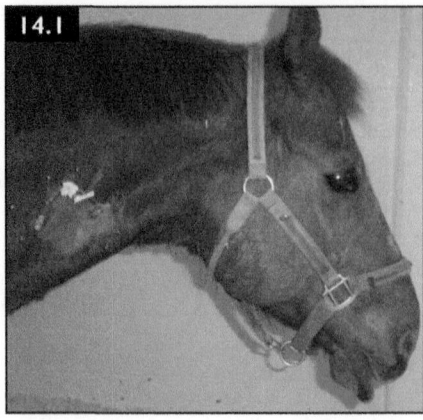

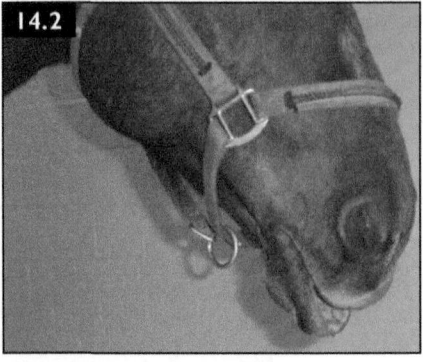

CASE 14 19-year-old Quarter Horse gelding presented for acute onset of paresis of the mandible and lower lip (**14.1, 14.2**). The horse was last seen to be normal 12 hours prior to presentation when turned out onto the pasture for the night. The owners first noticed the abnormality when they brought the gelding in that morning. The horse seemed unable to eat and drink, although he was attempting to do so. On examination the horse was quiet but responsive, with pink mucous membranes and a normal CRT of 1.5 seconds. He was normothermic (T = 99.8°F [37.7°C]), mildly tachycardic (HR = 48 bpm), and had a normal RR (20 bpm). Neurologic examination revealed a symmetric lower lip droop with decreased to absent sensation of the lower lip and gums adjacent to both first mandibular incisors. The horse was able to weakly appose the upper and lower incisor teeth. Tongue tone was normal. There was no obvious swelling or muscle atrophy evident on the head or neck. All other CN testing (e.g. palpebral reflex, sensation in the nostril) was within normal limits. The remainder of the physical and neurologic examinations was normal. When offered a soft bran mash, the horse was able to slowly prehend, masticate, and swallow.

1 What is the neuroanatomic lesion in this horse?
2 How might this lesion have occurred?
3 Although not affecting this horse, what important neurologic disease can produce these clinical signs as part of its constellation of possible symptoms?

CASE 15 A 2-month-old Warmblood foal presented for diarrhea of 3 days' duration, mild fever, and swollen hocks without lameness (**15.1**). There were no prior medical treatments and the 20 other foals on the farm were unaffected. Clinical examination revealed a bright and alert foal with normal HR and RR, normal thoracic auscultation, and a fever of 103°F (39.4°C). The diarrhea was frequent but relatively low volume. The hocks were swollen but no lameness was observed at the walk. The skin appeared normal and the foal did not appear to be clinically dehydrated. CBC revealed a WBC count of 7,500 cells/μl (7.5 × 10^9/l) with 5,300/μl (5.3 × 10^9/l) mature neutrophils, PCV of 24%, fibrinogen of 500 mg/dl (5.0 g/l), and TP of 5.0 g/dl (50 g/l). Serum biochemistries were unremarkable, except for slight hyponatremia (130 mEq/l) and hypokalemia (2.6 mEq/l).

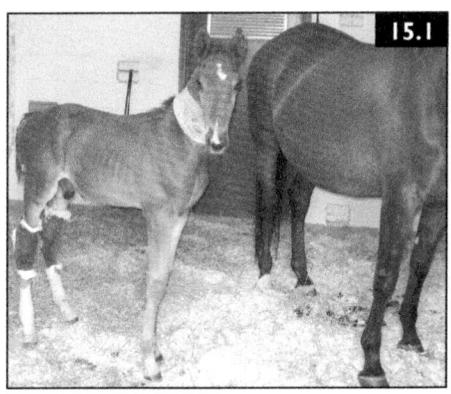

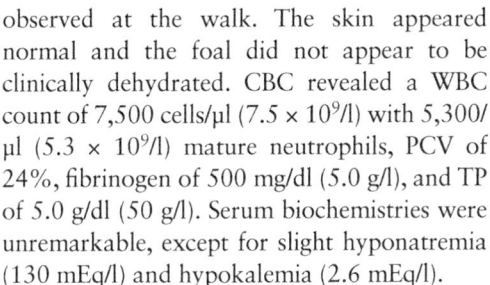

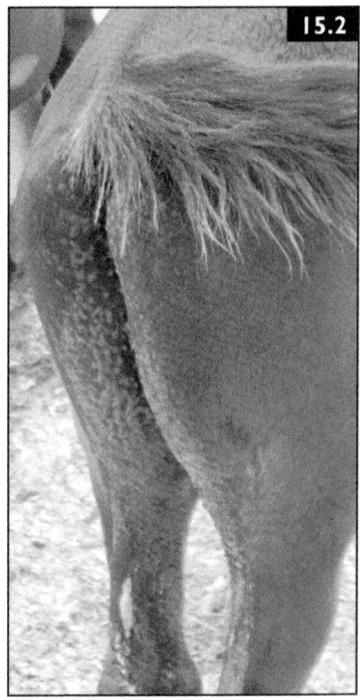

1 What are the differential diagnoses for diarrhea and fever in a 2-month-old foal?
2 What diagnostics should be performed?
3 Why might the hocks be swollen with little or no lameness?

The foal was treated with clarithromycin, metronidazole, rifampin, and omeprazole PO and IV fluid therapy; the diarrhea resolved within 6 days. In spite of very good nursing care, which included deeply bedding the stall, cleaning the perineal area with warm water several times per day, and bandaging the swollen hocks, the foal developed an unusual perineal dermatitis (**15.2**).

4 What could be the cause of the dermatitis?

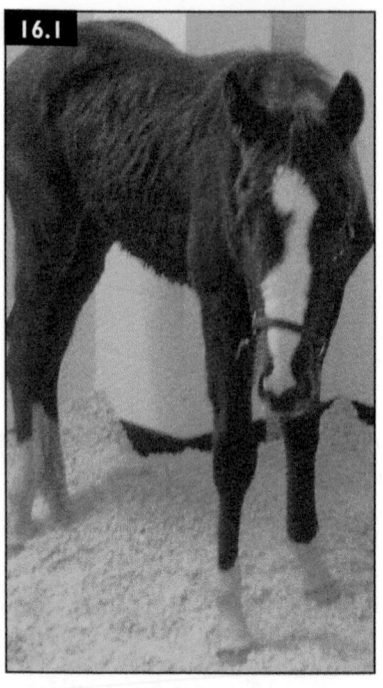

CASE 16 A 7-month-old Thoroughbred colt was treated with IV fluids on the farm because of anorexia and suspected gastric ulcers. The foal developed septic thrombophlebitis of the jugular vein and was treated with chloramphenicol PO for 3 days without improvement and then referred for evaluation. The most remarkable laboratory findings were azotemia (BUN = 66 mg/dl [23.6 mmol/l], Cr = 6.2 mg/dl [472.7 mmol/l]), hyponatremia (Na$^+$ = 121 mEq/l), hypochloremia (Cl$^-$ = 82 mEq/l), and an anion gap (28 mEq/l) with an L-lactate of 2.8 mmol/l. Urine SG was 1.014. The azotemia resolved over 72 hours with fluid therapy, but within 36 hours of hospital admission the colt spiked a fever of 103°F (39.4°C) and developed lameness in both forelimbs, with swelling of the right carpus and left fetlock (**16.1**).

1 What is the most likely reason for the lameness?
2 What diagnostic procedures could be performed to determine an etiologic cause?
3 What are the treatment options?
4 What would be the benefits of therapeutic drug monitoring in this case?

The foal seemed to be improving, but fell when trying to stand immediately following regional limb perfusion treatment and was acutely non-weight bearing in the right hindlimb. A large swelling developed in the caudal thigh region. Ultrasound of the medial aspect of the semimembranosus muscle revealed this image (**16.2**).

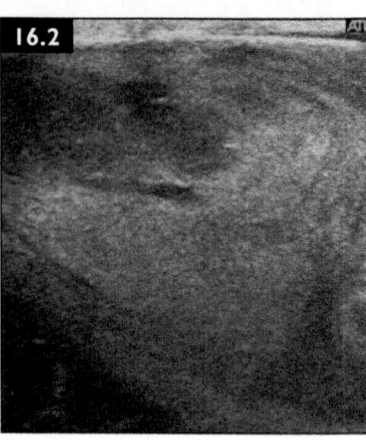

5 What are the differentials and likely cause for this rapid swelling and new lameness?

CASE 17 A 48-hour-old Quarter Horse colt presented for colic, lethargy, and disinterest in nursing. The owners reported that the foaling was unobserved but the foal appeared to be normal for the first day and was seen vigorously nursing. The owners had not witnessed the foal defecating or urinating, although they worked during the day and did not have time to observe him very closely.

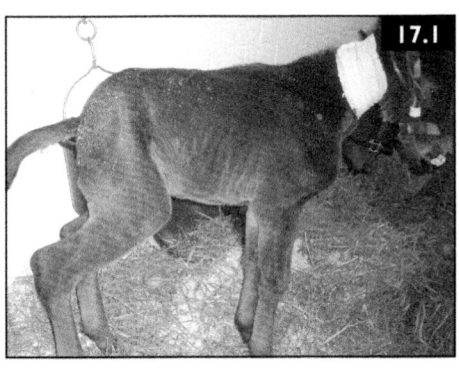

On presentation, the foal was depressed, normothermic (T = 101.9°F [38.8°C]), tachycardic (HR = 140 bpm), and tachypneic (RR = 60 bpm). Mucous membranes were pink but injected and slightly tacky with a CRT of 2.5 seconds. Moderate abdominal distension was present. When observed unrestrained, the colt showed tail flagging, straining, and restlessness (**17.1**). Transabdominal ultrasound revealed an increase in anechoic free abdominal fluid to a maximum depth of 5 cm (**17.2**). Serum biochemistry revealed hyponatremia (Na^+ = 118 mEq/l), hyperkalemia (K^+ = 6.6 mEq/l), hypochloremia (Cl^- = 85 mEq/l) and azotemia (BUN = 35 mg/dl [12.5 mmol/l]; Cr = 4.1 mg/dl [362.4 µmol/l].

1 What are the three most common causes of colic and depression in a neonatal foal?
2 Given the information provided above, what is the most likely etiology of the clinical signs in this 48-hour-old colt?
3 What diagnostic test should be performed next to help confirm the diagnosis?
4 What ECG abnormalities might be expected with hyperkalemia?
5 What initial medical treatments should be administered to a foal with this condition?

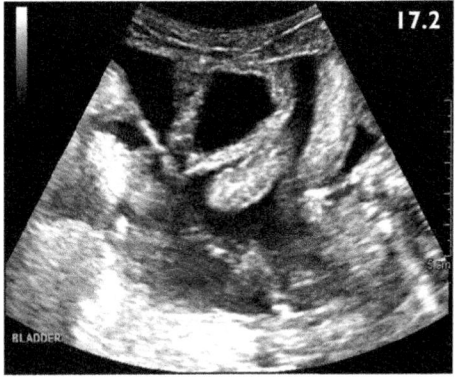

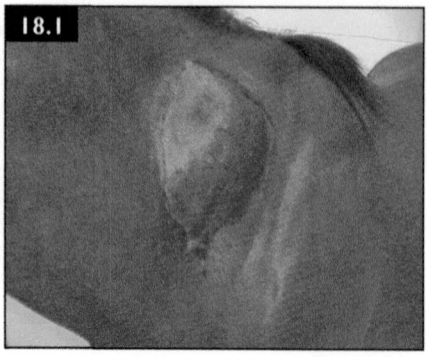

CASE 18 A 3-year-old Thoroughbred was examined because of a swelling on the left side of the neck (**18.1**). The swelling started one week after the horse was vaccinated with a modified-live intra-nasal strangles vaccine and at the same time an IM West Nile virus vaccine was given in the left neck. The swelling was believed to be at the same site as the West Nile vaccination. The swelling increased in size over 5 days and the horse became febrile (103.4°F [39.7°C]).

1 What is the most likely cause of the swelling?
2 What is the best treatment for this condition?
3 Should this horse be isolated?
4 How long does the organism survive in the environment?
5 Is there a way to determine if the horse is a chronic shedder?
6 Is there a way to confirm that the organism cultured from the abscess is the vaccine strain organism?

CASE 19 A 4-year-old Thoroughbred presented for a 3-day history of mild ataxia with severe and progressive weakness leading to recumbency (**19.1**). The horse was bright and alert with a normal rectal temperature and a HR of 54 bpm. Neurologic examination subjectively revealed decreased muscle tone of the limbs but normal tail tone, eyelid tone, and ability to eat. The muscle enzymes were increased (CK = 2,800 IU/l and AST = 780 IU/l) but there was no evidence of muscle atrophy. The serum electrolytes were normal except for a very mild hyponatremia and hypochloremia. Treatment with IV fluids to correct electrolyte abnormalities was performed for 48 hours with no improvement in the clinical condition.

1 Botulism was the initial tentative diagnosis, but some of the clinical findings were not characteristic of botulism. Which ones?
2 List other differentials for the skeletal muscle weakness in this horse.

CASE 20 A 36-hour-old Standardbred colt (**20.1**) was examined because of failure to actively nurse during the first day of life. Clinical examination was unremarkable other than mild weakness and a lack of affinity for the mare. Temperature was normal, HR and RR were increased (140 bpm and 40 bpm, respectively) but ausculta-

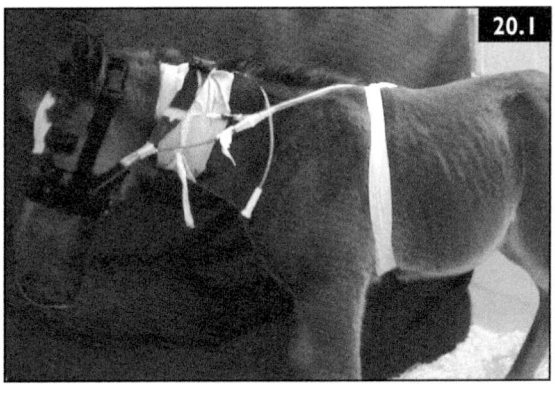

tion was noticeably abnormal. The foal was given a plasma transfusion because of a low IgG, was started on antibiotics (ceftiofur 5 mg/kg IV q8h), and received intranasal oxygen.

Abnormal CBC results and biochemistry findings were: WBCs (1,000/µl [1.0 × 10^9/l]): neutrophils (790 /µl [0.79 × 10^9/l]); bands (200/µl [0.2 × 10^9/l]); lymphocytes (100/µl [0.1 × 10^9/l]); platelets (82,000/µl (82 × 10^9/l); moderate toxic changes in neutrophils. Direct bilirubin was high (0.7 mg/dl [12.0 µmol/l]) with normal SDH (4 IU/l), low AST (185 IU/l), and normal GGT (20 IU/l).

1 What is the interpretation of the foal at this point?
2 The placenta weighed 12% of the foal's body weight. What is the interpretation of this information?

In spite of treatment and with the addition of IV fluids and flunixin meglumine, the foal developed progressive respiratory distress over the next 20 hours (PaO_2 progressively declined with a progressive increase in $PaCO_2$) and died suddenly. On necropsy examination, the most striking lesions were in the lungs and serosal surface of the small intestine (**20.2**). The enteric mucosa appeared normal on opening the bowel.

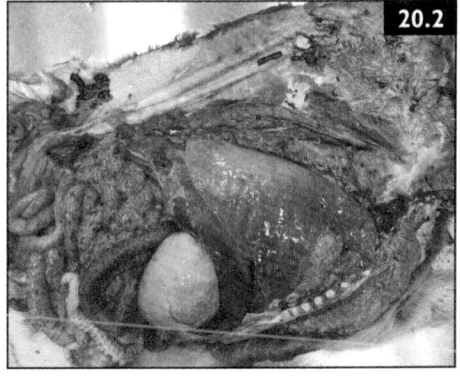

3 What is the interpretation of the rapid deterioration of the foal and the gross pathology findings?
4 What are the most likely differentials at this time?

CASE 21 A 13-year-old Morgan gelding was examined for fever and pruritus. Six days earlier, the horse had acutely developed pruritic hives on the neck, shoulders, and abdomen while being ridden. He appeared to recover without treatment but 2 days later developed swollen legs and sheath. He again developed generalized edema, pruritus, and generalized dandruff (**21.1**). Physical examination revealed a rectal temperature of 104°F (44.0°C) and tachycardia (HR = 54 bpm). The legs and ventral abdomen had pitting and painful edema. There was evidence of thickened skin and hair loss over the face, neck, and dorsum (**21.2**). There was no evidence of petechiation on the mucous membranes and no recent history of drug administration or infectious disease exposure.

1 What generalized type of disease does this horse have?
2 What are the differential diagnoses for this type of disorder?

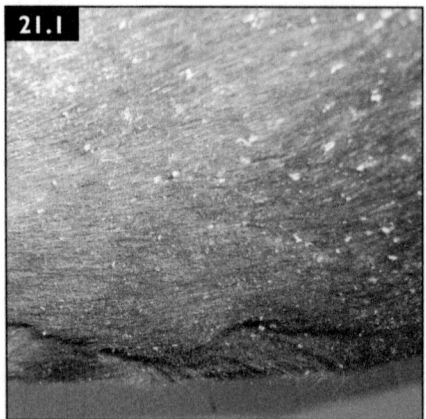
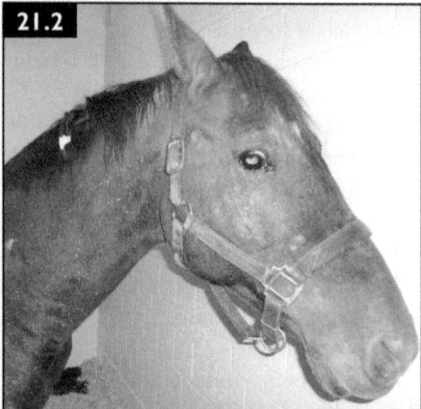

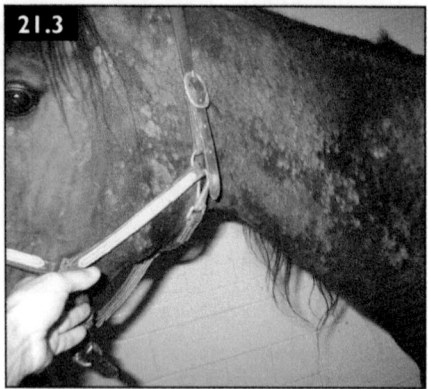

3 What tests could be run to confirm the diagnosis and identify initiating antigens?
4 What treatments should be recommended?
5 What is the primary concern with the drug administered in this case?
6 This horse (**21.3**) had another immune-mediated disease that often causes a generalized dermatitis and can appear similar to the horse in this case. What is this disease?

CASE 22 A 3-day-old Warmblood colt presented because of an acute onset of salivation, 'matted' eyes, difficulty nursing, and multifocal dermatitis over mainly the head and neck region (**22.1**). After an uncomplicated foaling, he had received adequate colostrum, had the umbilicus dipped with dilute chlorhexidine three times daily, and had received an injection of vitamin E and selenium (1 ml) in the hindlimb. The foal had seemed normal for the first 2 days of life. Examination on day 3 revealed normal temperature, HR, and RR. The only other abnormalities identified were ecchymotic hemorrhages and erosions in the mouth (**22.2**). The owners commented that the foal must have eaten something toxic.

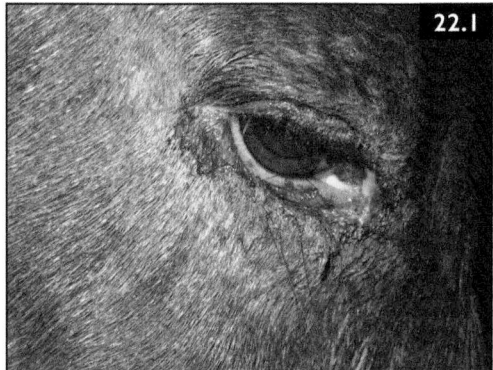

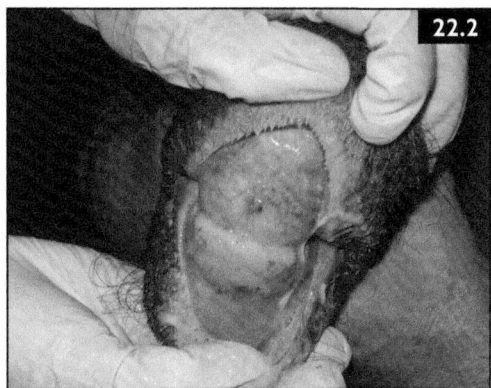

1 What is the diagnosis for these clinical signs in this foal?
2 What laboratory test should be recommended?
3 How should the foal be treated, and what complications could occur?
4 What recommendations should be made to the owner regarding the management of future foals from this mare?
5 What does the mean platelet volume (MPV) indicate?

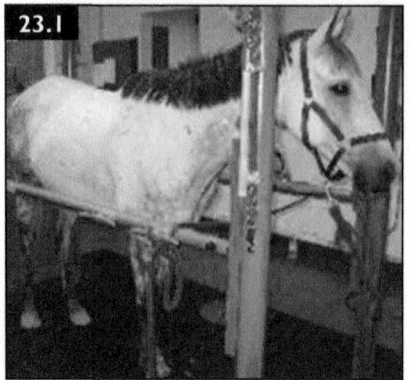

CASE 23 A 10-year-old Thoroughbred gelding used as a pleasure horse was examined because of a 3-week history of cough and bilateral mucopurulent nasal discharge with some food material at the nares (**23.1**). The coughing was most pronounced when the horse was eating. There had also been some weight loss and the horse was underweight. A 10-day course of trimethoprim-sulfa given PO had not produced any noticeable improvement. On clinical examination the horse was bright and alert and all vital signs were normal. There was a green (feed) and yellow discharge present at both nostrils. There were no gait abnormalities and cranial nerves II–VIII and XII tested normal. Auscultation of the chest revealed a few crackles and wheezes in the right ventral area. Thoracic ultrasonography was performed (**23.2**).

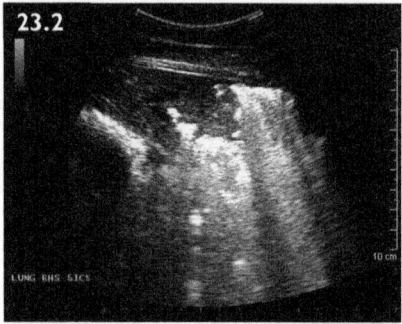

1 What are the major concerns as to the clinical conditions affecting this horse?
2 What does the thoracic ultrasound suggest?
3 What organisms are commonly associated with this type of pneumonia?

Endoscopy of the oropharynx revealed food material in the trachea and the soft palate became easily displaced (even without tranquilization). This lesion (**23.3**) was observed within the medial compartment of the right guttural pouch (GP).

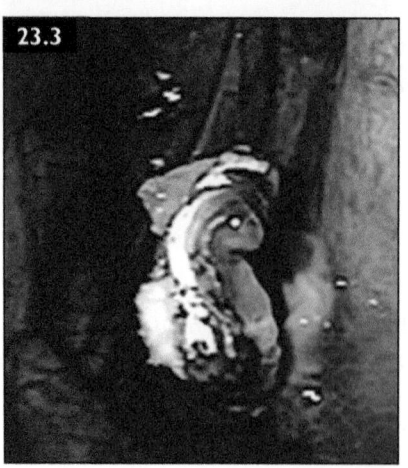

4 What is the most likely diagnosis, and how is it causing the clinical signs?
5 What are the treatment options for the guttural pouch lesion?

CASE 24 A 5-year-old Throughbred-Warmblood cross gelding presented for intermittent fever of several weeks (sometimes as high as 103.5°F [39.7°C]), weight loss of 6 months' duration, and development of bony swellings on the distal radii and tibias within the last 2 weeks. Physical examination revealed a normal HR with an increased RR (24 bpm). Auscultation of the chest was normal, there was no nasal discharge, and a cough could not be elicited. There was no enlargement of the peripheral lymph nodes. Mild circumlimbal edema was noted in both eyes with no evidence of uveitis. On palpation of the limbs, there was marked enlargement of the distal radial metaphyses (**24.1**), with lesser enlargement of the mid-left metacarpus, distal tibial metaphyses, and metatarsal metaphyses and diaphyses. The horse was neither painful on palpation of the limbs nor noticeably lame at a trot.

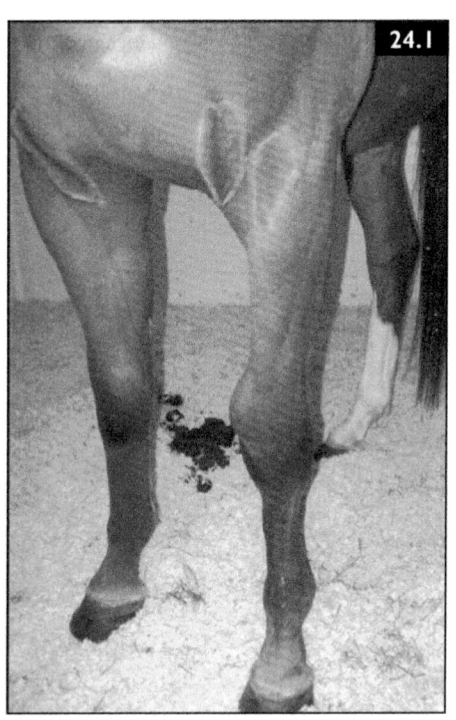

24.1

1 What would be the next diagnostic test?
2 What is the most likely diagnosis for the limb lesions?
3 With this differential diagnosis and observation of an increased RR, what should be done next?
4 What did this test show?
5 Multiple lung lesions, fever, weight loss, lymphopenia (1,100 cells/µl [1.1 × 10^9/l), and elevated fibrinogen (600 mg/dl [6.0 g/l]) might be suggestive of what pulmonary disease?
6 How could this diagnosis be confirmed?
7 How could this case be treated?

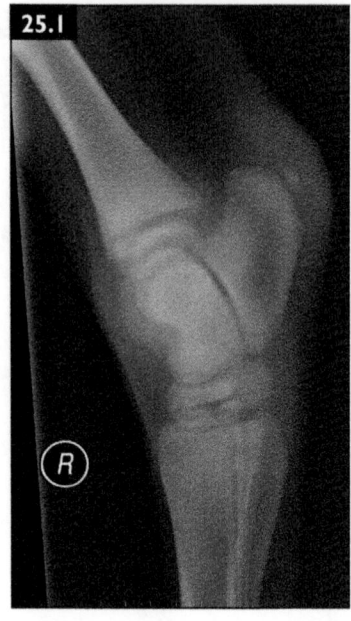

CASE 25 A 36-hour-old premature (319 days) Arabian colt was treated for pneumonia and sepsis resulting from failure of passive transfer. The foal had a marked leukopenia (600 cells/µl [0.6×10^9/l]), neutropenia (100 cells/µl [0.1×10^9/l]) with toxic changes, lymphopenia (200 cells/µl [0.2×10^9/l]), and thrombocytopenia (69,000 cells/µl [69×10^9/l]). Blood ACTH and cortisol levels were above the normal range, suggesting that the hypothalamus/pituitary/adrenal (HPA) axis had responded appropriately to the illness. Mild symmetrical distension of the carpal, tarsal, and fetlock joints was noted. By day 3, the foal was able to stand and did not demonstrate any lameness. A radiograph of the right tarsus was obtained (**25.1**). The left carpus was considered normal.

1 What does the radiograph demonstrate?

The foal responded well to medical treatment for the pneumonia and was discharged with instructions to limit exercise and be maintained on good footing at home. The foal was re-examined 1 month later for angular limb deformity. A radiograph of the right tarsus was taken at that time (**25.2**).

2 What is shown in the new radiograph?
3 What treatment should be recommended?

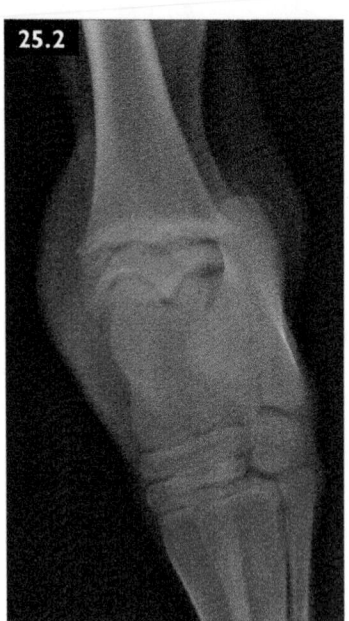

The foal was discharged and confined to a small dirt stall with instructions that the owners help him up hourly until he was comfortable doing so on his own. Treatment with omeprazole was continued. A cast change was recommended at every 2-week recheck. After two cast changes, the foal demonstrated improvement and was kept in a grass paddock for the rest of the year with

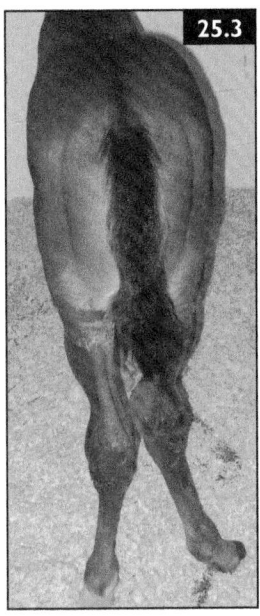
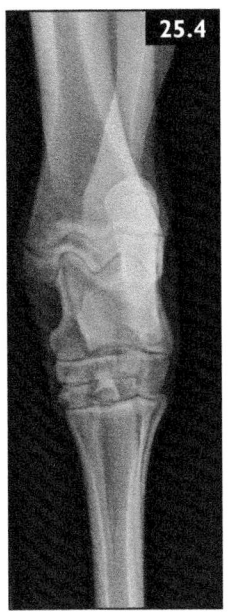
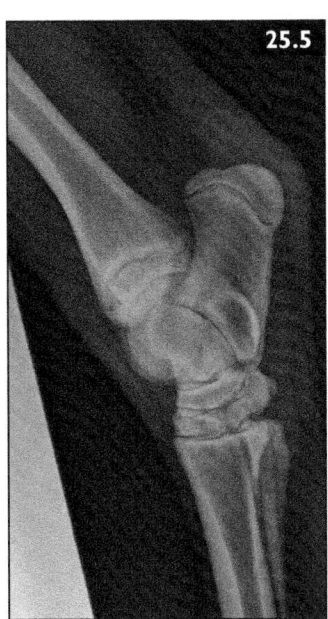

no further treatment. The foal was re-examined when he was 4 months old (**25.3**) and radiographs were taken of the hocks at that time (**25.4**, **25.5**). The foal was sound at a walk and trot.

4 Based on these radiographs, what is the most likely diagnosis, and what treatment should be recommended?

CASE 26 An 18-month-old Percheron gelding presented for evaluation of acute respiratory distress and dark colored urine (**26.1**). His paddock mate had shown similar symptoms in the previous 24 hours and was then found dead. On evaluation the horse was dull with pale/gray mucous membranes. He was tachycardic, tachypneic, and febrile. A stream of black urine was passed during the examination.

1 What diagnostic tests should be performed to determine the cause of the urine discoloration?
2 What is the likely cause of this problem?
3 What secondary problems should be considered when treating this case?

23

CASE 27 A 3-year-old Appaloosa gelding was initially examined on the farm because of a fever of 104°F (44.0°C) and mild pectoral edema without other signs (**27.1**). The horse was treated with flunixin meglumine (500 mg IV once) and a 5-day course of trimethoprim sulfa (20 mg/kg PO q12h). Febrile episodes without other signs continued and after 10 days the horse was treated with K$^+$ penicillin (22,000 IU/kg IV q6h), gentamicin (6.6 mg/kg IV q24h), and metronidazole (15 mg/kg PO q8h). Coggin's test for equine infectious anemia was negative. There were eight other adult horses on the farm with no new additions to the farm within the past 6 months. After 3 days of penicillin, gentamicin, and metronidazole treatment, the horse still had a fever of >103°F (39.4°C) and was referred for further evaluation. On initial hospital examination, the horse was bright and alert but had a temperature of 101.6°F (38.7°C), RR of 24 bpm, and HR of 56 bpm. A 2/6 diastolic murmur was heard on auscultation over the aortic area. The horse appeared thin (BCS 3/9) and the owners reported that he had lost weight and muscle mass over the past 2 weeks. Pitting, non-painful edema was noted over the sternal area (**27.2**) and prepuce.

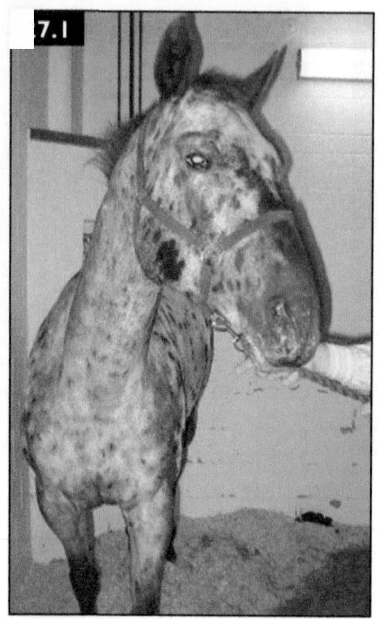

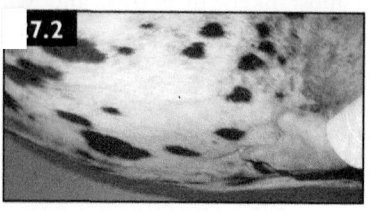

1 List the initial differentials for this horse.
2 What additional diagnostic procedures should be performed?

The horse was treated with flunixin meglumine (0.5 mg/kg IV q 12 h) for 1 day as symptomatic treatment of the fever. After 36 hours, the fever and tachycardia persisted. In the absence of clinical or laboratory findings supportive of a bacterial infection, an immune-mediated or viral myositis/myocarditis became the tentative diagnosis and the horse was given 20 mg dexamethasone IV. The horse became afebrile within 12 hours of the dexamethasone, but the tachycardia persisted (52–56 bpm) and the horse experienced an episode of significant tachycardia (>180 bpm) with some distress and jugular pulses within 1 hour of the dexamethasone injection. ECG was performed (**27.3**, paper speed = 50 mm/s).

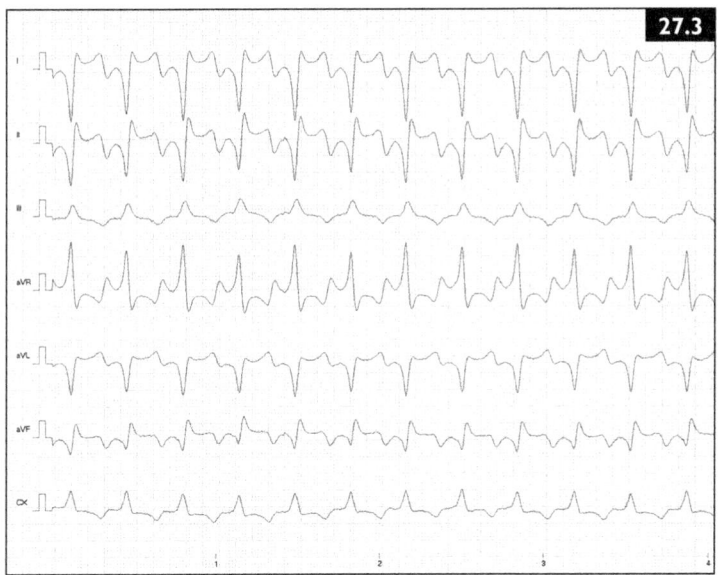

3 What is the abnormality on the ECG?
4 Does this arrhythmia need to be treated? If so, what treatments could be used?

Because of the strong suspicion of myocardial involvement, a sample was tested for cardiac troponin (cTnI), which was elevated at 2.27 ng/ml (2.27 µg/l). An echocardiogram was performed (**27.4**).

5 Interpret the echocardiogram.

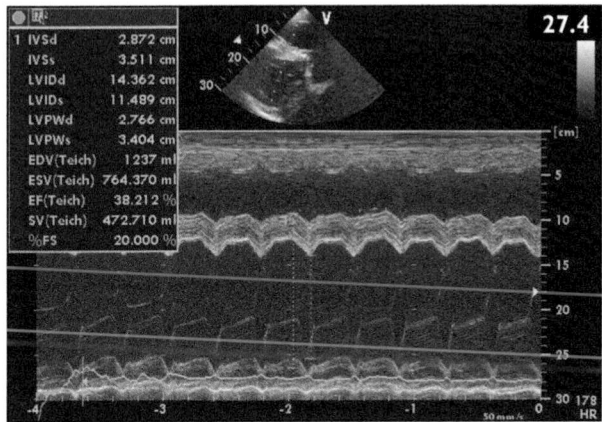

CASE 28 A 36-hour-old Warmblood filly was evaluated for failure to urinate since birth and mild colic. The foal received a 1 ml dose of flunixin meglumine in the muscle of the right hindlimb and was treated with sucralfate on the farm. On presentation, the filly was 'tail flagging' and appeared mildly colicky, but was seen to pass feces and urine normally. Rectal temperature was 101.6°F (38.7°C), HR was 116 bpm, and RR was 40 bpm. Mucous membranes appeared normal, there was no evidence of rib fractures on palpation, and chest auscultation was normal. The abdomen was not distended, auscultation revealed normal GI borborygmi, and the foal was still nursing, albeit less than 1 minute at a time. The umbilical area was normal on palpation. CBC and serum IgG were normal. Serum biochemistries were normal except for a mild hyponatremia (127 mEq/l), hypochloremia (92 mEq/l), hyperkalemia (4.9 mEq/l), and an increase in CK (582 U/l). Creatinine was 1.2 mg/dl (91.5 µmol/l); BUN was 22 mg/dl (7.85 mmol/l). The foal was sedated with 5 mg of diazepam IV and abdominal ultrasound performed. The urinary bladder appeared normal, the intestines were motile and normal, and all umbilical structures appeared normal. However, there was increased abdominal fluid that appeared more echogenic than normal (**28.1**). Based on the mild signs of colic, the unusual peritoneal fluid appearance on ultrasound, and the mild electrolyte abnormalities, abdominocentesis was performed using a teat cannula (**28.2**). The fluid had 17,200 WBCs/µl (17.2×10^9/l) (72% non-degenerate neutrophils, 27% macrophages, <1% lymphocytes). TP was 5.6 g/dl (56 g/l) and lipid droplets were seen in the macrophages.

1 What is the interpretation of the peritoneal fluid analysis, and what other test should be performed?
2 How should the effusion be treated?

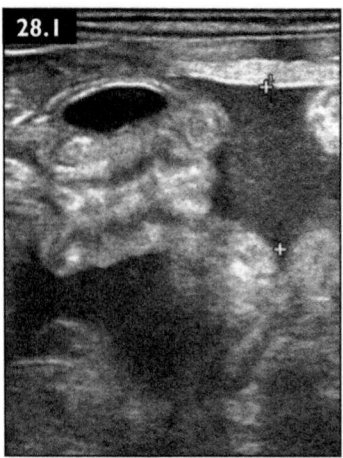
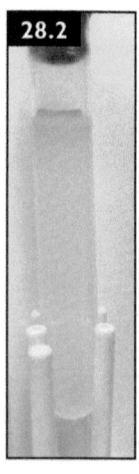

28.1 28.2

The next day the filly was lame in the right hindlimb and mild swelling was noted at the caudal thigh area (**28.3**). Ultrasound revealed extensive gas distension between the muscle planes. CBC revealed 1,200 immature neutrophils/µl (1.2×10^9/l) (band cells). The fibrinogen had risen from 200 mg/dl (92.0 g/l) at admission to 600 mg/dl (6.0 g/l) 18 hours later.

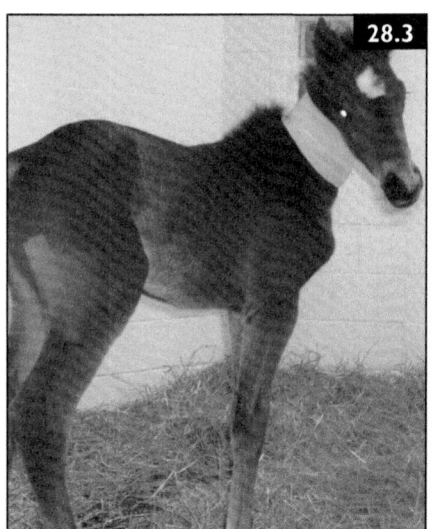

3 What would be the most likely diagnosis for the lameness?
4 Since there was no rupture of the bladder, how could the electrolyte abnormalities be explained in light of the new clinical sign?
5 What is the black hypoechoic structure in **28.1**?
6 How should the swollen leg be treated initially?

CASE 29 A 23-year-old Thoroughbred gelding had been dribbling urine for at least 3 weeks, but had been otherwise healthy. Clinical examination was unremarkable except for the dribbling urine and mild associated contact dermatitis on the hindlimbs (**29.1**). There were no neurologic deficits on gait analysis or reflex evaluation (CNs, anal), and the tail tone was normal. Rectal examination revealed a very large urinary bladder that could be manually expressed to void small amounts of urine. Ultrasound examination per rectum revealed large amounts of hyperechoic material in the ventrum of the bladder.

1 What is the tentative diagnosis?
2 What would the urinalysis and urine culture results likely show on the sample taken at the initial evaluation?
3 What would the urine culture results likely show on all subsequent samples?
4 What are the treatment options for this horse?
5 What are possible predisposing factors?

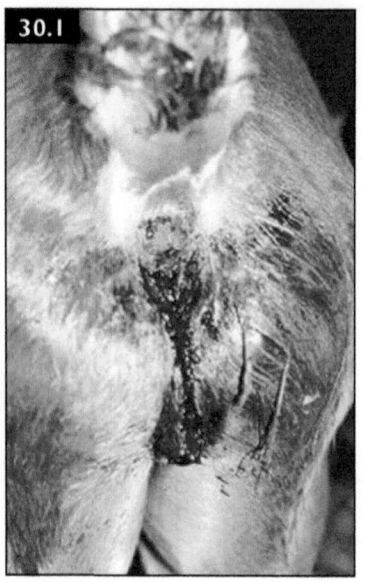

CASE 30 A 3-day-old Thoroughbred foal was examined because of an acute onset of hemorrhagic diarrhea (30.1), mild colic, and depression. The foal was treated for 1 day with bismuth subsalicylate and trimethoprim sulfa, but became more depressed. The foal stopped nursing but still had interest in the mare. Rectal temperature was 103°F (39.4°C), HR was 110 bpm, and RR was 28 bpm. The abdomen was slightly enlarged and GI borborygmi were nearly absent on auscultation. An ultrasonogram of the abdomen is shown (30.2). None of the other 26 foals on the farm were affected.

1 What are the most likely differential diagnoses for the diarrhea?
2 What should be the immediate treatment plan?

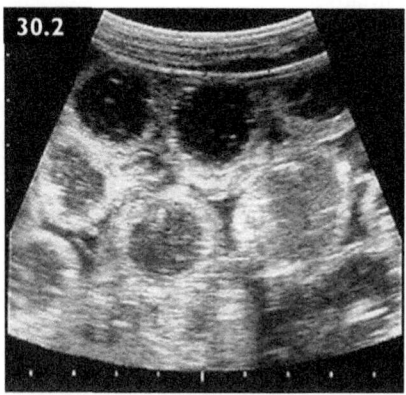

CBC revealed neutropenia (1,800 WBCs/μl [1.8 × 10⁹/l]), TP was 4.2 g/dl (42 g/l). Platelet count was normal. Toxic changes were noted in the neutrophils during cytologic examination. Biochemistry abnormalities included: Na = 118 mEq/l; Cl = 92 mEq/l; HCO_3 = 13 mEq/l; K = 1.7 mEq/l; BUN = 22 mg/l (7.85 mmol/l); Cr = 2.0 mg/l (176.8 μmol/l); glucose = 88 mg/l (4.88 mmol/l); pH = 7.27; PCO_2 = 28 mmHg (3.72 kPa).

3 After reviewing the laboratory findings, how can the results be interpreted, and what would be the treatment plan?
4 What complications can be associated with neonatal foal diarrhea, and how can these be prevented?

CASE 31 A 14-year-old Thoroughbred gelding sustained a 10-cm wire laceration over the left hock. The laceration was tended to by the owner, a veterinary student who had been caring for two hospitalized patients with catheter site infections caused by MRSA. One of the patients was being treated for pneumonia and then developed septic jugular vein thrombophlebitis (**31.1**). On hospital presentation, the injured horse was bright and alert with normal vital signs and 4/5 lameness in the left hindlimb. Exploration of the wound over the tarsocrural joint revealed that the long digital extensor tendon was partially torn, but the wound did not appear to communicate with the joint. The laceration was sutured and the leg was bandaged (**31.2**). The horse was started on antimicrobial treatment with PO trimethoprim sulfa and

phenylbutazone. The horse had not improved the following day, had a rectal temperature of 103.2°F (39.6°C), and distension of the tarsocrural joint. Arthrocentesis revealed 44,000 cells/μl (44.0×10^9/l) and a TP of 5.4 mg/dl (54 g/l). No bacteria were seen on cytology of the fluid, but the fluid was submitted for culture, which showed a moderate and pure growth of *S. aureus* sensitive only to enrofloxacin and vancomycin (i.e. MRSA).

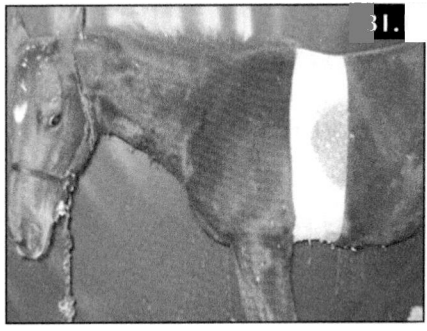

1 What should be done next?
2 How could the infection in this horse be explained?
3 What would be the treatment recommendations?
4 The horse became non-weight bearing on the leg (**31.3**). What would be a concern regarding the opposite hindlimb?
5 When considering how frequently vancomycin and enrofloxacin should be administered to this horse, what should be taken into account?

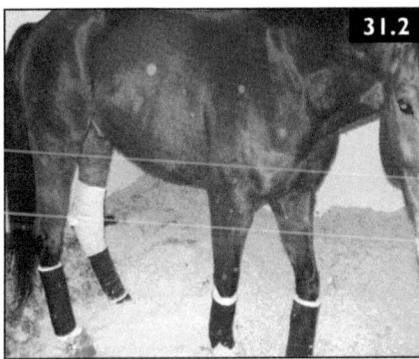

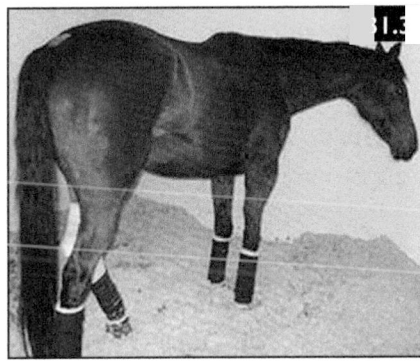

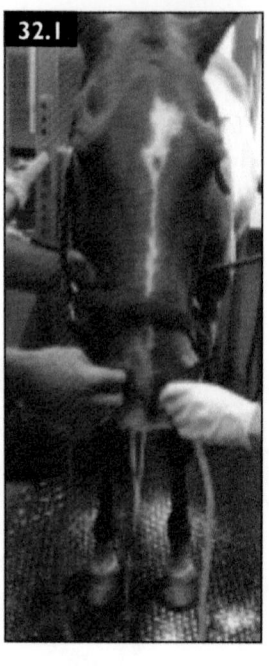

CASE 32 A 14-year-old Thoroughbred gelding used as a polo school horse was first noted on the farm to be in respiratory distress with a large amount of clear, foamy fluid, without feed or odor, running from both nostrils. HR was 88 bpm and RR was 80 bpm with increased abdominal effort. Although there was no feed in the nasal exudate, a nasogastric tube was passed after both xylazine and Buscopan® had been administered. The tube was passed into the stomach without resistance, and afterwards the nasal discharge diminished significantly. The horse was administered a single dose of furosemide IM. He appeared normal for 4 days and was even lightly worked on day 2. On day 4 there was an acute onset of nasal discharge (**32.1**) with a constant flow of a large volume of fluid from both nostrils and respiratory distress. No cardiac murmur was auscultated but crackles were noted on auscultation of the lungs. Ultrasound examination of the chest was performed. The image shown (**32.2**) is of the dorsal lung field, which was representative of the entire lung.

1 What is the tentative diagnosis?
2 What is the most likely cause of the acute left heart failure and cardiogenic pulmonary edema in this case? A necropsy specimen is shown (**32.3**).

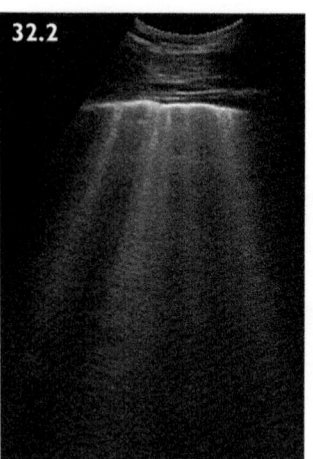

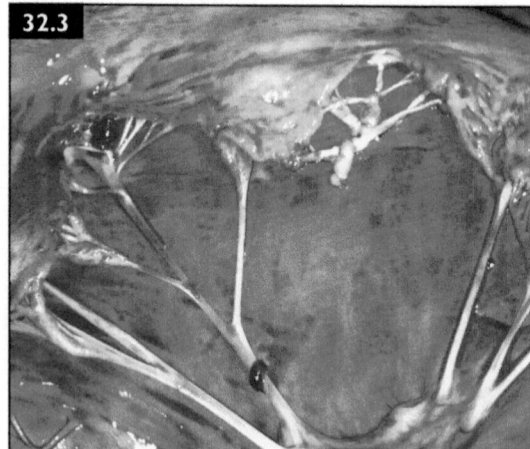

CASE 33 After a qualifying race, a 3-year-old Standardbred was immediately treated for stiffness and suspected rhabdomyolysis with a single dose of flunixin meglumine. The following day (a hot day), the horse raced and did poorly. Immediately after the race the horse was noted to be distressed, very stiff, and sweating profusely. Twelve hours later the horse was observed to be more depressed and stiff. On examination (**33.1**) the HR was 66 bpm, RR was 54 bpm, and the mucous membranes were a dark red color. Palpation of most large muscle groups revealed that they were firm, swollen, and painful to palpation. PCV = 80%; TS = 7.8 g/dl (78 g/l). Pertinent laboratory findings included: Na = 124 mEq/l; Cl = 79 mEq/l; K = 6.0 mEq/l; Ca (ionized) = 0.9 mmol/l; HCO_3 = 16 mEq/l; AST = 40,723 U/l; CK = >400,000 U/l; BUN = 44 mg/dl (15.7 mmol/l); Cr = 6.4 mg/dl (488 µmol/l). After initial fluid therapy, urination was not observed and abdominal palpation per rectum revealed only a small amount of urine in the bladder. Ultrasound examination of both kidneys was similar and an image of the left kidney is shown (**33.2**).

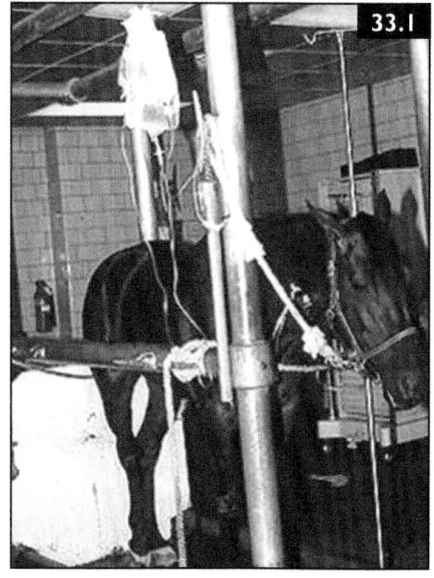

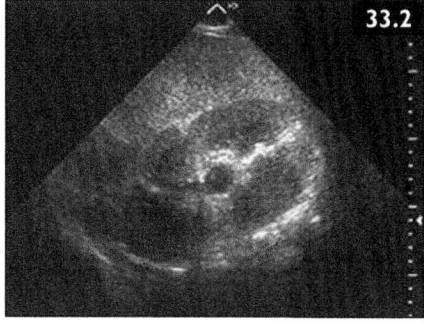

1 What immediate treatment should this horse receive?
2 What clinical and laboratory diagnosis is most likely responsible for the stiffness, tachycardia, and distress? Explain the reason for the laboratory findings provided.
3 Interpret the ultrasound findings.
4 What is a major concern at this time, and what monitoring and treatments should be used?

CASE 34 A 5-year-old Miniature horse mare foaled uneventfully 3 days prior to presentation. She had a healthy foal running at her side, constantly attempting to nurse. Since foaling, the mare had been lethargic and had shown signs of difficulty eating and swallowing. She wanted to eat but could not seem to chew or move food to the back of her mouth for swallowing. On presentation, her tongue was sticking out of the mouth (**34.1**). There were 37 other animals on the farm and none appeared similarly affected. Additional clinical examination findings revealed icterus (**34.2**), masseter muscles that were uniformly enlarged for a female horse (**34.3**), and tachycardia (HR = 64 bpm). During the examination, the mare passed urine that was grossly abnormal (dark). The urine was evaluated for myoglobin (positive occult blood) and a complete biochemistry and CBC performed. Abnormal results included: WBCs = 3,700/µl (3.7 × 10^9/l); neutrophils = 2,100/µl (2.1 × 10^9/l); PCV = 43%; TP = 7.2 g/dl (72 g/l); pH = 7.29; HCO$_3$ = 20 mEq/l; lactate = 3.5 mmol/l; CK = 57,040 U/l; AST = 13,030 U/l; SDH = 362 U/l; GLDH = 280 U/l; ALP = 812 U/l; GGT = 77 U/l; glucose = 212 mg/dl (11.7 mmol/l); ammonia = 110 µg/dl (78.5 µmol/l); total bilirubin = 6.2 mg/dl (106.0 µmol/l) [direct 0.7 mg/dl, 12.0 µmol/l; indirect 5.5 mg/dl, 94.1 µmol/l]).

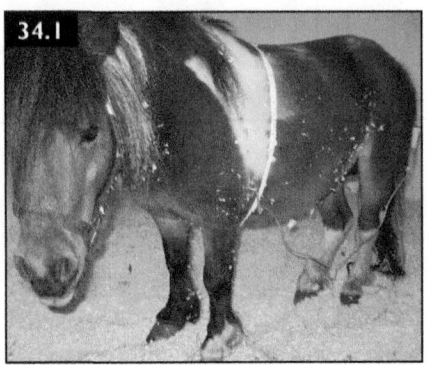

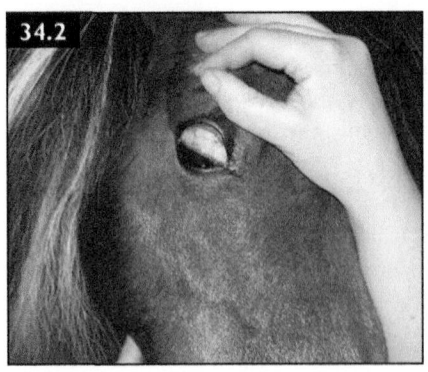

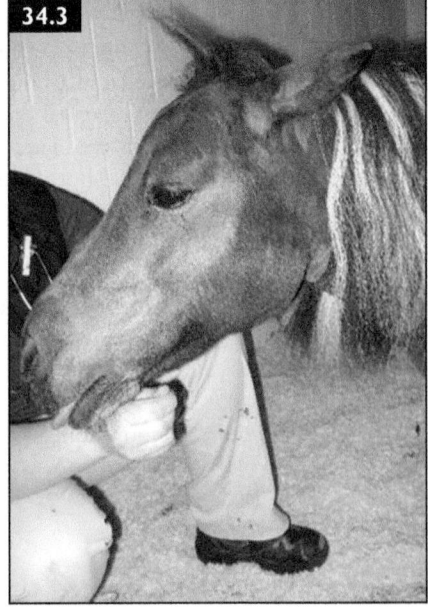

1 The primary diagnosis was not obvious, but what complication should be a concern since this is a Miniature horse that is unable to eat properly and has a 3-day-old foal by its side?

2 The spun down plasma is shown (34.4). What is this an indication of?

3 What is the diagnosis based on the labwork?

4 How should the hyperlipemia be treated?

5 Since there was strong evidence that a myopathy might have been the primary disorder, what additional diagnostic test should be performed?

CASE 35 A 17-year-old pony used as a show jumper developed an upper respiratory flutter during exercise for 6 months prior to presentation. In the last 2 weeks there had been black, fetid, unilateral (left) nasal discharge (35.1). The pony was rubbing its nose and causing black discoloration of the walls (35.2). The pony was otherwise healthy based on clinical examination.

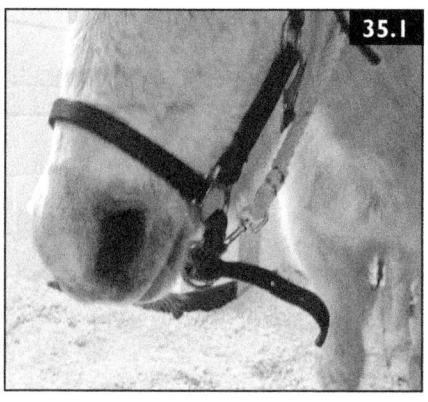
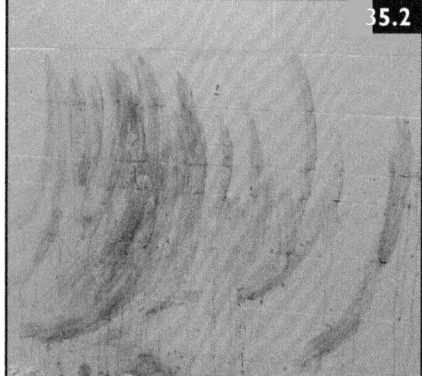

1 What would be the next diagnostic procedure?

2 What is the likely cause of the black nasal discharge?

3 What are the treatment options?

4 What is the most common anatomic site for this condition?

CASE 36 A 22-year-old mixed breed mare was found in the field with food contents in both nares. Additional history revealed that the mare had choked three times within the past year. The referring veterinarian had unsuccessfully tried to relieve the choke 18 hours prior to presentation. Overnight tranquilization with acepromazine had also failed to relieve the choke. The owners had given the horse a 'handful' of hay just prior to hospital admission to be certain she was still choked (**36.1**). On clinical examination, the horse was depressed and in some respiratory distress. HR was 84 bpm, RR was 56 bpm, and the gums were dark red and congested with a toxic line (**36.2**). Crackles and fluid could be heard on auscultation of the lungs. The choke was relieved after administering xylazine and further flushing with water via a nasoesophageal tube. A thoracic radiograph was obtained (**36.3**) and the esophagus was examined endoscopically (**36.4**).

Common and often successful management of equine esophageal obstruction ('choke') includes tranquilization, removing all feed and water from the stall, and attempts to relieve the obstruction via a nasoesophageal tube.

1 What historical information in this case makes it not surprising that tranquilization treatment for choke was unsuccessful?
2 What are the greatest medical concerns regarding complications of the choke in this horse?

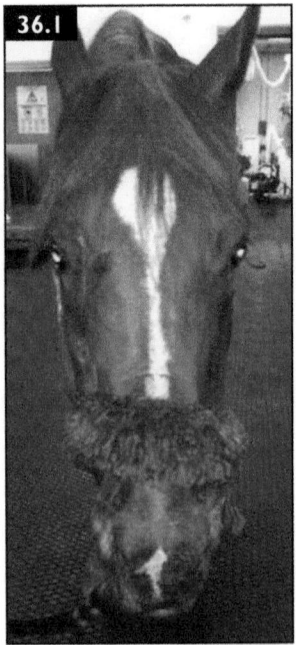

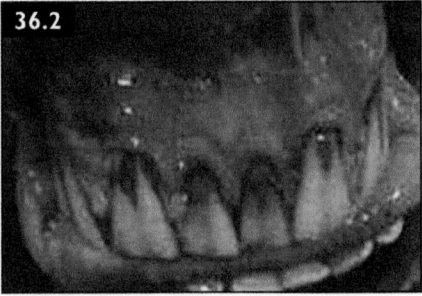

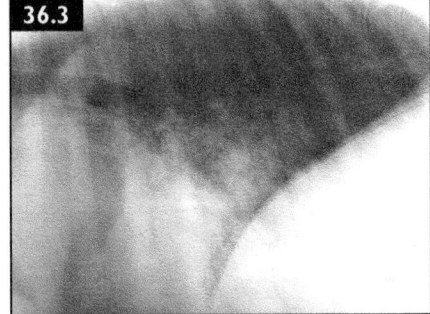

3 What is the radiographic interpretation?
4 Describe the endoscopic findings from the cardia region of the esophagus.
5 In addition to immediately removing all feed and water, administering a tranquilizer to relax the esophagus and lower the horse's head, and flushing, what additional parasympatholytic drug could be considered in a horse with choke?
6 What is the predominant anion in equine saliva?

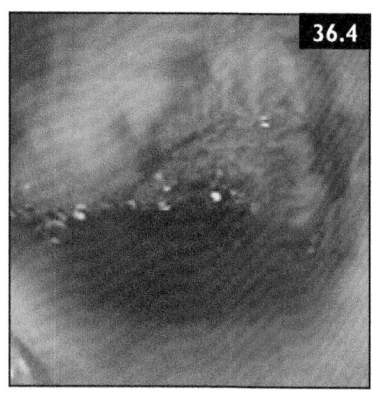

CASE 37 An 8-year-old Warmblood gelding was purchased from Europe 1 year prior to presentation. The only abnormality noted at a prepurchase examination was left-sided Horner's syndrome. There was no reported investigation into the reason for the Horner's syndrome at that time and it was assumed to be caused by a perivascular injection. The horse competed successfully at a high level for 1 year after arriving in North America. On the day of presentation, the horse had appeared normal when loaded onto a van and there was no report of trauma during the 10-hour trip. On arrival, the following abnormalities were noted: severe head tilt to the left, severe balance deficits, a mild soft tissue swelling just caudal to the left guttural pouch, and noticeable sweating on the left side of the face and proximal neck (**37.1**).

1 What is the differential diagnosis for the neurologic signs in this case?
2 What should the diagnostic plan include (in order of importance)?
3 What are the most common causes of Horner's syndrome (mild miosis, enophthalmos, hyperemic mucous membranes, sweating of the face and cranial neck, mild ptosis) in the horse?
4 Can the lesion identified explain the clinical signs?

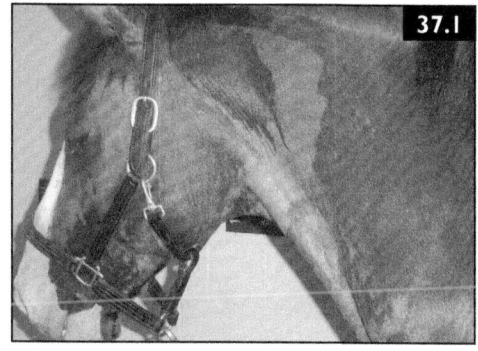

5 Since there was soft tissue swelling caudal to the guttural pouch and clinical evidence (i.e. the vestibular disease) of pathology of the tympanic bulla, what additional procedure could be performed?

CASE 38 A 13-year-old Thoroughbred mare at pasture was found acutely blind. There had been a 3-week history of severe uveitis in the right eye and muscle wasting over the epaxial region for several weeks. The uveitis had been treated with topical 1% atropine 2–3 times daily, topical 1% dexamethasone in combination with neomycin and polymixin B 4–6 times daily, and 1.1 mg/kg flunixin meglumine PO once a day for 7 days. In spite of treatment, the eye did not improve and there were no menace, pupillary, or dazzle responses in the eye. A large yellow fibrin clot was present in the anterior chamber of the right eye (**38.1**) but there was no fluorescein stain uptake. A lighter yellow clot and an intensely miotic pupil were found in the more recently affected left eye (**38.2**). The posterior segment could not be well visualized and the horse had a normal palpebral reflex but no menace response in the left eye. The horse was euthanized because of the rapid progression of the ocular disease. Aqueous and vitreous samples were collected following euthanasia. Both the right and the left aqueous humor samples were of increased neutrophilic cellularity (4,365and 4,907 cells/µl [4.37 and 4.91 × 10^9/l], respectively) and no infectious agents were identified. The cellularity of the left vitreous humor was much lower (10 cells/µl [0.01 × 10^9/l]) but was still mildly elevated (**38.3**).

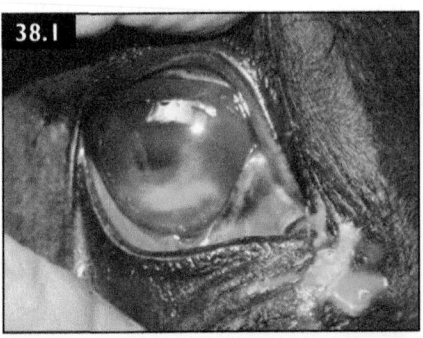

1 What would be potential causes for the acute progressive bilateral uveitis?
2 What is the interpretation of the cytology of the vitreous?
3 What diagnostic test should be performed next?
4 How could the muscle wasting over the horse's top-line be explained?

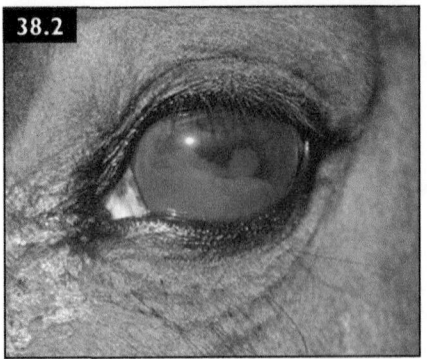

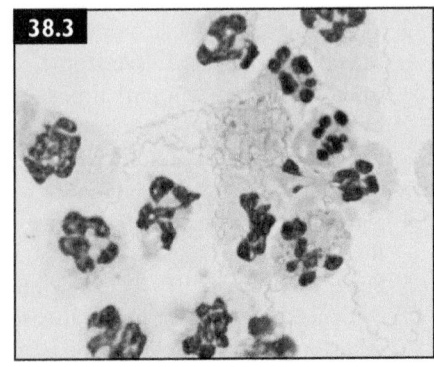

CASE 39 A 10-year-old Standardbred mare, 9 months pregnant, was examined for progressive enlargement of the abdomen and a plaque of ventral edema (**39.1**). The udder was in the normal anatomic position. Rectal palpation revealed an enlarged, fluid-filled uterus, making it difficult to palpate the fetus. The fetus was alive and on transabdominal ultrasound examination had a HR of 100 bpm.

1 What are the differential diagnoses for this mare?

2 After 24 hours, the mare became recumbent due to the progressively enlarging uterus. What would be the treatment options?

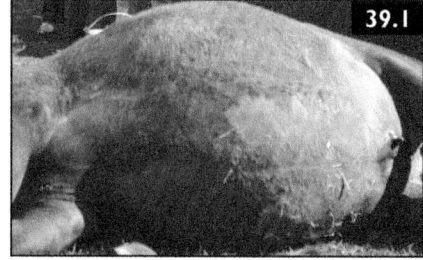

Although an attempt was made to control the speed at which the allantoic fluid was removed, the mare developed hypotensive shock after 80 or more liters were removed.

3 What is the mechanism of hypotensive shock, and what would be the treatment?
4 What would be a probable uterine complication following delivery?
5 In less severe cases than the mare in this case, what can be done to help prevent abdominal wall injury/hernia?

CASE 40 A 3-year-old Standardbred gelding was examined by a referring veterinarian for a decline in performance without any clinical abnormalities on routine examination, CBC, or serum biochemistry. Endoscopic examination of the airway within 60 minutes of training revealed a pool of mucus in the trachea (**40.1**). Cytologic examination of the fluid revealed mild neutrophilic inflammation (approximately 25% neutrophils and a mixture of macrophages and lymphocytes) with large amounts

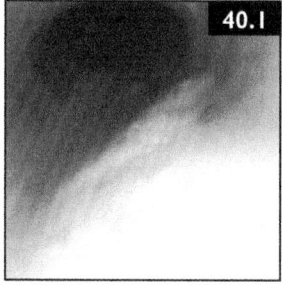

of thick mucus. The horse was referred 5 days later for bronchoalveolar lavage (BAL), which revealed the following: 4% neutrophils, <1% mast cells and eosinophils, 54% macrophages, and 41% lymphocytes.

1 Interpret the BAL results.
2 What is the most likely diagnosis?
3 What would be the preferred treatment(s) for this horse?

CASE 41 A 5-year-old Friesian mare was initially examined because of an acute onset of anorexia and a fever of 103.6°F (39.8°C). Clinical examination revealed bounding pulses, HR of 80 bpm with an intermittent cardiac arrhythmia, no auscultable murmurs, and a frequent cough. She was treated with antibiotics and flunixin meglumine for 5 days. Evaluation on day 6 found the mare to be afebrile, partially anorexic, and having bounding pulses with runs of tachycardia and arrhythmia. The manure had also become less formed than normal. She was up to date on vaccines but had not been tested for equine infectious anemia (EIA). The mare was referred for more testing to determine the cause of the clinical signs. The variable tachycardia (60–80 bpm) and irregular cardiac rhythm were noted along with pale mucous membranes. Mild, non-painful pitting edema was noted on her ventrum. The jugular veins appeared mildly distended, but there was no evidence of jugular pulses. However, when the veins were held off near the head and manually compressed caudally to remove residual blood, a refill was noted in the most ventral aspect of the vein after 2 minutes of the bilateral proximal jugular occlusion. No cardiac murmurs were auscultated and the heart was easy to auscultate. ECG (**41.1**) was performed. A complete chemistry profile was normal except for an elevated creatinine (3.4 mg/dl [300.5 µmol/l]), BUN (66 mg/dl [23.6 mmol/l]), and L-lactate (7.22 mmol/l).

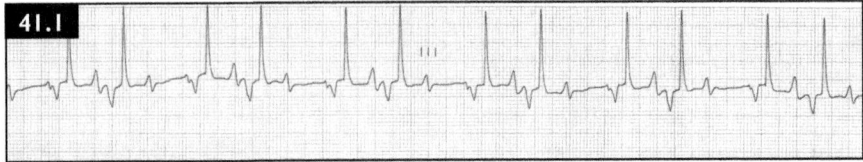

41.1

1 What differentials should be included at this time?
2 What blood test would aid in confirming the main differential?
3 Interpret the ECG. What other test should be performed?
4 How could EIA be ruled out?

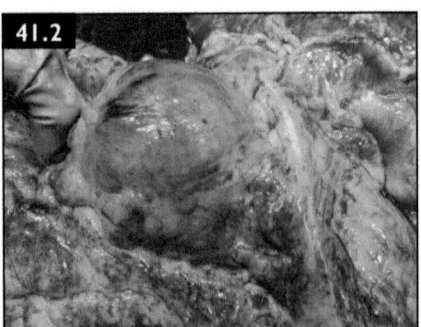

41.2

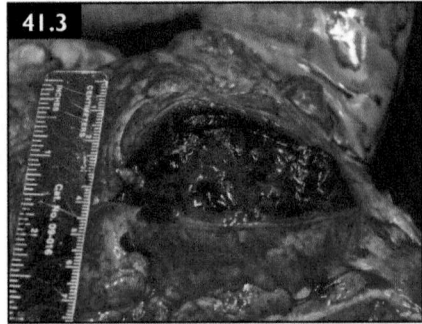

41.3

5 What are the laboratory abnormalities most likely caused by?

6 What would be a reasonable treatment plan?

7 The mare was euthanized and several encapsulated hematomas were found at the base of the heart (**41.2, 41.3**). What is the most likely diagnosis?

CASE 42 A 10-year-old Warmblood mare presented for multiple dermal plaques that had developed over the right masseter muscle following removal of a tick from the same site 3 months earlier (**42.1**). The lesions had expanded in the past 2 weeks prior to presentation but were not painful to palpation. The lesions were believed to be dermal without subcutaneous extension. The remainder of the physical examination was normal. Skin biopsy suggested cutaneous lymphoma (**42.2**). ELISA titer and Western blot for *Borrelia* spp. were very high and strongly positive, respectively.

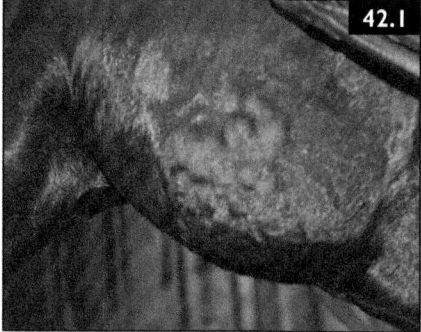

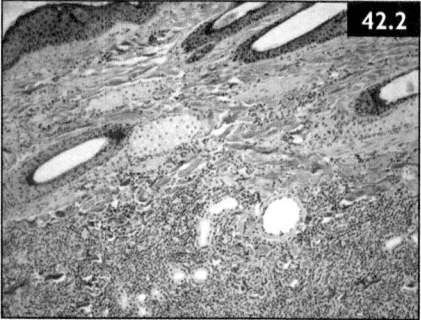

1 What other test could be performed to determine if *Borrelia* might be associated with the disease?

2 What treatment is recommended for Lyme disease?

Re-evaluation of the biopsy determined that the skin lesion was pseudolymphoma, a form of cutaneous Lyme disease also seen in humans. Horses are not known to have the common erythema macula lesions seen in many humans with early Lyme disease. Differentation between pseudolymphoma and cutaneous lymphoma can be difficult. The depth of the lesion in this case was not characteristic of cutaneous lymphoma in horses.

3 How is this different from cutaneous lymphoma?

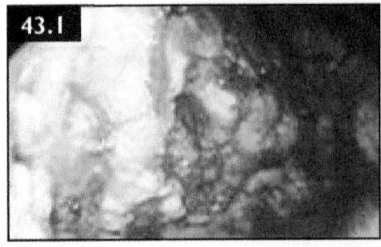

CASE 43 A 22-year-old Arabian mare presented for a 2-month history of weight loss, mild colic after eating, and decreased appetite. She also appeared to have been choked for a short time (<10 minutes), based on the owner having noted head and neck extension and excessive salivation. She had been treated with omeprazole for 4 weeks. Initially there seemed to be some improvement, but the signs then progressed. Routine physical examination was normal and CBC and biochemistries were unremarkable except for a PCV of 23%, low albumin (1.9 g/dl [19 g/l]), increased globulins (4.5 g/dl [45 g/l]), and a serum calcium of 13.9 mg/dl (3.48 mmol/l). Gastroscopy revealed this image (43.1).

1 What would be the next diagnostic procedure(s)?
2 What is the tentative diagnosis based on the history, laboratory findings, and clinical and gastroscopic examination?
3 What other procedures might allow a diagnosis if gastroscopy is not available?

CASE 44 A 6-year-old Warmblood mare was examined for coronary band lesions in all four limbs (44.1). The lesions had been present for 4 months and were thought to be a fungal infection. They had been treated with diluted betadine solution topically and ketoconazole orally. The horse did not appear lame and clinical examination was normal except for the coronary band lesions. The coronary bands were very hyperkeratotic with some hyperemia (44.2).

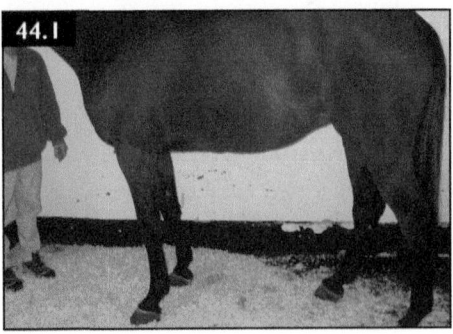

1 What would be the primary differential diagnosis in this case?
2 How could the diagnosis be confirmed?
3 What are the possible treatments?

CASE 45 A 5-year-old Standardbred mare and her 2-day-old colt presented because of milk observed in the foal's nostrils after nursing. Clinical examination of the foal was normal except for a 'gurgling' tracheal sound as the foal nursed and milk appearing at the nose following nursing (**45.1**). Oral examination and endoscopic examination of the pharynx and upper airway did not reveal any abnormalities except that milk was seen in the trachea.

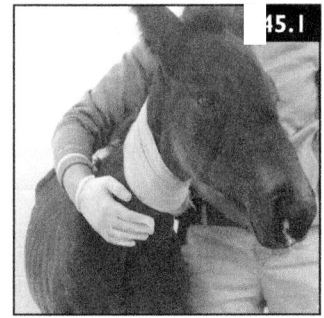

1 What are possible causes for the mild-to-moderate dysphagia in this foal?
2 How should this foal be treated?

While the foal was being treated, the mare (on day 3) developed a fever (103.2°F, [39.6°C]), a swollen left side of the udder (**45.2**), and a positive California mastitis test (CMT) in the milk from the left side (**45.3**).

3 What further treatment and diagnostics should be recommended for the mare?

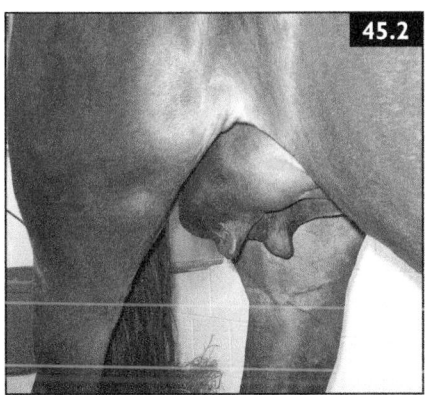
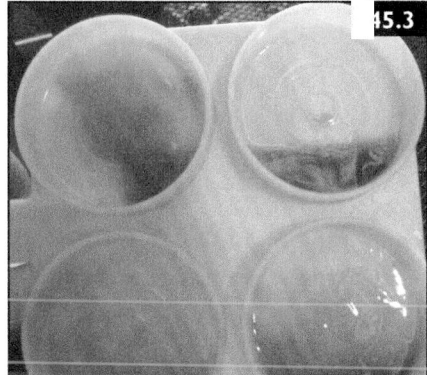

CASE 46

1 List the clinical findings that support a diagnosis of infected ulcerative keratitis (melting corneal ulcer) (**46.1**).

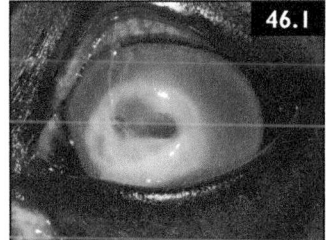

CASE 47 A 13-year-old Thoroughbred mare was examined by the referring veterinarian because of weight loss, lethargy, fever, and conjunctival erythema (**47.1**). Biochemical profile revealed elevation in liver enzymes (GGT, AST, SDH) and both direct and indirect bilirubin concentrations. A diagnosis of cholangiohepatitis was made and the horse was treated with enrofloxacin, pentoxifylline, milk thistle, and

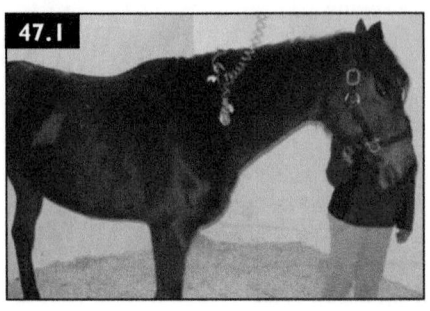

amino acids with only marginal improvement. The horse was then referred for more complete diagnostic testing.

On presentation, the mare was dull, febrile (103.2°F [39.6°C]), icteric, and in poor body condition. Bilateral conjunctivitis was noted. Pertinent laboratory findings included: PCV = 65%; segmented neutrophils = 14,200/µl (14.2 × 10^9/l) with no bands; lymphocytes = 800/µl (0.8 × 10^9/l); fibrinogen = 300 mg/dl (3.0 g/l); potassium = 2.2 mEq/l; BUN = 10 mg/dl (3.57 mmol/l); TP = 4.9 g/dl (49 g/l); globulins = 1.6 g/dl (16 g/l); AST = 2,117 U/l; SDH = 84 U/l; GGT = 19,121 U/l; total bilirubin = 7.8 mg/dl (133.4 µmol/l); direct bilirubin = 1.3 mg/dl (22.2 µmol/l); iron = 384 µg/dl (68.7 µmol/l) (94% saturation); bile acids = 145 µmol/l.

1 What is the explanation for the swollen and red conjunctiva in both eyes?
2 Based on this information, does it seem reasonable that the horse has chronic bacterial cholangitis? Why?
3 What laboratory findings are not consistent with chronic bacterial cholangitis?
4 What further diagnostic test(s) should be performed?
5 Since the PT and PTT were prolonged, should a liver biopsy be avoided?
6 Undoubtedly the horse had hepatitis, which was presumably bacterial in origin, but what predisposing factors should be considered based on the laboratory testing?

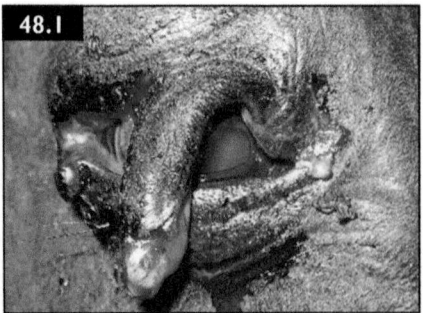

CASE 48 A 13-year-old Arabian gelding presents acutely with ocular trauma (**48.1**).

1 What diagnostics and therapy are indicated?

CASE 49 A 12-year-old Quarter Horse gelding was examined because of a 2-week history of mild dysphagia (some food and water intermittently refluxed out of the nares) and episodes of airway noise with respiratory distress. The owner had noticed some progressive stumbling during riding over the past 6 months. All vital signs were normal on physical examination but the horse seemed dull and depressed at times, especially when not being handled. He swallowed hay normally but some water refluxed from the nostrils when drinking. CN examination revealed normal vision, normal pupillary response to light, no strabismus, normal head carriage, and normal palpebral response.

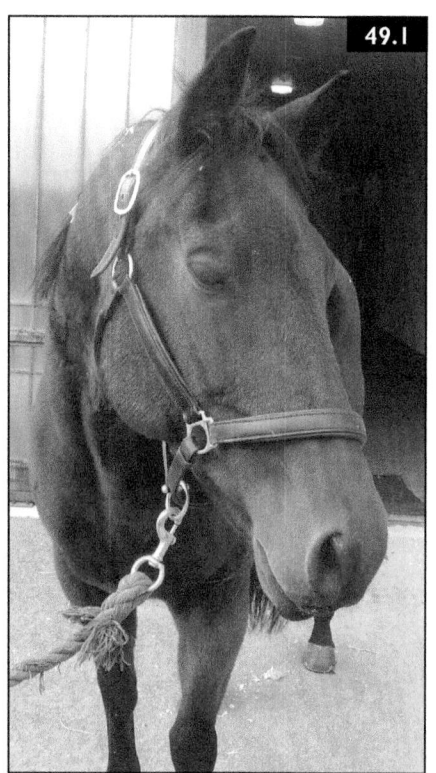

Abnormalities were found on examination of the masseter muscles (**49.1**) and endoscopy of the larynx (**49.2**). The horse notably failed to cough during endoscopy of the larynx and trachea. Symmetrical ataxia (general proprioceptive deficits) was observed in all four limbs. Other procedures at the initial evaluation included endoscopy of the guttural pouch (normal), measurement of serum selenium, vitamin E, and lead (all normal), serum antibody testing for EPM (negative), and serum biochemistry (normal).

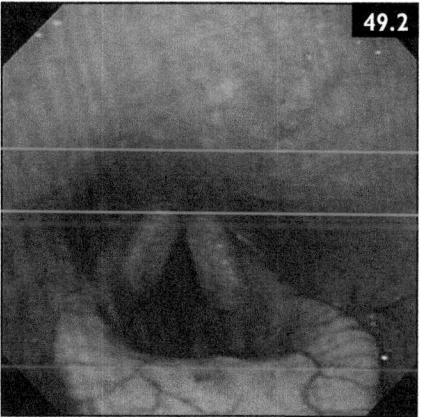

1 Where is the lesion(s) in the nervous system?
2 What would be the next diagnostic procedure?

CASE 50 A 10-month-old Arabian gelding presented with a history of intermittent bleeding from his castration wounds (50.1). The animal had been castrated standing 5 days previously, and had intermittently dripped blood from the wounds since then. In the 24 hours prior to presentation, he had become very dull, lethargic,

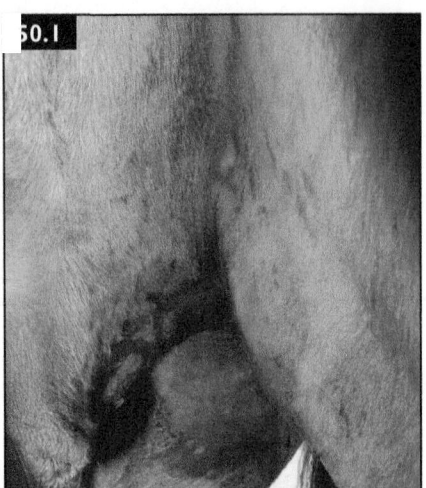

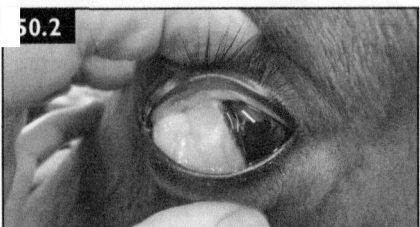

and anorexic, and was lying down most of the time in sternal or lateral recumbency. On examination, the gelding was dull, with a HR of 80 bpm, RR of 24 bpm, and rectal temperature of 98.4°F (36.9 °C). The mucous membranes were pale (50.2), but there were no signs of petechial or ecchymotic hemorrhages. Abnormalities of the initial hematology included a low PCV (13%) and hypoproteinemia (3.4 g/dl [34 g/l]). Prothrombin time (13 sec), activated partial thromboplastin time (42 sec), and platelet count (90,000/μl [90 × 10⁹/l]) were all within normal limits. Abdominal ultrasonography revealed a large volume of peritoneal fluid of mixed echogenicity with a swirling appearance (50.3). Abdominal paracentesis yielded a heavily blood-stained fluid (50.4) with a PCV of 30% and TP of 4.0 g/dl (48 g/l); the fluid did not clot.

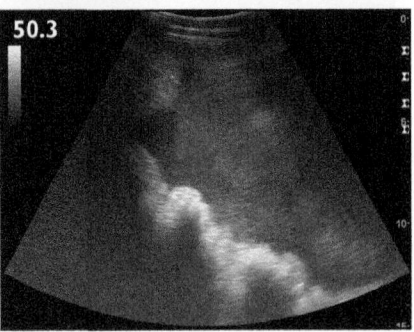

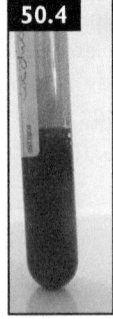

1 What is the most likely cause of the hemoab-domen and peripheral anemia?

2 How should this case be managed?

3 If it is decided to administer a blood transfusion, how much blood should be collected and administered?

CASE 51 A 3-year-old Thoroughbred colt developed an acute swelling just caudal to the point of the hock 1 day after a race. The colt was lame on the leg and had a fever of 104°F (40.0°C). The swelling over the hock became progressively more painful over the first 12 hours and the horse bore little weight on the limb. Although no noticeable 'break' in the skin was identified, there was concern about the rapidly progressive infection that did occur (**51.1**) and surgical drainage of the area was performed under anesthesia. The horse's fever quickly abated after surgery.

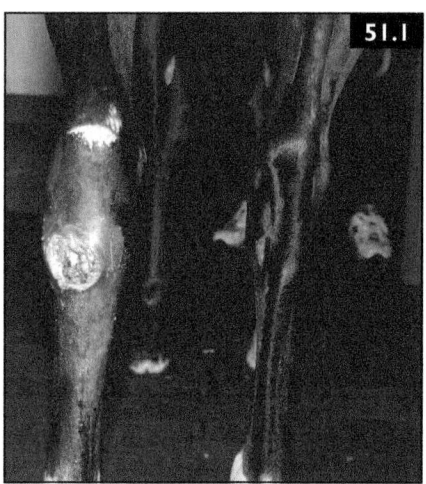

1 What caused this disease?
2 What organism is most commonly associated with this injury, and what antimicrobial should be selected?
3 Is it better to give the selected antibiotic as a large dose once a day or split the dose every 12 hours?
4 What supportive care should the horse receive?
5 What is the prognosis?
6 What other complication might occur in the opposite limb if the horse remains non-weight bearing on the diseased limb for a prolonged period?

CASE 52 A 2-year-old pony filly had a 48-hour history of dullness and inappetence, followed by the development of profuse watery diarrhea in the 12 hours prior to presentation. The owners had noticed a large red structure appearing at the anus (**52.1**) and have requested your advice.

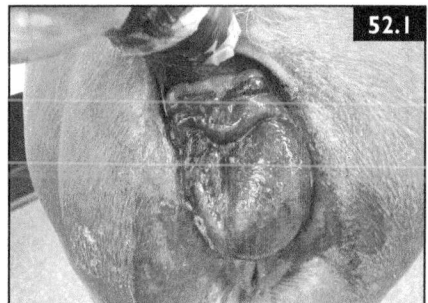

1 What is this structure?
2 What are the different classifications of this disease?
3 What are the common causes of this disease?
4 How should this condition be treated?

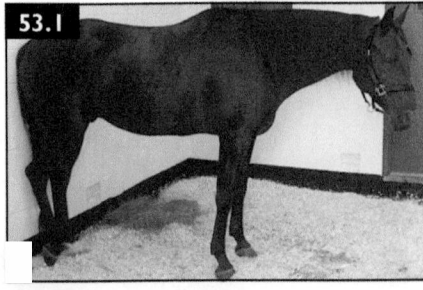

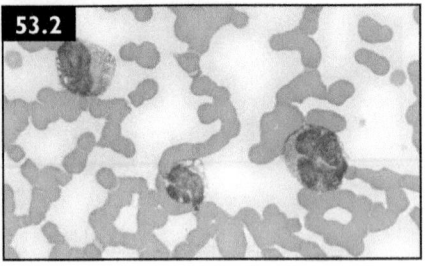

CASE 53 A 5-year-old Quarter Horse (**53.1**) was examined at the referral hospital because of a 12-day history of fever (up to 107°F [41.7°C]) and mild lethargy. The horse was originally treated with oxytetracycline because of an initial suspicion of anaplasmosis, but a lack of response led to a secondary diagnosis of acute viral infection. The fever persisted in spite of no obvious reason identified on thorough clinical examination. Chemistry panel was unremarkable except for high plasma iron of 355 µg/dl (63.5 µmol/l). CBC had the following: PCV = 35%; MCV = 62%; RDW = 24.4%; nucleated RBCs = 8/100 RBCs; WBCs = 8,200/µl (8.2 × 10^9/l); no segmented PMNs; immature/band PMNs = 10/µl (0.1 × 10^9/l); lymphocytes = 5,200/µl (5.2 × 10^9/l); other WBCs that could not be differentiated = 2,800/µl (2.8 × 10^9/l); platelets = 19,000/µl (19.0 × 10^9/l); TP = 8.2 g/dl (82 g/l); fibrinogen = 600 mg/dl (6.0 g/l). A blood smear is shown (**53.2**).

1 What is the interpretation of the blood smear?
2 What additional procedure(s) could be performed to help confirm the diagnosis?

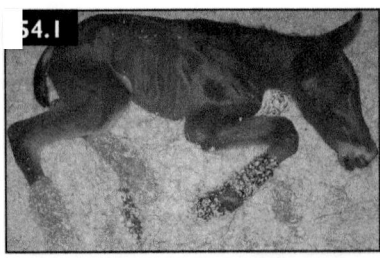

CASE 54 A 12-hour-old crossbred filly was unable to stand (**54.1**). The foal was full-term and the birth was uneventful. The foal had tried to stand on her own but did not appear to be strong enough to get up. The owner had milked the mare and tried to offer it to the foal in a bottle, but the foal appeared unable to swallow. On examination, the foal was dull and weak. When assisted to stand, she was unable to take weight and collapsed. HR was 110 bpm, RR was 32 bpm. Auscultation of the heart and lungs revealed no specific abnormalities. CK and AST concentrations were markedly elevated (66,000 IU/l and 1,422 IU/l, respectively).

1 What is the most likely diagnosis?
2 What is the etiology and pathogenesis of this condition?

CASE 55 A 5-week-old Standardbred colt was examined because of increased respiratory effort, fever (104.5°F [40.3°C]), severely swollen tarsal and stifle joints (**55.1**) with no noticeable lameness, and green discoloration of the eyes (**55.2**).

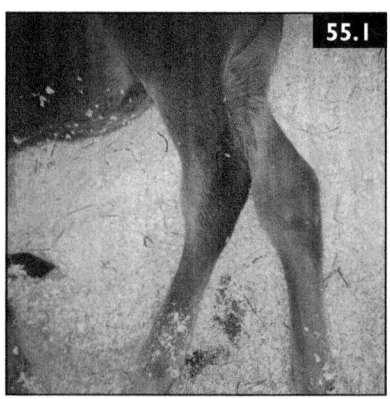 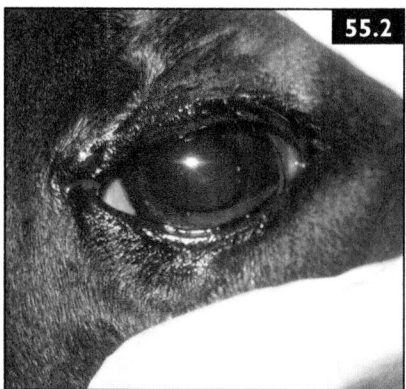

1 Based on this information, what is the most likely etiologic diagnosis?
2 How could the diagnosis be supported and then confirmed?

The foal did not respond to appropriate therapy with clarithromycin and rifampicin. One week after therapy was begun, the foal had not improved and was maintaining rectal temperatures as high as 104.0°F (40.0°C).

3 List some possible explanations for the poor response to seemingly appropriate treatments.

CASE 56 4-year-old Andalusian mare is referred because she has developed multiple skin nodules, mainly over the ventral abdomen, over the previous 12 months. Most of these nodules are wart-like and 1–2 cm in diameter, but one had enlarged (4 cm diameter) and ulcerated (**56.1**).

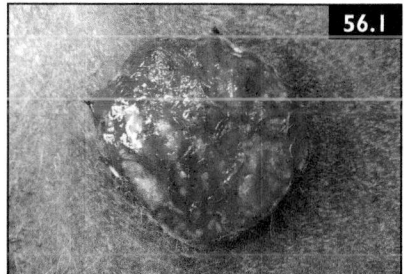

1 What is the most likely diagnosis?
2 What is the etiology of this condition?
3 What treatment options are available?

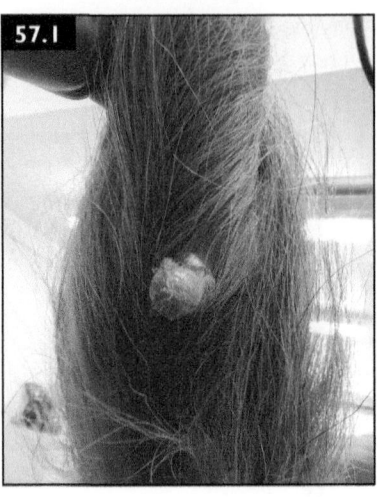

CASE 57 A 14-year-old chestnut Warmblood gelding presented with a mass on the ventral tail (**57.1**) that had grown rapidly over the previous 4 weeks. The mass had the typical appearance of a skin tumor. There were no other identifiable skin masses and the horse appeared to be fit and well, apart from this lesion.

1 What possible tumor types should be considered?
2 What should be the next course of action?
3 Anaplastic malignant melanoma was diagnosed. What further advice or treatment should be recommended?

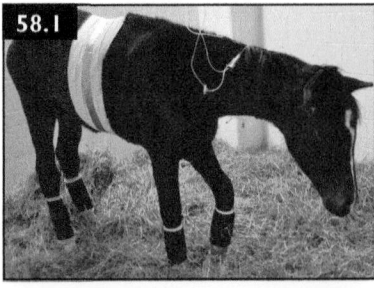

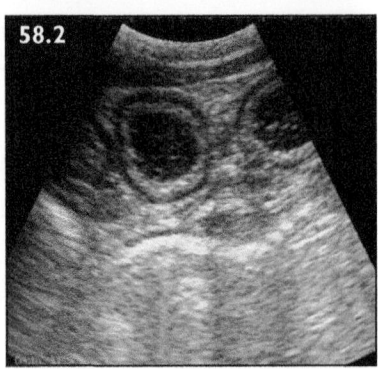

CASE 58 A 6-month-old Thoroughbred filly presented with weight loss of approximately 150 lbs (67.5 kg) over 10 days, diarrhea, and mild pitting edema of her ventral abdomen and distal limbs (**58.1**). Thickened loops of small intestine were seen on ultrasound (**58.2**) and mild pleural roughening was noted in the thorax. No free abdominal fluid was seen and the large intestine, cecum, spleen, liver, and kidneys all appeared normal. On presentation, blood work abnormalities included hypoproteinemia (TS = 2.8 g/dl [28 g/l]), dehydration, hyponatremia, hypochloremia, hypokalemia, and a mild left shift without a neutrophilia.

1 List some differential diagnoses for infectious causes that could be responsible for the diarrhea in this filly.
2 Based on the hypoproteinemia, history, physical examination, and ultrasound findings, what is the most likely diagnosis?
3 How should this infection be treated?

CASE 59 A 3-week-old Paint filly was examined because of mild colic signs, slightly loose manure, and fever of 102.7°F (39.3°C). Clinical examination was relatively normal except for enlargement of the umbilical stump on palpation. Ultrasound examination of the umbilical structures was performed (**59.1**). An ultrasonogram from a normal foal with both umbilical arteries and the urinary bladder identified is shown (**59.2**).

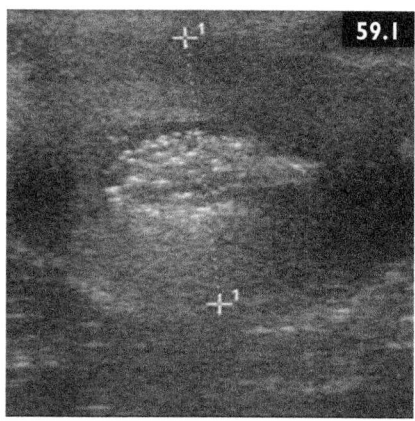

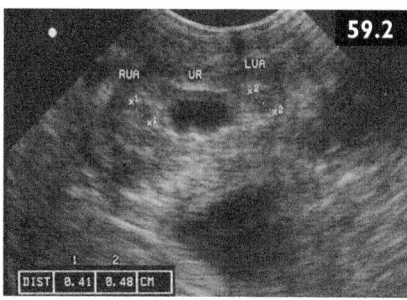

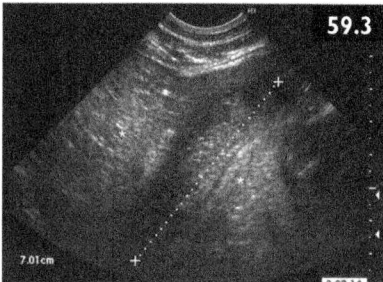

1 What is the most likely diagnosis?
2 What is the recommended treatment for this foal?
3 What is a common and serious complication with this condition in foals?
4 An ultrasound image of the cranioventral abdomen of another foal with fever is shown (**59.3**). What is the mixed echogenic structure coursing within the liver (identified by the broken line)?

CASE 60 A 2-year-old crossbred gelding presented with protrusion and swelling of the penis (**60.1**).

1 What is this condition?
2 What are the possible causes of this condition?
3 How should this condition be treated?

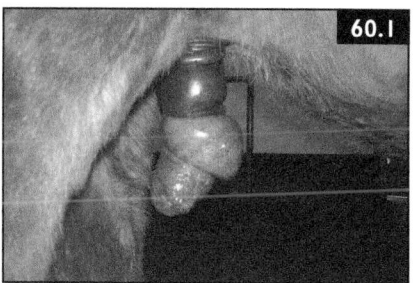

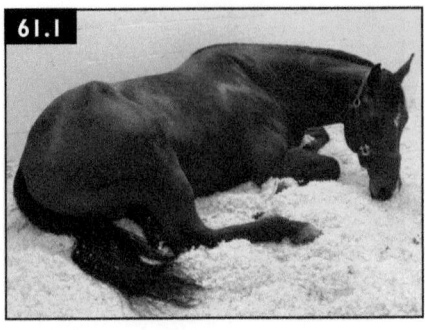

CASE 61 A 4-year-old Quarter Horse gelding was examined because of a 7-day history of mild colic that was responsive to 500 mg flunixin meglumine daily. The horse was passing normal manure and had a fair appetite. There was no history of fever and no abdominal distension had been noted. The horse was purchased 3 months prior from a desert area in Nevada. Clinical examination revealed normal temperature, pulse, respiration, and auscultation of both the thorax and abdomen with good GI borborygmi. The only unusual finding on auscultation of the most cranial ventral abdomen was a 'gushing' sound not heard elsewhere in the abdomen. Abdominal palpation per rectum was normal, with normal-appearing feces. Following examination, the horse lay down and looked back at its flank (**61.1**).

1 What type of disease process (i.e. strangulation, complete obstruction, partial obstruction) is causing the colic?
2 What initial diagnostic tests should be run?
3 What other diagnostic test(s) could be performed?
4 What is the interpretation of the diagnostics?
5 What is the preferred treatment for the type of colic diagnosed?
6 What are the most common risk factors for colic in horses in sandy areas?
7 Although ultrasound examination was normal in this case, what sonographic abnormality might be seen that would be supportive of the colic type diagnosed?

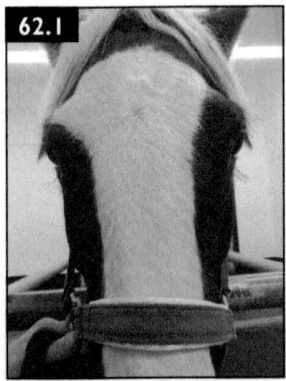

CASE 62 The frontal view of the face of a 4-year-old gelding suspected to be affected by subacute equine grass sickness is shown (**62.1**).

1 What clinical feature can be seen that would be compatible with this diagnosis?
2 What other common clinical signs would be expected in subacute grass sickness?
3 What abnormalities would be detected on rectal examination?

CASE 63 A 14-year-old Standardbred stallion was being treated for a corneal ulcer in the left eye with topical antibiotics and atropine and IM administration of flunixin meglumine in the neck. Seventy-two hours after the start of treatment, the neck was swollen at the injection site and the horse had a temperature of 104°F (40.0°C). The horse was then treated with IM procaine penicillin and IV phenylbutazone. The next day, the neck was more swollen and painful. On presentation, rectal temperature was 101°F (38.3°C), HR was 80 bpm, RR was 40 bpm, and the left side of the neck was very swollen and painful (**63.1**). PCV was 42%, TS were 5.5 g/dl (55 g/l), and creatinine was 2.2 mg/dl (167 µmol/l).

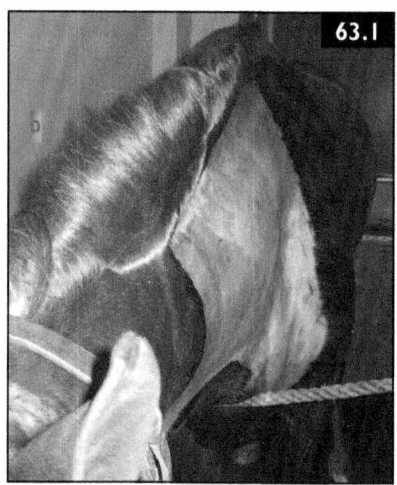

1 What procedures should be performed to help diagnose the cause of the swollen neck?
2 What further treatment should be provided following diagnostics?
3 Should the standard dose of 22,000 IU/kg K^+ penicillin IV q6h be used in this case or should a larger dose be used in light of the life-threatening possibility of this disease?

CASE 64 A 12-year-old Thoroughbred cross gelding suffered a small wound to the left nostril 4 weeks ago. Initially the wound started to heal by secondary intention, but subsequently developed into a small ulcerated nodule that intermittently discharges small volumes of purulent material (**64.1**). On palpation the nodule was firm and appeared painful.

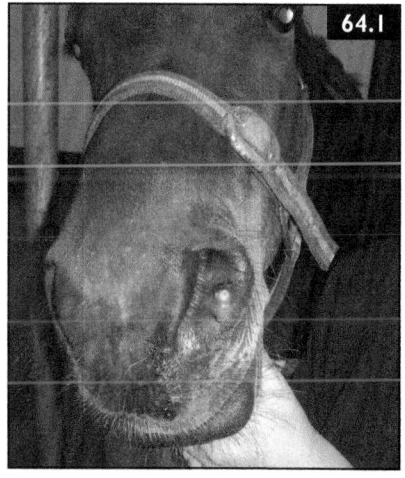

1 What is the tentative diagnosis and differential diagnoses?
2 What is the pathogenesis of this condition?
3 How should this condition be treated?

CASE 65 A 12-year-old crossbred mare presented 48 hours after parturition. The birth was unremarkable, and the foal had been normal. The mare had been becoming progressively depressed since the birth and had stopped passing feces. In the 24 hours prior to presentation, she started showing colic signs that were responsive to analgesic drugs until the last few hours. On examination the mare was severely depressed (**65.1**). She had severely congested mucous membranes with a toxic ring around the dental margins (**65.2**). The vaginal mucosa was bruised and congested (**65.3**). HR was 120 bpm and RR 24 bpm. Auscultation of the abdomen revealed absent GI borborygmi and abdominal ultrasound revealed excessive peritoneal fluid. Abdominal paracentesis yielded grossly dark red/brown fluid (**65.4**); cytologic examination revealed RBCs, neutrophils, abnormal particulates, and protozoa.

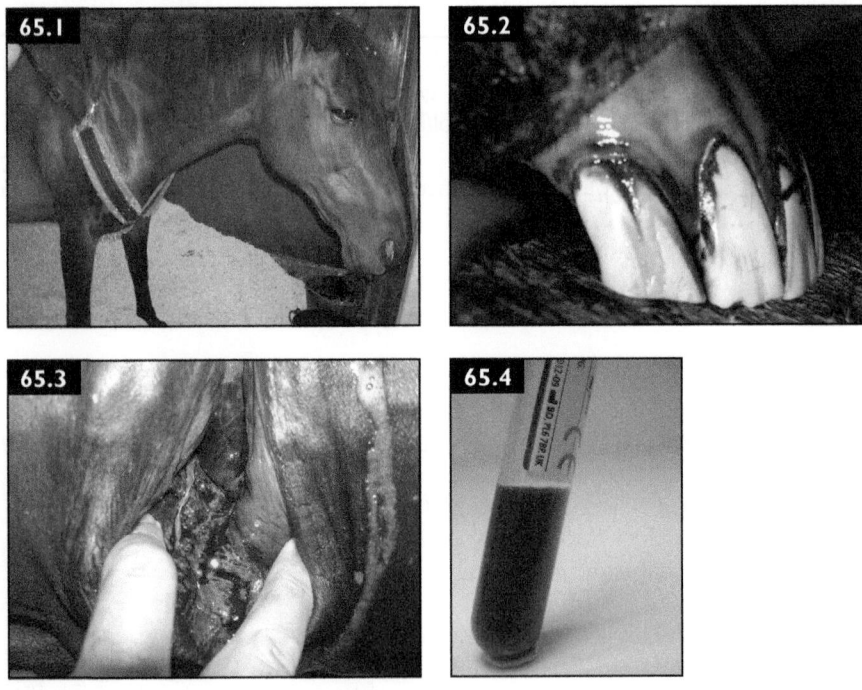

1 What is the cause of this horse's severe shock?
2 What possible underlying causes are suspected?

CASE 66 A 7-year-old Thoroughbred stallion was examined for intermittent colic of 4 years' duration (**66.1**). Colic episodes were generally mild but in the past 2 months had become more frequent in occurrence. Three years earlier, a course of steroid treatment had been attempted because of recurrent colic and loose manure, but a response to the treatment could not be documented. On examination, the stallion was under-

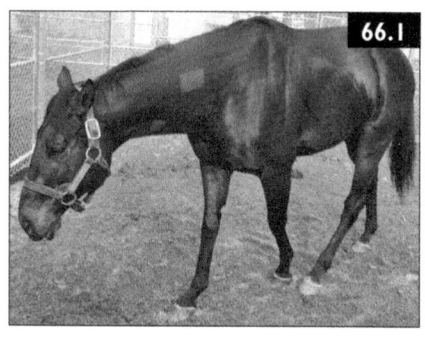

weight but had normal temperature, a mildly elevated HR (48–52 bpm), and normal respiration. Intestinal sounds were within normal limits and the manure was slightly more liquid than normal. Abdominal palpation per rectum revealed notably thickened small intestine and passage of a nasogastric tube caused spontaneous reflux of 8 liters of gastric contents. Laboratory findings were normal except for a mildly decreased TP (5.2 g/dl [52 g/l]) and mildly increased plasma lactate (1.75 mmol/l). Ultrasound examination of the abdomen revealed diffuse markedly thickened small intestine (**66.2**, duodenum; **66.3**, right flank; **66.4**, left flank). Peritoneal fluid analysis, including measurement of lactate, was normal.

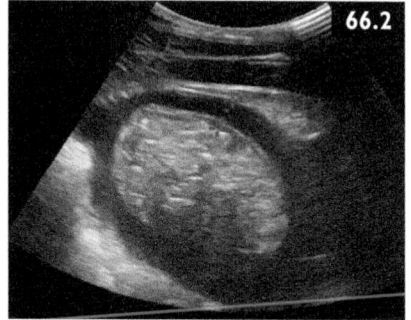

1 Interpret the abdominal ultrasound.
2 What diagnostics should be performed next?
3 How should this horse be treated?

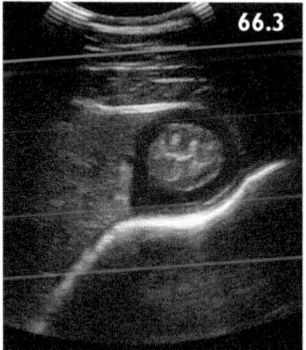

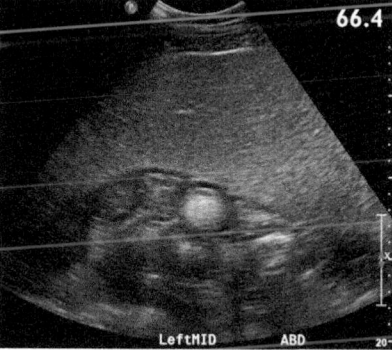

CASE 67 A 3-year-old Thoroughbred stallion was found in his stall acutely ataxic in all four limbs. His left carpus had been injected with hyaluronic acid 1 week earlier to treat the carpal injury that had ended his racing career. The ataxia was graded as 4/5 and the colt was administered IV dexamethasone, DMSO, and thiamine and PO vitamin E. The next day the colt was slightly improved and referred for further examination. Clinical examination revealed a bright and alert colt with 3 to 4/5 ataxia, proprioceptive deficits in all four limbs,

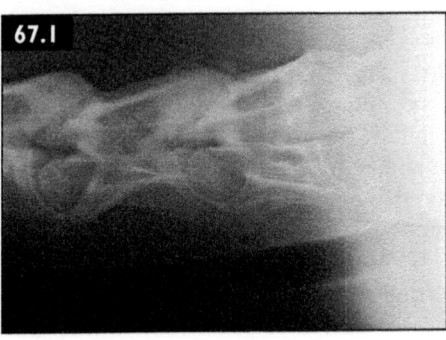

and some evidence of cervical pain on turning in either direction. The stallion would commonly be seen with his limbs placed in awkward positions. The forelimbs would 'buckle' when the colt lowered his head to eat. All CN functions appeared normal and the colt was urinating and defecating normally. A lateral cervical radiograph was obtained (**67.1**).

1 What would be the most likely neuroanatomic location for a lesion?
2 What is the interpretation of the radiograph?
3 What is the most likely diagnosis based on the radiograph and the clinical picture?

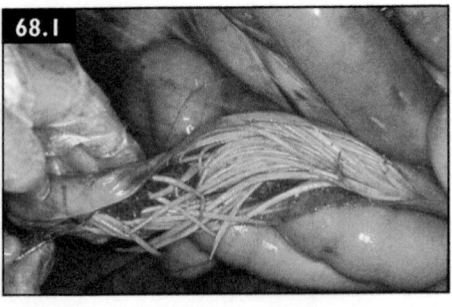

CASE 68 A postmortem examination is carried out on a 9-month-old pony colt that was euthanized following a severe episode of colic (**68.1**). The colt came from a pony stud farm that has approximately 20 foals grazing with their dams on permanent horse pastures. The owners lost another foal with colic approximately 3 weeks earlier (but no postmortem examination was undertaken in that case). All foals on the farm were routinely administered anthelmintic drugs (ivermectin or moxidectin) every 3 months.

1 What is the diagnosis?
2 How is the occurrence of this disease explained in foals that are routinely administered anthelmintic drugs?
3 Describe the life cycle of this parasite.
4 What other clinical signs can infestation with this parasite cause?

CASE 69 A 13-year-old Warmblood gelding presented with stubborn hives that had been occurring for over 1 year. The hives were raised without hair loss, not weepy or crusty, and there was associated pruritus when they were present (**69.1**). The hives would develop at many times during the year (spring, summer, fall, and deep winter) and a pattern could not be identified. The horse had been on multiple types of feed without any positive results. He had been treated with two different antihistamines: hydroxazine and cetirizine. The hydroxazine seemed to have been efficacious, but the hives continued to develop occasionally. The horse had also been treated with prednisolone with some response but the owners were concerned about laminitis; as soon as the prednisolone was discontinued the hives returned. CBC and chemistry values were normal and the horse was otherwise clinically normal. No other horses in the barn were affected.

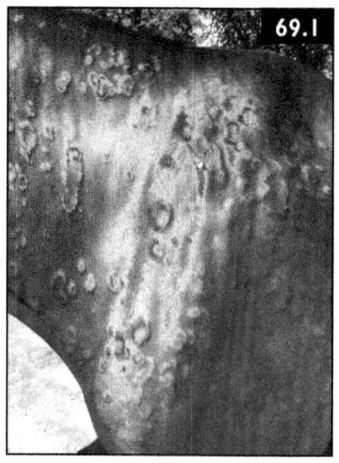

1 Based on the history and appearance of the lesions, what is the most likely diagnosis?
2 What procedure(s) should be performed to help confirm the diagnosis?
3 What treatments should be recommended for this horse?

CASE 70 A 12-year-old Quarter Horse was examined because of acute swelling of the hindlimb following a rabies vaccination. A diagnosis of *Clostridium perfringens* myositis was made. A surgical incision into the infected site was made to allow drainage, and IV penicillin and flunixin meglumine therapy was initiated. The horse appeared to be improving until day 4, at which time he became anorexic and developed a fever (104°F [44.0°C]) and watery manure.

1 Based on the history, what are the additional differentials?
2 What diagnostics should be performed to determine the cause?
3 What treatment should be recommended for the colitis?
4 What are the most common life-threatening complications associated with any inflammatory colitis in the adult horse?
5 What is the single most important treatment that should be incorporated to help prevent this complication?
6 The horse stabilized but continued to have very watery diarrhea and a poor appetite, and the plasma protein decreased (but remained low normal). Is there any other treatment that could be highly effective in treating the diarrhea?

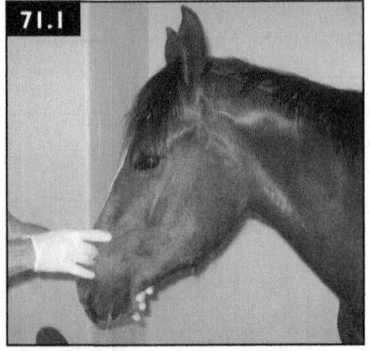

CASE 71 A 10-year-old Thoroughbred mare, used as a polo horse, had a swelling on the left mandible for 1 month that then developed a draining tract (**71.1, 71.2**). The horse appeared otherwise healthy, maintained a good appetite, and had a BCS of 5/9. CBC showed no abnormalities.

1 What are the differential diagnoses in this case?
2 What diagnostic procedures should be considered?
3 What is the radiographic diagnosis (**71.3**)?

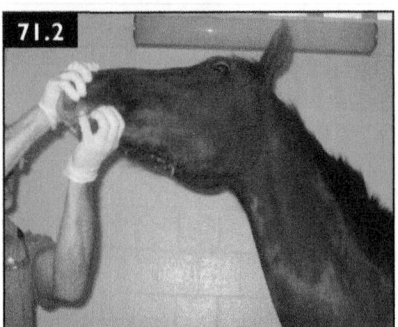

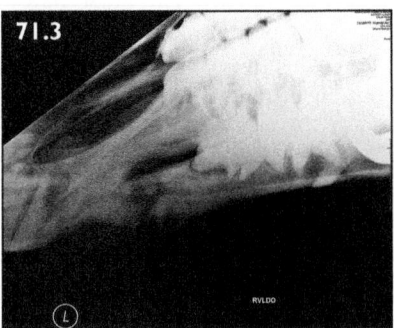

CASE 72 A 10-year-old Miniature Horse mare presented for evaluation of sudden onset of stiffness and extreme distress (**72.1**). She was overconditioned (BCS 6/9) with moderate abdominal distension. After questioning, the owners reported that she could be pregnant due to the escape of a stallion on their property 11 months earlier. Evaluation revealed trismus, generalized muscle tremors, a spastic hypermetric (goose-stepping) gait involving the hindlimbs, and extreme respiratory distress. The mare collapsed once in the stall and was more relaxed when recumbent.

1 What are the three most likely differential diagnoses?
2 Given the high index of suspicion, what would be the first treatment?
3 What else should be taken into consideration with this case, given the signalment and history?

CASE 73 A 17-year-old Appaloosa gelding (**73.1**) was examined for a 4-week history of lethargy and intermittent low-grade abdominal pain. Physical examination was unremarkable. Routine hematology and serum biochemistry revealed a mild neutrophilia (12,000 cells/µl [12 × 10⁹/l]) and hyperfibrinogenemia (980 mg/dl; 9.8 g/l). Abdominal paracentesis yielded slightly turbid peritoneal fluid with a total nucleated cell count of 44,000 cells/µl (44 × 10⁹/l) (80% PMNs, 20% mesothelial cells) and TP concentration of 3.5 g/dl (35 g/l). Abdominal palpation per rectum revealed an ill-defined, painful mass in the mid-abdomen. Transabdominal ultrasound was performed (**73.2**, image from the right cranioventral abdomen).

1 What is the diagnosis?
2 What is the most likely cause?
3 What other diagnostics should be recommended?
4 What treatment options should be considered?

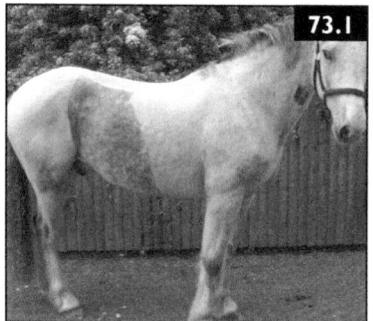
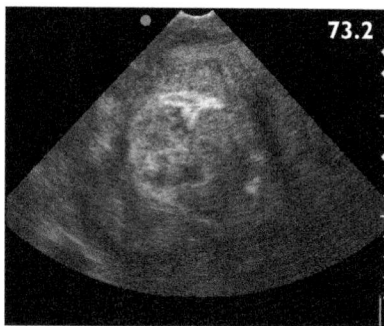

CASE 74 A 2-day-old 140-pound (63 kg) Thoroughbred colt presented for evaluation of weakness and increased respiratory effort (**74.1**). There was no history of dystocia and the foal had good transfer of passive immunity with an IgG >800 mg/dl (>8 g/l). Clinical examination revealed tachycardia (HR = 130 bpm), decreased lung sounds bilaterally, pale mucous membranes, and a large soft/painful swelling over the left thorax.

1 What are the most likely differential diagnoses?
2 What diagnostic tests should be chosen first?
3 What treatment options should be discussed with the owner?

CASE 75 A 13-year-old Quarter Horse mare used as a hunter-jumper was examined for severe lameness in her right hindlimb, localized to the stifle. The mare had been diagnosed with a stifle abnormality and the stifle was injected with corticosteroids 9 days prior to the development of the lameness. The lameness was moderately responsive to oral phenylbutazone and topical antiphlogistine therapy,

but after 3 days the mare was referred for follow-up diagnostics (**75.1**). The mare's rectal temperature was 101.5°F (38.3.6°C) and the HR was 48 bpm. On palpation of the limb, the stifle joint was painful and swollen, and crepitus was present. The swelling was believed to be due to joint effusion.

Aspiration of joint fluid after sterile preparation of the femoropatellar joint was performed using an 18 gauge, 1½ inch needle and twitch restraint. Fluid analysis showed: color = light yellow; turbidity = slightly cloudy; TP = 5.3 g/dl (53 g/l); nucleated cells = 21,600/µl (21.6 × 10^9/l); RBCs = 3,700/µl (3.7 × 10^{12}/l); estimated viscosity = decreased. Direct and sediment smears were examined, showing markedly increased cellularity consisting mostly of leukocytes with low numbers of RBCs on a pink granular background. The leukocytes were composed of 86% non-degenerate neutrophils, 8% macrophages, and 6% small lymphocytes. Despite careful searching on both Wright's- and Gram-stained slides, no infectious agents were identified.

1 What diagnostic procedure(s) should be performed next?
2 What is the interpretation of the synovial fluid analysis and cytology?
3 What else should be done with the synovial fluid?
4 If there could be a reasonable chance of sepsis in this case, and the owner wants the horse to return to performance soundness, should arthroscopic surgery and flushing the joint, in addition to local and systemic antibiotics, be recommended?

Culture results found on enrichment 3 days later revealed a pure growth of a yeast species (*Cyberlindnera* sp., which may include *Candida* sp.). The joint was then aspirated a second time for culture and a pure growth of the same organism was found.

5 What are the options for therapy in light of the culture results?

CASE 76 A 13-year-old Clydesdale gelding was examined because of a 1-year history of waxing and waning lameness of the hindlimbs that prevented him from competing as a member of a driving team. On clinical examination, multiple crusts were noted on the skin of the distal one-third of all four limbs (**76.1, 76.2**). The limbs at these sites were painful to palpation with the left hind being the most severely affected. That limb had markedly thickened skin from the pastern to the coronary band. The only abnormalities noted on CBC and biochemistry profile were an elevated fibrinogen (400 mg/dl [4 g/l]), low iron (50 µg/dl [8 µmol/l]), and increased globulins (5.5 g/dl [55 g/l]).

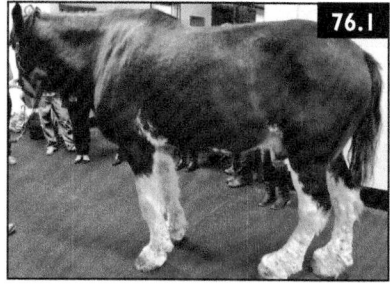

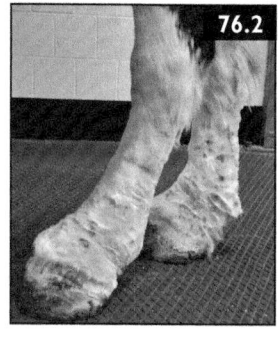

1 What is the most likely explanation for the high fibrinogen, low serum iron, and elevated serum globulins?
2 What is the most likely diagnosis in this case?
3 How can this condition be treated?

CASE 77 A 12-year-old Quarter Horse gelding (**77.1**) presented for evaluation of weight loss, lethargy, variable appetite, and cutaneous nodules/ulcers of several months' duration. Clinical evaluation revealed a BCS of 2.5/9 with thickening/ hyperemia and small, firm, intradermal nodules of the muzzle and lips. Gingival hyperplasia and ulceration of the mucocutaneous junctions were apparent. There were diffuse, firm intradermal nodules present on the skin of the neck, back, and flanks. The submandibular lymph nodes were enlarged.

1 What five diagnostic tests should be performed next?
2 List the most likely differential diagnoses.
3 What advice should the owner be given regarding the likely response to treatment?

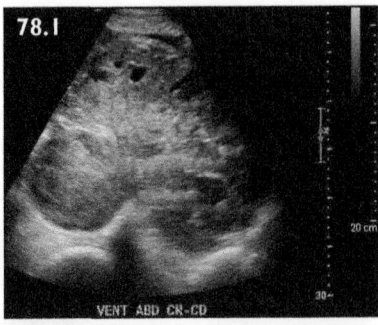

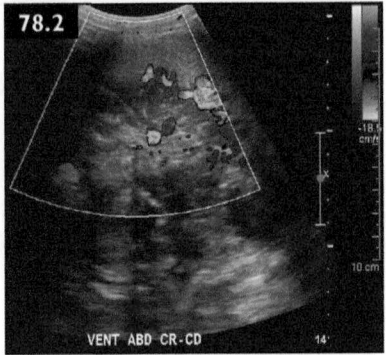

CASE 78 An 18-year-old Thoroughbred mare used for pleasure riding was examined because of a 6-day history of colic that had been responsive to treatment with flunixin meglumine. The mare had continued to eat and pass some manure, although both appetite and manure quantity were decreased from normal. The owner reported that the mare had difficulty maintaining a normal body condition for the previous 2 years in spite of dental prophylaxis, deworming, and good nutrition. CBC, fibrinogen, and biochemistry were normal. On presentation, the mare was bright and alert with a BCS of 3/9. Physical examination was otherwise normal, abdominal palpation per rectum was unremarkable, PCV was 40%, and TS was 6.6 g/dl (66 g/l). Ultrasound of the ventral abdomen revealed the following (**78.1**). The mass appeared to move with the colon and was believed to be attached to the colon wall. Doppler examination (**78.2**) showed that many of the large hypoechoic regions seen on ultrasound were blood vessels.

1 List some differentials for the chronic colic and weight loss in the face of an otherwise normal physical examination and normal laboratory testing?
2 What other procedures should be performed to arrive at a diagnosis?

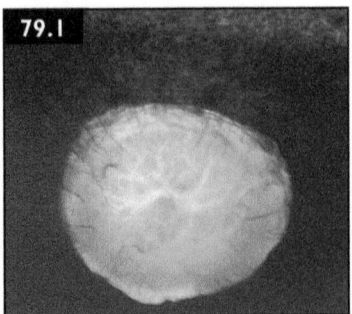

CASE 79
1 Describe the fundic examination abnormalities in this horse (**79.1**).

CASE 80 A 20-year-old Thoroughbred mare delivered a large colt (128 lb [58 kg]) with veterinary assistance (chains on the foal's forelimbs and vigorous pulling by two people). The mare was administered 500 mg of flunixin meglumine within 1 hour of foaling and passing the entire placenta. Aside from moderate bruising in the vagina, the mare appeared clinically normal. Twenty-four hours after foaling, she was noted to be depressed, had a fever of 103.6°F (39.8°C), and was dribbling urine (**80.1**).

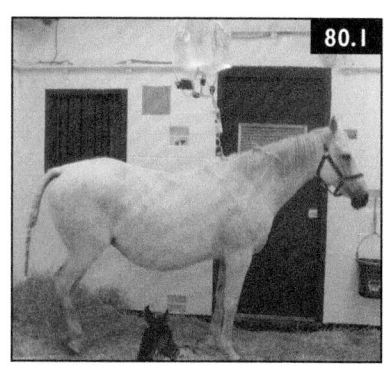

1 What would be the likely differentials for these clinical signs?

Transabdominal ultrasound was performed (**80.2**).

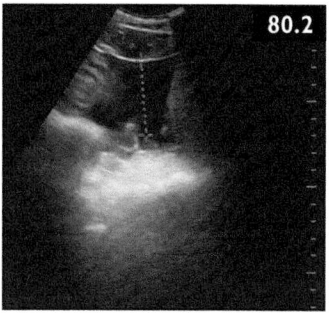

2 What other diagnostic procedures could be performed?
3 What would the peritoneal fluid analysis in a normal post-foaling mare be expected to show?
4 How should this mare be treated?

CASE 81 A 6-year-old Clydesdale broodmare presented with a 2-week history of fever, poor appetite, and lethargy (**81.1**). Severe ventral and pectoral edema had developed in the 48 hours prior to presentation. The mare had a history of early embryonic loss (EEL) and premature delivery of her last pregnancy. Examination revealed muffled/quiet heart sounds with a HR of 120 bpm. Jugular vein distension and jugular pulses were present bilaterally. Severe pectoral and ventral edema was present.

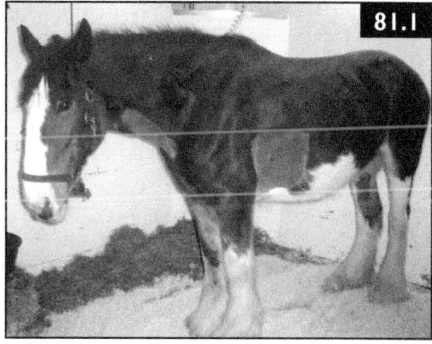

1 What other questions should the owner be asked?
2 What diagnostic tests should be performed initially?
3 What are the most likely pathogens associated with this problem?

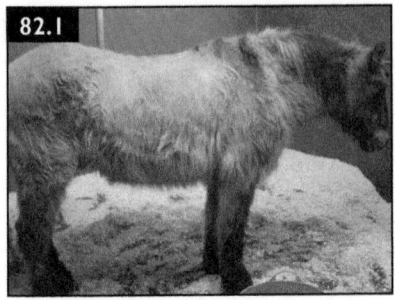

CASE 82 The owner of a 20-year-old pony gelding sought advice regarding the development of an abnormal haircoat that had become apparent over the last 2 years (82.1). The pony was retired and appeared well in other respects. The owner had read on the internet about Cushing's disease and was concerned that the pony could be affected by this disease and that it might predispose him to other problems.

1 What is equine Cushing's disease, and what clinical signs are commonly associated with it?
2 How can the diagnosis be confirmed?
3 Should this pony be treated, and if so, how?

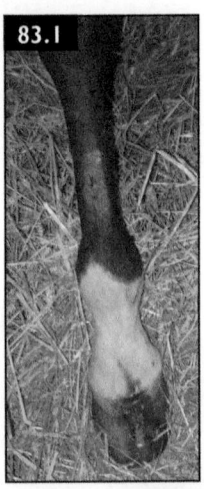

CASE 83 An 8-year-old Thoroughbred stallion had a progressive lameness in the left hindlimb over a 5-week period. The lameness was considered to be coming from the foot based on response to hoof testers and an abaxial (basisesamoid) nerve block. The foot was soaked and the sole explored to try (unsuccessfully) to find an abscess. A small lesion had erupted at the coronary band but there was no improvement in the lameness (83.1). On referral, the horse was bright and alert with normal vital parameters and 4/5 lameness in the left hindlimb. The left rear hoof appeared normal, had normal digital pulsation, no difference in temperature on palpation compared with the other feet, and was painful on application of hoof testers to the sole over the toe region. A radiograph of the foot was obtained (83.2).

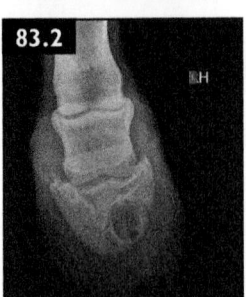

1 Interpret the radiograph.
2 What is the most likely diagnosis?
3 What would be the preferred treatment?
4 Because of the severity of the lameness in the left hindlimb, what would be a major concern for the right hindlimb?
5 What could be done to help prevent this condition?

CASE 84 A 13-year-old Irish Sport Horse gelding presented with a 2-day history of sudden onset dyspnea. The horse had been in full work (eventing) until the onset of dyspnea. On examination the horse was depressed, tachypneic (RR 40 bpm), tachycardic (HR 80 bpm) and pyrexic (temperature 102.6°F [39.2 °C]). Mucous membranes were pale and slightly cyanotic. Auscultation of the chest revealed dullness ventrally and increased lung sounds with mild wheezing dorsally (bilaterally). Thoracic ultrasonography (**84.1**) and radiography (**84.2**) showed a bilateral pleural effusion. A chest drain on the right side yielded heavily blood-stained fluid (**84.3**) with a nucleated cell count of 1,100/µl (1.1 × 10⁹/l), PCV of 2%, and TP concentration

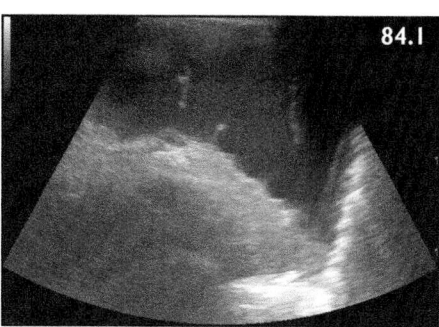

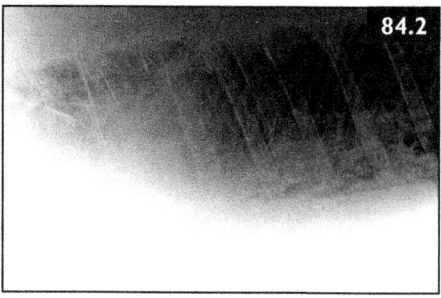

of 2.0 g/dl (20 g/l). Cytologic examination of the pleural fluid showed numerous RBCs and moderate numbers of reactive mesothelial cells. Hematology and biochemistry revealed a PCV of 49%, lymphopenia (1,000 cells/µl [1.0 × 10⁹/l]), hyperfibrinogenemia (1,200 mg/dl [12 g/l]), elevated SAA (3,680 µg/ml), hypoalbuminemia (2.8 g/dl [28 g/l]), and hypokalemia (serum K 2.5 mmol/l).

1 What is the most likely diagnosis?
2 How should this horse be further evaluated?

CASE 85 A 16-year-old mixed breed gelding presents with a 1-month history of corneal ulceration that was initially treated with topical antibiotic/steroid medications; the cornea is now fluorescein negative (**85.1**).

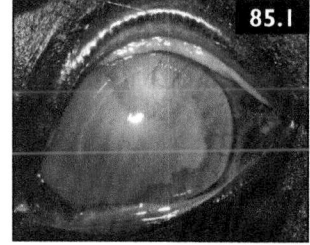

1 What is the diagnosis?

CASE 86 A 4-month-old Quarter Horse colt was examined initially because of a mandibular abscess from which a heavy growth of *Actinobacillus* sp. was grown on culture. The abscess was drained and the colt returned to normal. Two months later the colt was febrile and had a nasal discharge that responded to 2 weeks of trimethoprim sulfa. A CBC/fibrinogen performed at the end of the treatment was normal.

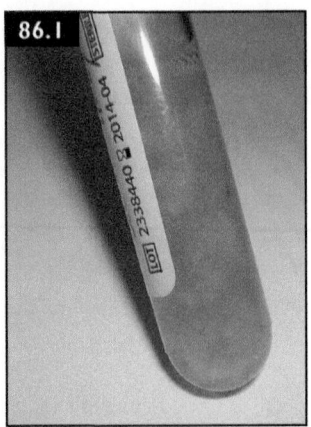

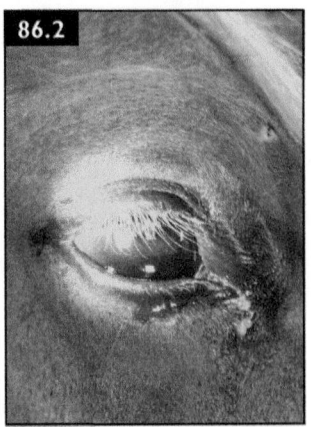

One month later, the colt experienced another episode of nasal discharge, cough, and fever. The colt was treated for pneumonia for 1 week with procaine penicillin IM followed by 30 days of doxycycline PO and appeared to have a clinical response. A transtracheal wash was performed during one of the episodes of nasal discharge and a purulent exudate was recovered (**86.1**), which had a heavy growth of *Pasteurella* spp. and *Streptococcus zooepidemicus* on culture. CBC at the end of the doxycycline treatment was normal. One month later the nasal discharge and fever recurred and the horse developed acute hyphema in both eyes, which was so severe that he did not appear visual (**86.2**). On examination there was cloudiness of the right cornea, but the cornea did not take up fluorescein stain. The horse did not appear to have ocular pain. The eyelids were subjectively believed to be thicker than normal. The horse was treated topically with an ophthalmic preparation of dexamethasone and atropine and with doxycycline PO. Treatment was continued for 16 days with only mild improvement in the hyphema, but the fever and respiratory signs resolved. The CBC/fibrinogen and platelet count were again normal.

There were two clinical issues in this case: (1) recurrent respiratory infections starting at 6 months of age; (2) more recent thickening of the eyelids and acute onset of persistent bilateral hyphema.

1 List some differentials for the recurrent bacterial respiratory disease and fever that responded to antibiotics but recurred.
2 What diagnostic test or procedure should be performed next?
3 List some differentials for the bilateral hyphema and mildly thickened eyelids.

CASE 87 A 1-day-old Warmblood colt presented because of an inability to stand on his own. With help, the colt could stand enough to nurse. The mare had a retained placenta for 10 hours, and when it was passed it was noted to be discolored and weighed 12% of the foals body weight (11% or less is normal). The foal was depressed, HR was 140 bpm, and RR was 32 bpm. Auscultation of the chest was normal except for a systolic murmur heard on the left, which was believed to be due to a patent ductus arteriosus. Rectal temperature was 100.2°F (37.9°C) and the peripheral pulses felt weak; mean blood pressure was 70 mmHg. Serum IgG was 400–800 mg/dl (4.0–8.0 g/l) on the point of care SNAP test; normal values for this test are considered to be >800 mg/dl (8 g/l). Point of care testing found an oxygen saturation of 99%, PCV 46%, and TS 6.7 g/dl (67 g/l), the latter two suggesting dehydration. Results of the I-STAT revealed hyponatremia (123 mEq/l),

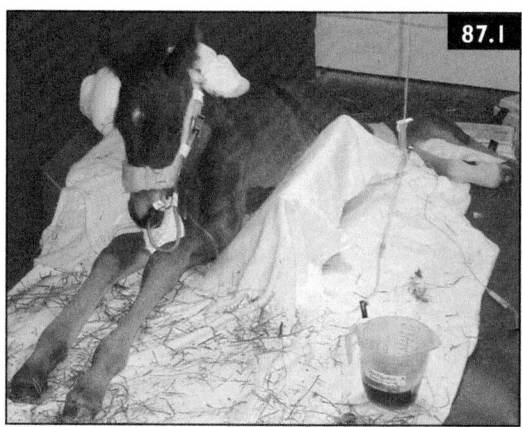

hypochloremia (92 mEq/l), hyperkalemia (7.9 mEq/l), lactate l6.07 mmol/l, BUN 36 mg/dl (12.85 mmol/l), creatinine 9.1 mg/dl (694 mmol/l), and glucose 233 mg/dl (12.9 mmol/l). After a catheter was placed in the jugular vein, 1 liter of 0.9% NaCl was administered and a urinary catheter was placed to monitor urine output (87.1). The urine was grossly discolored.

1 What is the most likely cause of the increased creatinine in this colt?
2 The dark colored urine was occult blood positive on urine dipstick examination. What are the possible causes for this finding?
3 Do these urinalysis and laboratory findings help explain the electrolyte abnormalities and elevated creatinine?
4 The mare and foal are from the northeastern US, so what would be a suspected nutritional cause for the severe myopathy in this case?

CASE 88

1 What are the most common causes of entropion in foals?
2 What treatment options are available?

CASE 89 A 20-year-old Connemara mare presented with a history of sudden onset of facial swelling, protrusion of the tongue, dysphagia, and drooling saliva. The mare had previously been diagnosed with pituitary pars intermedia dysfunction (equine Cushing's disease) and had been receiving pergolide treatment for approximately 1 year. Procaine penicillin and flunixin meglumine had been administered 24 hours prior to examination. At initial presentation, the mare was severely dyspneic with stridor and stertor; a temporary tracheotomy was immediately performed, which completely relieved the dyspnea. The mare had swelling of the head, worse on the right side; the tongue was swollen and partially protruded from the mouth (**89.1**). Oral examination, undertaken following sedation, revealed diffuse swelling of the tongue with linear ulceration of the ventral surface on the right side; no foreign bodies were identified. Radiographs of the head were obtained (**89.2, 89.3**). The mare was treated with antibiotics (penicillin and gentamicin IV), flunixin meglumine, and dexamethasone. The following day, the head swelling reduced and a more detailed oral examination was performed. This revealed necrotic tissue surrounding the linear ulceration on the ventral tongue (**89.4**). The mare was treated with a 5-day course of IV antibiotics and then discharged home with oral doxycycline. On re-examination 1 week later, purulent discharge was draining from the right side of the face. The head was re-radiographed (**89.5**).

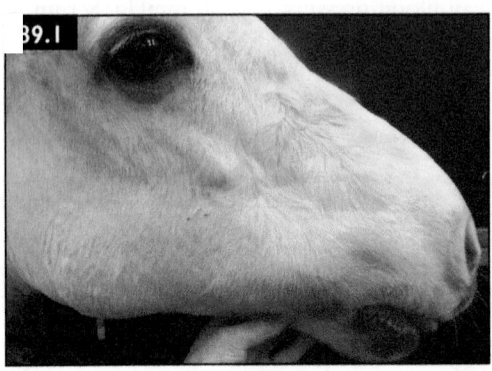

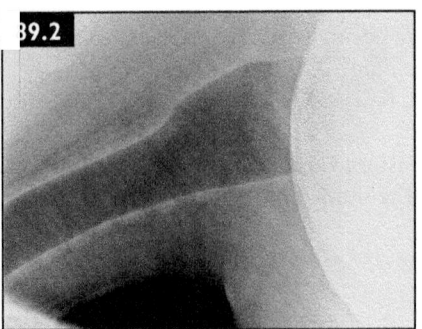

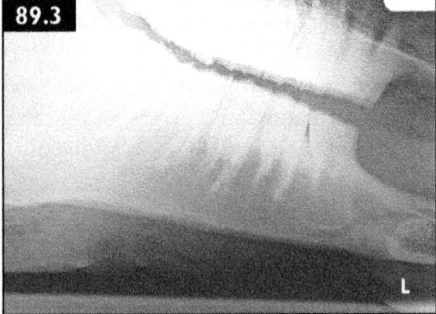

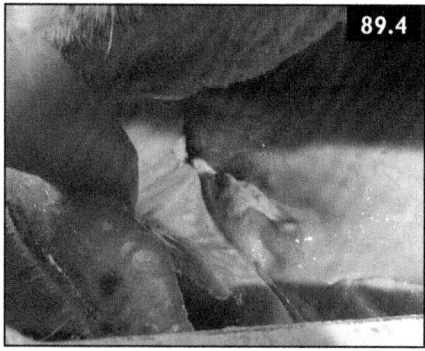

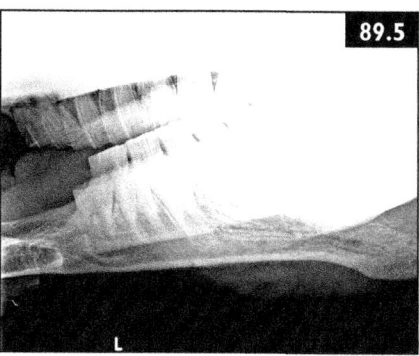

1 What conditions commonly result in acute swelling of the head?
2 What abnormalities can be seen on the first set of radiographs?
3 What abnormalities can be seen on the second set of radiographs?
4 What condition is now suspected?

CASE 90 A 3-day-old Thoroughbred colt presented with a history of depression, weakness, and unwillingness to suck from the dam. The foal's mucous membranes had become yellow (**90.1**). The mare had several previous foals without any problems. On examination the foal was weak, tachycardic (HR = 98 bpm) and tachypneic (RR = 30 bpm). Mucous membranes were profoundly icteric and the urine was discolored red. Hematology showed a PCV of 14% and leukocytosis (WBCs = 13,800/µl [13.8 × 10^9/l]); biochemistry revealed hyperbilirubinemia (total bilirubin 9.4 mg/dl [160 µmol/l]). The plasma was discolored red.

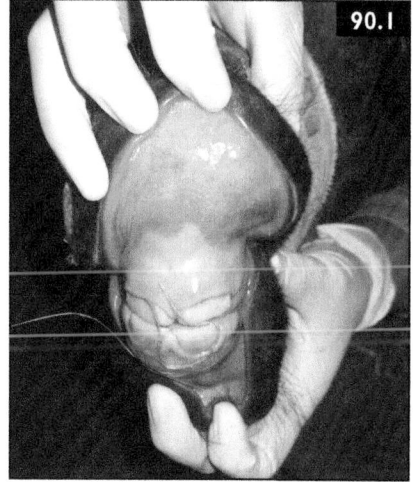

1 What is the most likely diagnosis?
2 What is the cause of this disease?
3 How should this foal be treated?

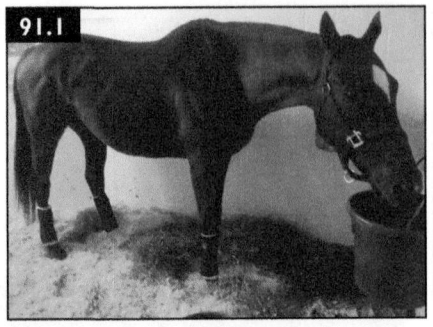

CASE 91 A 13-year-old Thoroughbred gelding (**91.1**) was examined because of suspected colic. The owner had been trail riding the horse when he became acutely uncomfortable, pawing, and wanting to lie down. The horse had not had any illness since the owner had him (4 years) and he was properly vaccinated. Clinical examination revealed a mildly uncomfortable horse that was pawing and kicking, and with a HR of 160 bpm. A nasogastric tube was passed and 6 liters of gastric reflux was obtained. Abdominal palpation per rectum and ultrasound examination of the abdomen were normal, hives were present on the abdomen, and a prominent jugular pulsation was noted extending half-way up the neck. PCV (48%) and blood lactate (5.4 mmol/l) were elevated. Electrolytes and peritoneal fluid analysis were normal. ECG (**91.2**) was performed 2 hours later. By that time the HR had decreased to 60 bpm, the jugular pulses had resolved, and the horse was comfortable. The ECG was abnormal. Cardiac troponin I (cTnI) was measured (4.2 ng/ml [4.2 µg/l]). Since the horse appeared to be clinically improved and the HR had decreased, IV fluids were administered overnight. The horse also had a good appetite. The following morning, his HR suddenly increased from 60 bpm to 100 bpm. A repeat cTnI was 5.4 ng/ml (5.4 µg/l).

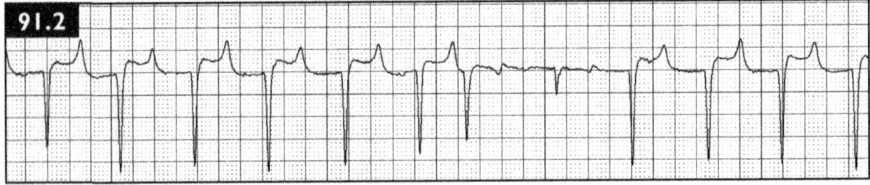

1 Should an abdominal disorder be further pursued (i.e. repeat abdominocentesis, exploratory laparotomy) as the most likely diagnosis, or should another organ system disorder be investigated?
2 Is the increase in cTnI clinically important?
3 What treatments should be considered?

CASE 92 A 22-year-old Holsteiner gelding was examined because of a 3-day history of acute onset of left hindlimb swelling. A small injury to the central part of the coronary band had been noted 3 days earlier and had been cleaned with betadine ointment; at that time the horse was not lame. The swelling had developed quickly, spreading from the coronary band to the stifle, and was noticeably warm and painful to palpation throughout the limb. The horse had received flunixin meglumine twice daily, ceftiofur once daily, been walked, and the leg had been cold-hosed and wrapped without improvement. A single treatment of dexamethasone (16 mg) and furosemide (600 mg) had resulted in some improvement. On day 2 serum was noted to be 'oozing' from the skin from the coronary band to the hock. Lameness had been marked since day 1 and had progressed to a 4+/5, 'toe-touching' lameness. The horse was referred for more intensive work-up and treatment (**92.1**). On presentation, the HR was 52 bpm and the rectal temperature was 102.4°F (39.1°C).

1 What is the most likely diagnosis based on the clinical signs?
2 What diagnostic test should be performed to help establish the diagnosis?
3 What are the etiologic causes of the condition in this horse, and which one is most likely here?
4 How should this horse be treated?
5 Due to the non-weight bearing of this limb, what concern is there for the opposite limb, and how might this be prevented?

CASE 93 A 16-year-old crossbred gelding presented with a history of swelling of all four legs with serum oozing from the skin of the affected areas (**93.1**). There was also ventral abdominal subcutaneous edema and swelling of the prepuce (**93.2**). The horse was inappetent and reluctant to move. T = 101.7°F (38.7°C); HR = 70 bpm; RR = 40 bpm. The horse had a history of a strangles infection 4 weeks previously.

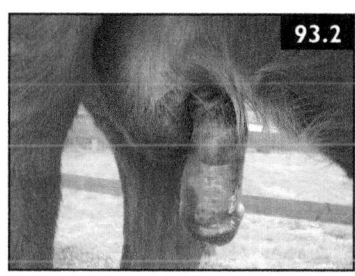

1 What is the most likely diagnosis?
2 How could the diagnosis be confirmed?
3 What treatments are recommended?

CASE 94 A 16-year-old gray Thoroughbred cross gelding presented with multiple dermal nodules up to 4 cm in diameter in the skin of the ventral tail (**94.1**) and in the perineal region (**94.2**). These lesions had been present for several years and had been slowly enlarging, but did not previously cause any problems to the horse. However, several of the nodules had recently become ulcerated and the flies were irritating them.

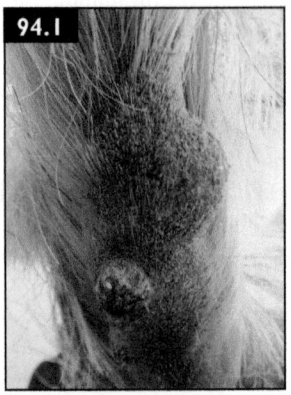

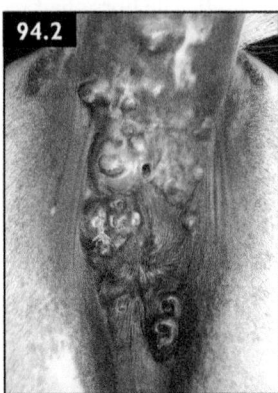

1 What is the likely diagnosis?
2 What are the different forms of this disease?
3 How can these lesions be treated?

CASE 95 A 5-year-old pregnant mare presented with signs of marked depression. The mare was in the last trimester of pregnancy and had been losing weight over the last 6 weeks. On examination, the mare appeared profoundly depressed. HR was 60 bpm and mucous membranes were congested. She had a poor appetite, but could prehend food, but then failed to chew or swallow it (**95.1**). Blood samples showed an abnormal opacity of the plasma (**95.2**).

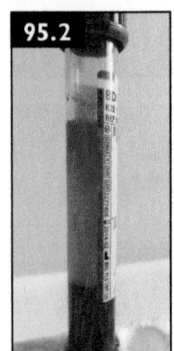

1 What is the cause of the opaque plasma?
2 What is the pathogenesis of this condition?
3 How should this pony be treated?

CASE 96 A 12-year-old gray Thoroughbred mare was examined because of a progressive history over 8 weeks of asymmetric muscle atrophy in the gluteal region (**96.1**) and mild hindlimb ataxia. The horse had been observed rubbing the perineal region for several weeks prior to the appearance of the gluteal muscle atrophy. The horse had become fecally incontinent and urine scalding of the hindlimbs was observed. On examination, the tail was flaccid and paralysed, as was the anus, and there was a zone of analgesia in the perianal and perineal regions (**96.2**).

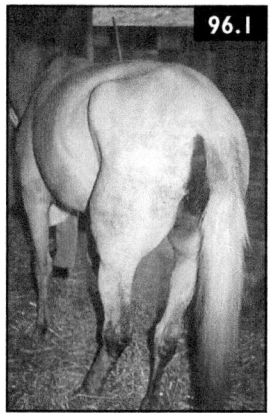
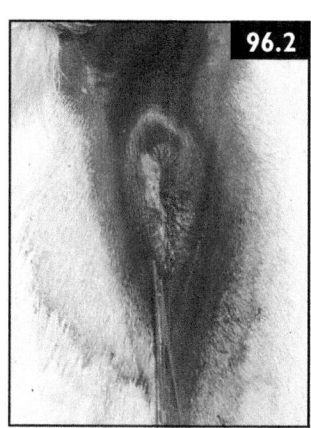

1 What is the most likely diagnosis, and what are the differential diagnoses?
2 What other clinical signs can be associated with this disease?
3 How should this horse be treated?

CASE 97 A 12-hour-old colt started to show colic signs (**97.1**). The foal appeared normal at birth, stood within 1 hour, and started to suckle from the mare soon afterwards. He behaved normally for the first 8 hours but then became dull and 'off suck' and subsequently started to show signs of abdominal pain and straining. Digital palpation per rectum revealed no significant abnormalities. A barium enema radiographic study was performed (**97.2**).

1 What is the likely diagnosis?
2 What other procedures can be used to help confirm the diagnosis?
3 How should this condition be treated?

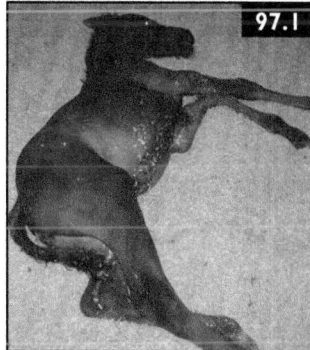
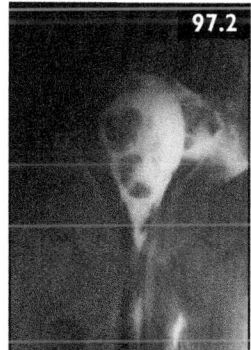

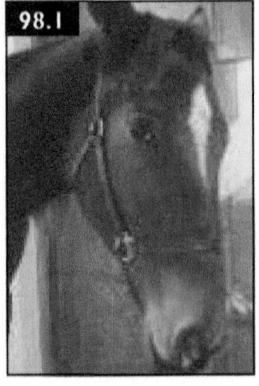

CASE 98 An 8-month-old Thoroughbred colt was examined because of an acute onset of presumed dysphagia. He was noted to have severe coughing and seemed to be unable to swallow hay when fed. A nasogastric tube was passed successfully to rule out esophageal choke. The colt was resentful of palpation of the intermandibular area and larynx, but there were no visible external abnormalities. Rectal temperature was 103.1°F (39.5°C), HR was 52 bpm, and RR was 20 bpm. Auscultation of the lungs was within normal limits. A mild bilateral nasal discharge was noted (**98.1**). On further questioning, it was discovered that several of the weanlings on the farm were coughing and had fevers. The colt was offered water but did not attempt to drink. A thorough oral examination and endoscopic examination (**98.2–98.5**) was performed. The oral examination was normal.

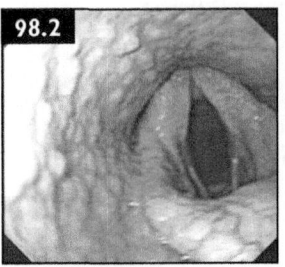 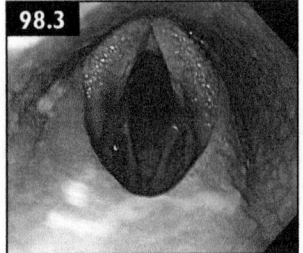 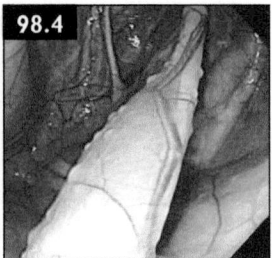

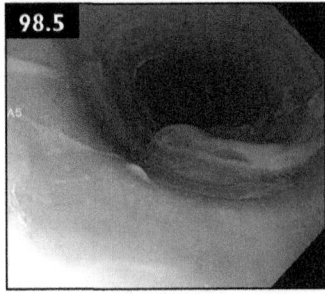

1 What do the endoscopic images show?
2 Image **98.4** is from endoscopy of the left guttural pouch. What abnormality is observed on the mucosa overlying the stylohyoid bone?
3 Image **98.5** shows endoscopy of the trachea. Interpret this image.
4 What sample should be collected to further identify the cause of this colt's clinical signs?
5 How should this weanling be treated?

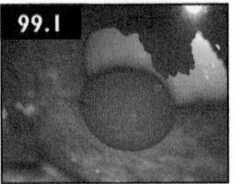

CASE 99 A 17-year-old Quarter Horse gelding, used as a cutting horse, presents with a 6-month history of poor performance on the left side (**99.1**).

1 What is the recommended treatment?

CASE 100 A 2-year-old pony gelding presented with a 48-hour history of weight loss, inappetence, diarrhea (**100.1**), and ventral abdominal and preputial edema (**100.2**). The pony lived out at permanent pasture in a group of approximately 20 ponies of various ages (none of the other ponies were ill). A fecal egg count yielded a worm egg count of 75 strongyle epg. Examination of a sample of feces mixed with water revealed numerous fine nematode worms (**100.3**).

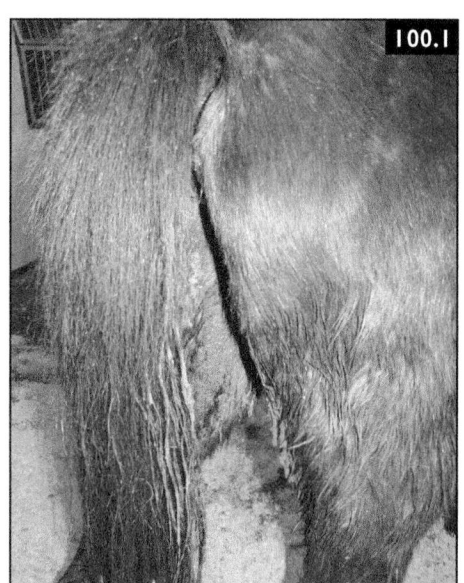

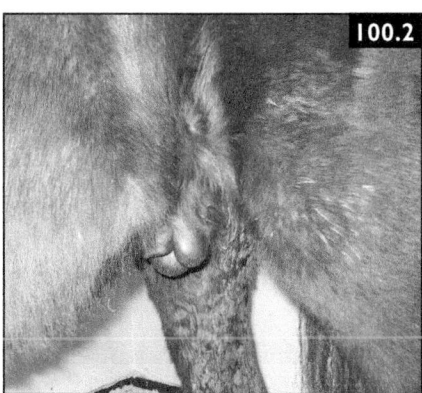

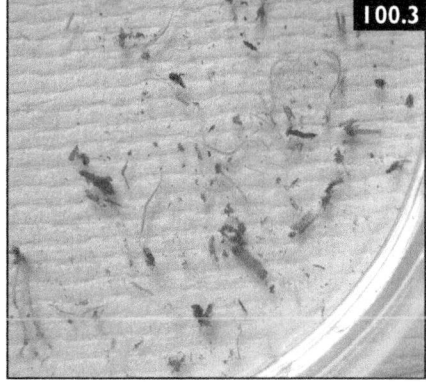

1 What is the most likely type of worm identified in the fecal preparation?
2 How does the presence of these worms in the feces relate to the fecal egg count of 75 epg?
3 Describe the life cycle of this parasite and how the condition 'acute larval cyathostominosis' occurs.

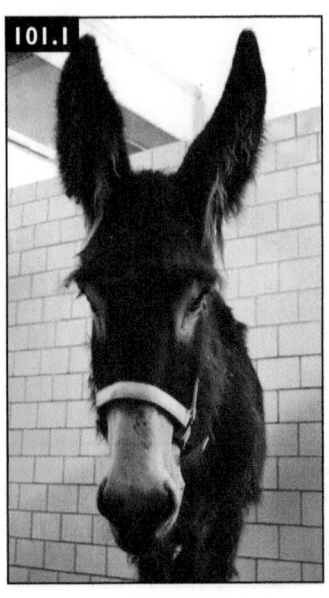

CASE 101 A 3-year-old female donkey (**101.1**) was examined because of miliary dermatitis (**101.2**) throughout most of her body; she also had perilimbal scleral nodules (**101.3**). Several other young adult donkeys of both sexes in the herd had similar lesions. On clinical examination, the donkey was underweight (BCS 3/9) with miliary dermatitis over most of her body and nodular lesions in both the sclera and nose. The nodules were neither painful nor pruritic. CBC and chemistry were normal and fecal flotation was negative. Since lesions were observed in the nose, the donkey was sedated with 0.5 mg/kg xylazine for endoscopic examination. As soon as the donkey was sedated, and before the endoscope was passed, the donkey had a 1–2 minute episode of sneezing. On endoscopic examination, nodules similar to those in the skin and nares were observed throughout the pharynx and larynx and down the proximal one-third of the trachea (**101.4, 101.5**).

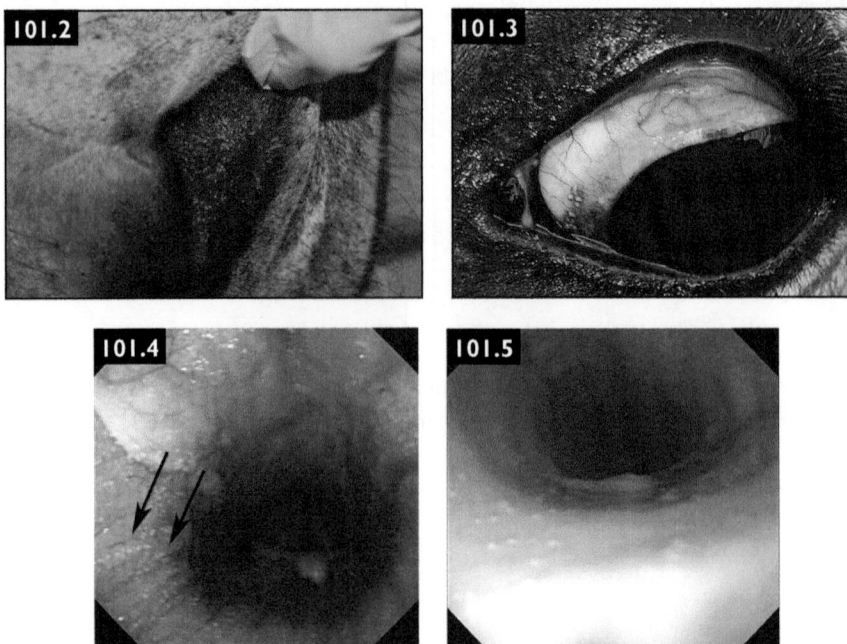

1 What is the most likely cause of the nodular lesions in the skin and mucous membranes?
2 What procedures or test could be performed to confirm the diagnosis?
3 How is the disease spread?
4 How should this disease be treated?
5 Why did the donkey sneeze immediately after the sedation?

CASE 102 An 11-year-old Arabian gelding was evaluated because of a 5-month history of stiffness and lameness in all four limbs that could not easily be localized. The horse also had a history of falling to his knees during training and more recently was noted to have lost muscle mass over his topline in spite of continued training. Clinical examination revealed significant atrophy over his topline and rump (**102.1**). There was no detectable effusion in any palpable joint and pain could not be detected on palpation of the back. The hind hooves had evidence of wear from dragging the toe and his tail was positioned to the right of the midline after being jogged (**102.2**). Dragging of the hind feet was occasionally noted during neurologic examination and the horse stumbled in all four limbs when walked up and down a hill. Lameness examination including flexion tests was unremarkable.

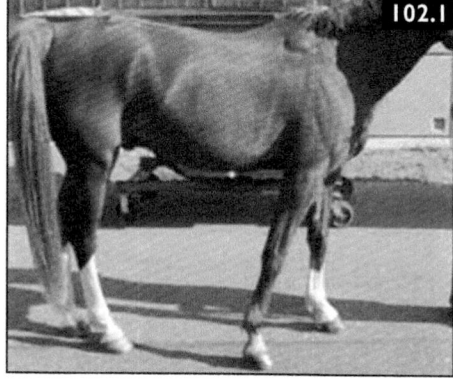

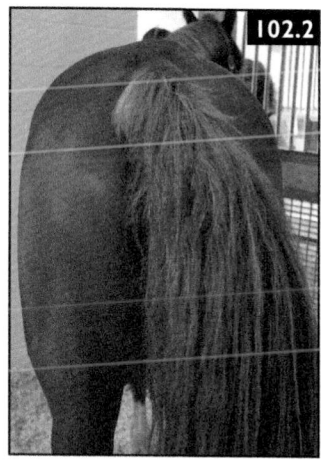

1 What diagnostic procedures should be recommended in the hope of determining the cause of the clinical abnormalities?
2 What is the preferred way to use serology for the diagnosis of EPM or neuroborreliosis?
3 What would be an appropriate treatment for this disease?
4 Why would doxycycline PO not be an appropriate treatment in this case?
5 Holding the tail to one side is a common finding with pain in what area?

CASE 103 A 15-year-old Arabian cross gelding was examined because of subcutaneous swellings of the limbs, neck, and body (**103.1**, **103.2**). The horse had presented 1 week earlier with diffuse swelling of the right hindlimb. Cellulitis was diagnosed and the horse was started on a course of procaine penicillin IM. In the 24 hours prior to presentation, the limb swelling worsened and the horse developed swellings elsewhere on its body. He became quiet and inappetent. A swelling developed at the site of an IM injection on the morning of your examination, and he subsequently developed mild bilateral epistaxis. On physical examination, the horse appeared depressed and had petechial and ecchymotic hemorrhages on the oral (**103.3**) and nasal (**103.4**) mucous membranes. There was no evidence of epistaxis at the time. Routine blood analysis revealed a profound thrombocytopenia (platelets = 10,000/µl

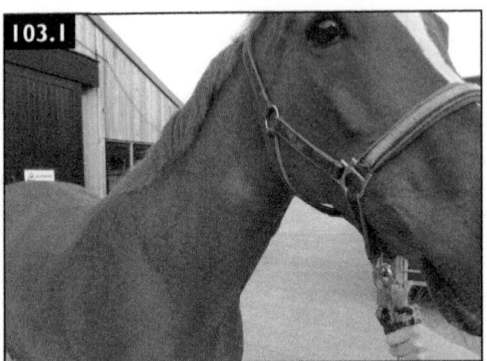

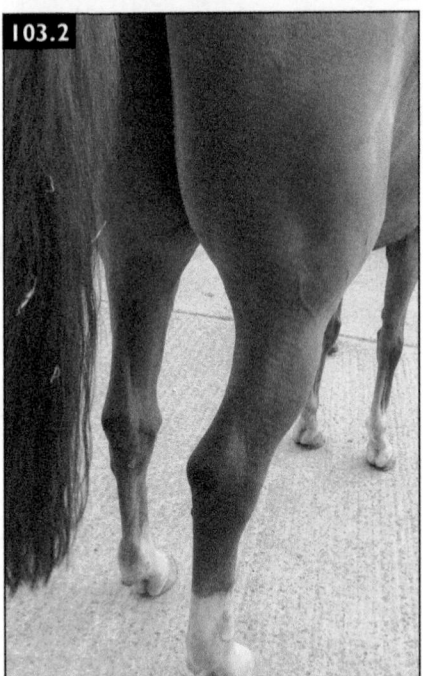

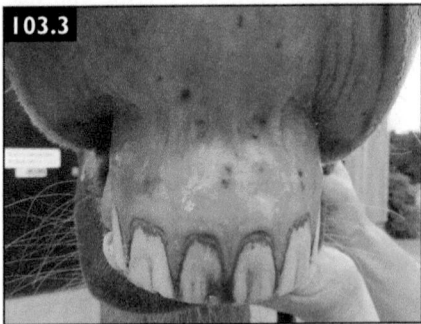

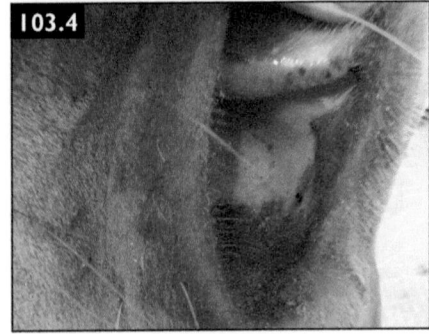

[10 x 10^9/l]) on an EDTA blood sample) and mild hyperfibrinogenemia (470 mg/dl [4.7 g/l]). You suspect thrombocytopenia.

1 What further tests are needed to confirm the suspicion of thrombocytopenia?
2 What are the important causes of thrombocytopenia in adult horses?
3 How should this horse be treated?

CASE 104 A 21-year-old Thoroughbred gelding presented for subcutaneous emphysema of the head and neck that had been present for 1 day. The owner reported that the horse was stiff, reluctant to walk, and that his head and neck were swollen and puffy. He was still eating, but much slower than normal. The only abnormal findings on clinical examination were diffuse subcutaneous emphysema, HR of 52 bpm, and RR of 30 bpm. Marked generalized subcutaneous emphysema was most pronounced around the throatlatch, shoulder, and neck region, but extended caudally and reached ventrally as far as the carpal and tarsal joints (**104.1, 104.2**). No external swellings, lacerations, or puncture wounds were identified. The clip marks seen in the photos were performed after hospital admission.

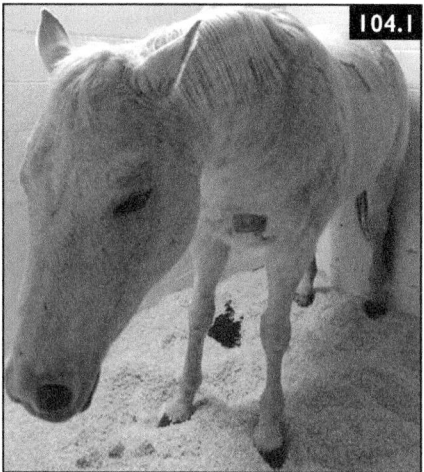
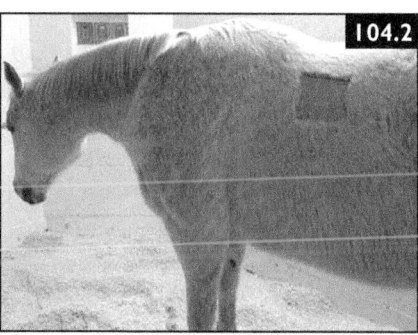

1 What are the differential diagnoses for diffuse subcutaneous emphysema of the neck and shoulder region?
2 What diagnostic procedure(s) should be performed next?
3 What clinical findings would help rule out a ruptured esophagus?
4 How should this case be treated?

CASE 105 A 23-year-old Arabian mare presented for a decreased appetite and dropping feed from her mouth. The problem had become progressively worse over the course of approximately 6 months. On examination of the oral cavity, there was a foul odor and pain on palpation of the incisors. Incisors 103, 203, 303, and 403 were deviated in a rostral direction but no teeth were palpably loose. There was obvious mineralization at the apex of the crown of each incisor and the gingiva was severely inflamed with evidence of some recession (**105.1**). This gingival inflammation contributed to advanced periodontal disease. Intraoral radiographs were obtained (**105.2, 105.3**).

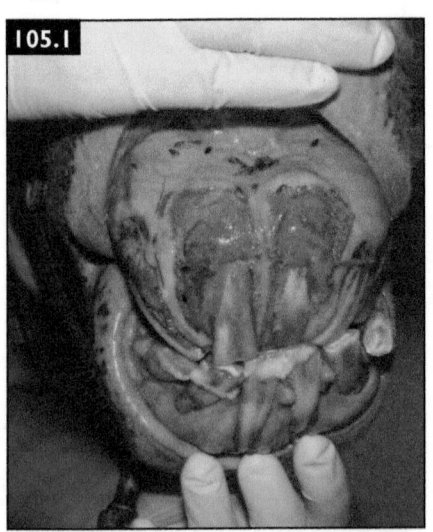

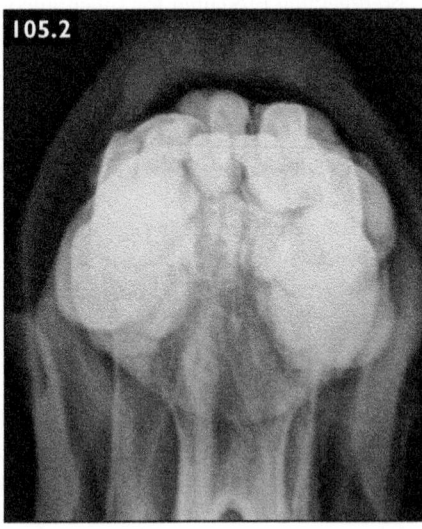

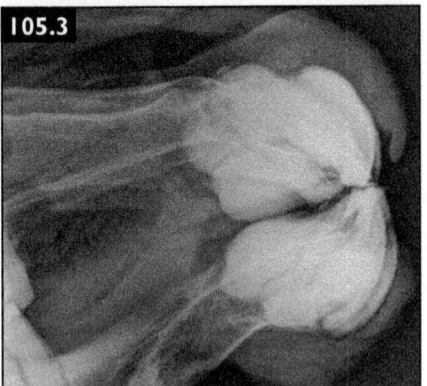

1 What is the interpretation of the radiographs?
2 Based on the history, physical examination, and radiographic findings, what is the most likely diagnosis for this case?
3 How should this horse be treated?

CASE 106 A 2-year-old Quarter Horse filly (**106.1**) from New York had a 2-day history of lethargy, limb edema, and fever (103.8°F [39.9°C]). The horse was treated with ceftiofur on day 1 for a presumed streptococcal infection, but on day 2 the fever and leg edema had increased and the mucous membranes were bright red. Examination on day 3 revealed mildy icteric membranes (**106.2**) and further depression in appetite. HR was 44 bpm, RR was 34 bpm, and the lungs and heart auscultated normally. There was no nasal discharge or cough. This was the only horse on the farm with these signs.

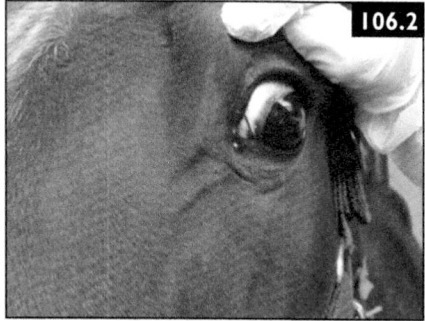

1 List some differentials for the clinical signs found in this horse.

A CBC was performed and the most abnormal findings were leukopenia (WBCs = 3,600/µl [3.6 × 10⁹/l]), neutropenia (2,200 cells/µl [2.2 × 10⁹/l]), lymphopenia (1,300 cells/µl [1.3 × 10⁹/l]), and thrombocytopenia (platelet = 37,000/µl [37.0 × 10⁹/l]). PCV was marginally decreased at 27% and fibrinogen was increased at 600 mg/dl (6.0g/l). This was observed on a buffy coat smear (**106.3**).

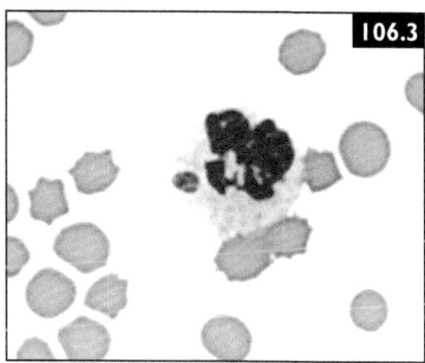

2 What is the tentative diagnosis?
3 What is the preferred treatment?
4 How quickly can the clinical and laboratory findings be expected to improve?
5 What additional clinical signs might occur with this disease?
6 What is the best (most sensitive) method to confirm the diagnosis during this febrile stage?
7 Would seroconversion be expected after 3 days of fever (likely 7 days after infection)?

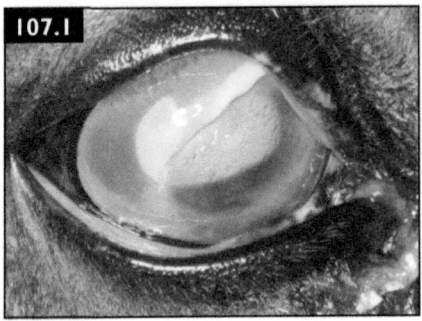

CASE 107 A 20-year-old Quarter Horse gelding presents with a 1-week history of blepharospasm and mucopurulent ocular discharge (**107.1**). Cytology reveals numerous gram-positive cocci.

1 What medical therapy is indicated?

CASE 108 A 3-year-old Arabian cross gelding presented with severe weakness that led to recumbency and an inability to get up. The horse had been living out permanently at pasture with other horses, and was found by the owner in the field 24 hours prior to presentation. He was depressed, sweating, and reluctant to walk

and, when brought into the stable, he appeared very weak but was alert and responsive when stimulated, and had a good appetite. On examination, the horse was recumbent and unable to rise (**108.1**). He was tachycardic (HR = 64 bpm), mildly tachypneic (RR = 20 bpm), and normothermic. Routine hematology and biochemistry revealed a very marked increase in muscle enzymes

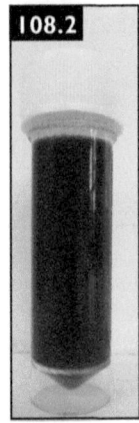

(AST = 11,591 IU/l; CK = 546,150 IU/l), a high lactate (7.3 mmol/l), high urea (27.9 mg/dl [10.0 mmol/l]), high glucose (192.8 mg/dl [10.7 mmol/l]), and raised inflammatory markers (fibrinogen = 680 mg/dl [6.8 g/l]; SAA = 954 mg/l). Abdominal palpation per rectum revealed a full bladder, which yielded thick, brown-discolored urine when catheterized (**108.2**). Urinalysis was consistent with hemoglobinuruia or myoglobinuria. Transabdominal ultrasound revealed a sluggish, mildly dilated small intestine and an increase in peritoneal fluid. Abdominocentesis yielded normal appearing peritoneal fluid.

1 What is the diagnosis?
2 What is the cause of this condition?

CASE 109 A 4-month-old Standardbred colt was examined because of an acute onset of abdominal pain. The foal was from a large broodmare farm and all of the foals 3 months of age or older had been dewormed for the first time with ivermectin the day before the onset of clinical signs. The foal's HR was 112 bpm, RR was 32 bpm, and rectal temperature was 101.2°F (38.4°C). The foal was sedated with 0.5 mg/kg xylazine so that the colic workup could be more easily completed. Passage of a nasogastric tube prompted 2 liters of spontaneous green-colored gastric reflux. Transabdominal ultrasound examination revealed numerous loops of thin-walled, amotile, and distended (5–8 cm diameter) small intestine. Two 'close-up' images of a loop of distended small intestine are shown (**109.1, 109.2**).

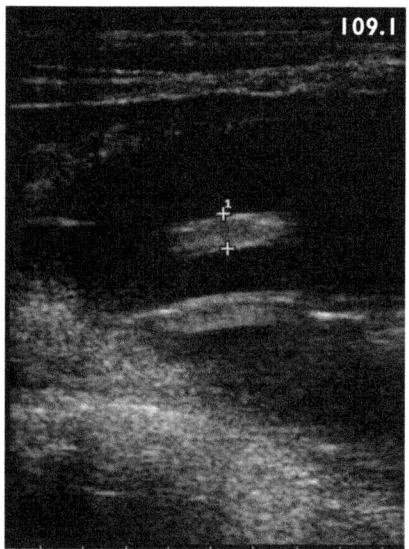 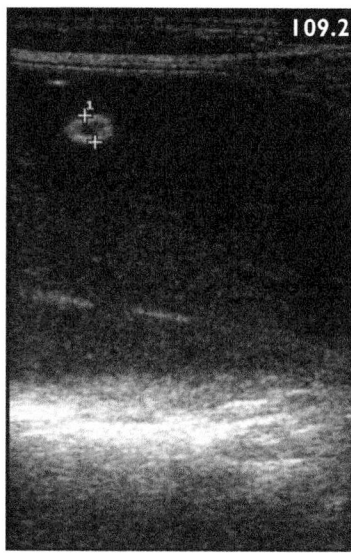

1 What is the tentative diagnosis based on the above information?
2 What are the hyperechoic areas marked (+) in the ultrasonograms?

The foal was anesthetized and an exploratory laparotomy performed. Most of the small intestine was markedly distended, with evidence of an intraluminal obstruction.

3 What is the greatest concern regarding complications that may occur within the next several days to months following this surgery?
4 What anthelmintic would have been a better and safer treatment in this foal?
5 What is the prepatent period for *P. equorum*?

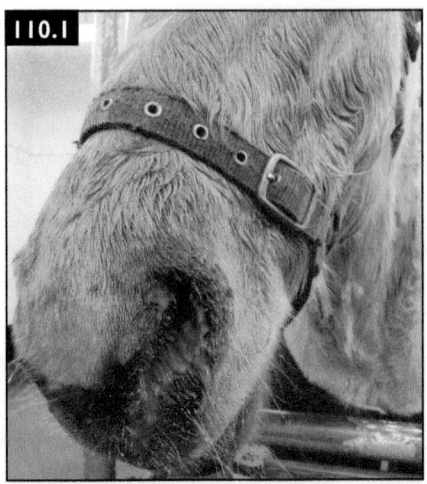

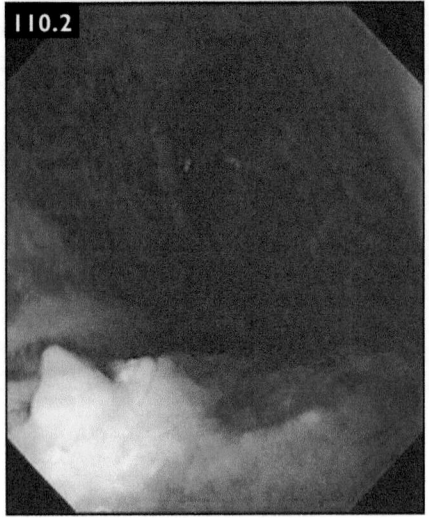

CASE 110 A 22-year-old Arabian cross mare was presented for evaluation of a sudden onset of profuse bilateral serosanguineous nasal discharge of 2 days' duration that had rapidly become thick, mucopurulent, and malodorous (**110.1**). The mare was depressed, inappetent, and dysphagic (quidding food). She had been diagnosed with equine Cushing's disease approximately 2 years earlier and was currently being treated with 0.5 mg pergolide PO once a day. The mare was in poor body condition and had stertorous breathing. The submandibular lymph nodes were mildly enlarged. She was mildly tachycardic (HR = 52 bpm) and mildly tachypneic (RR = 24 bpm). Temperature was mildly elevated (102.2°F [39.0°C]). Auscultation of the heart and lungs was unremarkable. Oral examination revealed a focal ulcer, approximately 1 cm in diameter, on the dorsal surface of the tongue. Endoscopic examination of the upper airways revealed severe bilateral purulent rhinitis with diphtheritic membranes (**110.2**); the lesions involved the mucosa of the entire nasal cavities and extended into the nasopharynx. The larynx, guttural pouches, and trachea appeared grossly normal. Esophagoscopy and gastroscopy were unremarkable. Routine hematology and serum biochemistry revealed leukopenia (WBCs = 2,300/µl [2.3 × 10^9/l]) and lymphopenia (lymphocytes 600 cells/µl [0.6 × 10^9/l]). Hypersegmented, slightly toxic ring-form neutrophils, occasional bands, occasional vacuolated monocytes, and some large reactive lymphocytes were seen on the blood smear. There was thrombocytopenia (platelets = 64,000/µl [64 × 10^9/l], hyperfibrinogenemia (83 mg/dl [8.3 g/l]) and increased serum amyloid A (1,360 mg/l). Other serum biochemical parameters were within normal reference intervals.

1 What is the tentative diagnosis?
2 What further diagnostic tests should be considered?
3 What is the significance of the large reactive lymphocytes identified on the blood smear?

CASE 111 A colleague asked you to perform a postmortem examination on a 12-year-old Thoroughbred gelding that had an 8-week history of depression, inappetence, rapid weight loss, intermittent pyrexia, and shifting leg lameness. Laboratory evaluations had revealed anemia, hyperproteinemia, hyperfibrinogenemia, and leukocytosis with a mature neutrophilia. The horse had been receiving treatment for about 1 week with broad-spectrum antibiotics and phenylbutazone, but deteriorated suddenly, collapsed, and died. Autopsy identified lesions on the aortic (**111.1**) and left atrioventricular (**111.2**) heart valves and well-defined focal discolored areas in the spleen and kidneys (**111.3**).

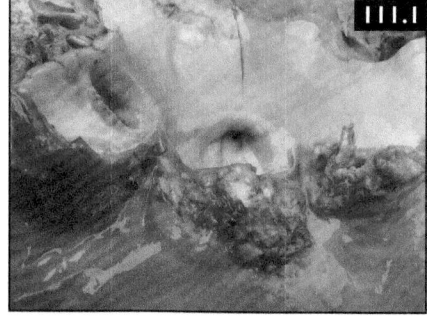

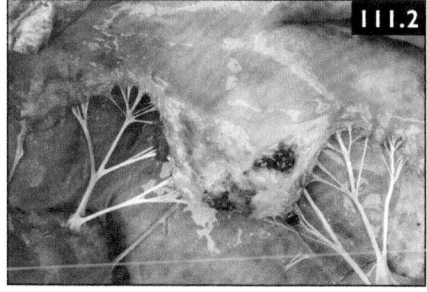

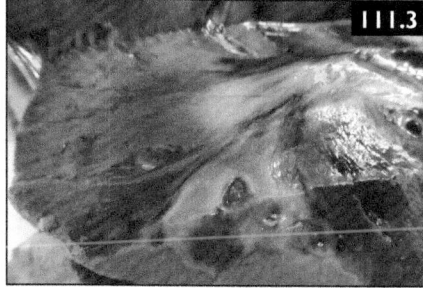

1 What is the provisional diagnosis?
2 What is the underlying cause of this disease?
3 How could this condition have been diagnosed during life?

CASE 112 A 6-month-old pony filly presented with an 18-hour history of persistent colic that was unresponsive to routine administration of analgesic drugs. No feces had been passed since the onset of colic signs. On initial presentation, the filly was showing signs of moderate pain and had a HR of 68 bpm. Intestinal borborygmi were absent and the mucous membranes were congested. Abdominal ultrasonography showed evidence of small intestinal distension. On the basis of this history and the clinical findings, an exploratory celiotomy was undertaken. At surgery, a cecocecal intussusception was identified (**112.1**) and the affected area of the cecal apex was resected. Over the next 3 weeks, following recovery from surgery, the foal remained dull/somnolent and inappetent, with persistent tachycardia (HR = 60–68 bpm). Feces were produced intermittently, but were small and dry (**112.2**). The foal lost weight and would stand with all four legs close together under her body (**112.3**). She intermittently showed muscle fasciculations of the triceps and gluteal muscles. There was ptosis of both eyes (**112.4**) and intermittent generalized sweating. Hematology and serum biochemistry showed no significant abnormalities.

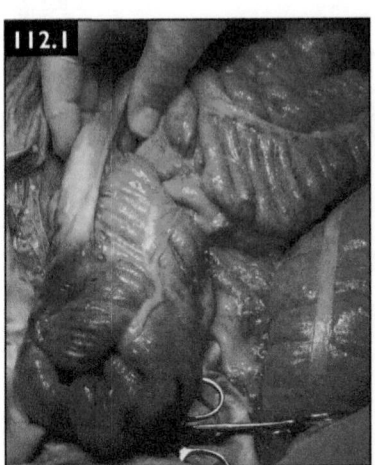

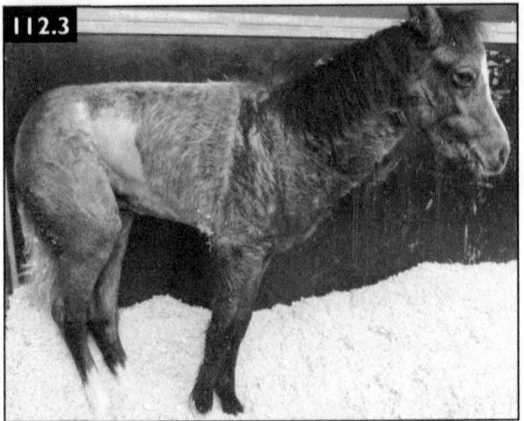

1 What condition should be suspected?
2 How does the cecocecal intussusception fit in with this disease?
3 What clinical tests could be performed to help confirm the provisional diagnosis?

CASE 113 A 10-year-old Cob mare had been treated for acute colitis and diarrhea. It was not possible to determine the cause – fecal cultures for enteric pathogens, including *Salmonella* spp., and toxin analysis for *Clostridium perfringens* and *Clostridium difficile*, have all returned negative results. Treatment included IV isotonic fluids, plasma, flunixin meglumine, pentoxyfylline, and di-tri-octahedral smectite. After 48 hours of treatment the mare deteriorated; her HR increased to 100 bpm and her mucous membranes became brick red with a toxic ring. Urine production decreased, despite increasing her fluid therapy rate, and she became azotemic. After a few hours she collapsed and died. Postmortem examination revealed a large volume of mildly turbid peritoneal fluid. There was diffuse congestion and mural edema of the entire cecum and large colon, focal petechial and ecchymotic hemorrhages on the serosal surfaces of the entire intestinal tract, and a focal area of intense discoloration of the colon wall close to the pelvic flexure (**113.1, 113.2**).

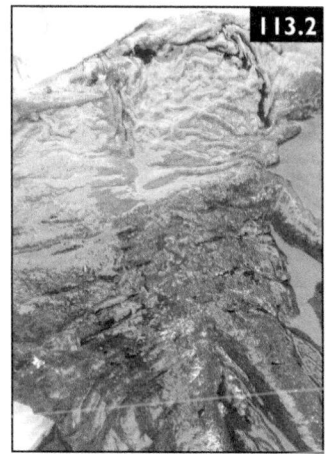

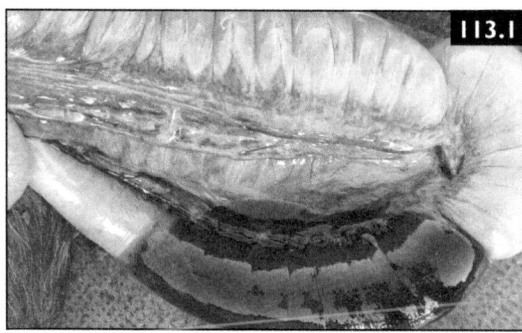

1 How do you account for the large volume of peritoneal fluid?
2 What is the cause of the focal severe discoloration of the large colon?

CASE 114 A 16-year-old Friesian mare presents with a 2-year history of an enlarging corpora nigra of the right eye (**114.1**) and decreased vision on the right side.

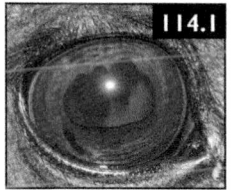

1 What are the differential diagnoses?
2 What diagnostic test is indicated to confirm the diagnosis?

CASE 115 A 15-year-old Arabian gelding presented with a 48-hour history of a mild right unilateral serosanguineous nasal discharge. The referring veterinary surgeon had performed an endoscopic examination, which revealed a small amount of blood at the right guttural pouch (GP) ostium and mild collapse of the dorsal nasopharynx. Attempted passage of the endoscope into the pouches was unsuccessful. Each GP was infused with dilute iodine solution via a catheter passed *per nasum*. Within 4 hours, the pony had developed depression and inappetence, with stertorous breathing and a profuse bilateral brown serous nasal discharge. Flunixin meglumine was administered prior to referral.

On examination, the pony was dyspneic with a bilateral foamy brown nasal discharge. HR and RR were 44 bpm and 20 bpm, respectively. The mucous membranes were congested and moist. Auscultation of the thorax revealed referred upper respiratory tract noise. The results of serum biochemistry and routine hematology (including a platelet count) were within normal limits. Endoscopic examination revealed severe pharyngitis, laryngitis, and tracheitis (**115.1, 115.2**). The dorsal nasopharynx was inflamed and collapsed, leaving a narrowed airway. Arytenoid movement appeared normal. The severe nasopharyngeal inflammation made the GP ostia difficult to visualize, and it was not possible to pass the endoscope into either eustachian tube. There was some pooling of brown mucoid material within the trachea. The pony was dysphagic, and it was not possible to perform esophagoscopy.

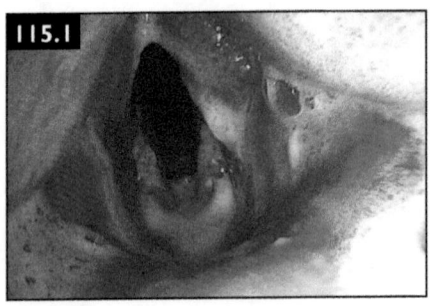 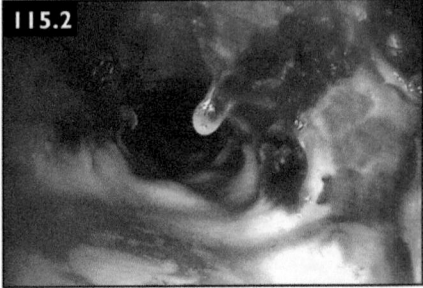

1 What is suspected?
2 What is the prognosis?

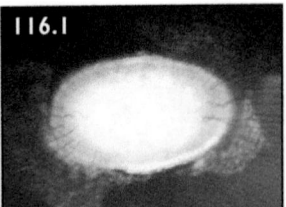

CASE 116 The fundic examination of a 1-year-old Haflinger filly with no visual deficits is shown (**116.1**).

1 Describe the lesion and provide possible etiologies.

CASE 117 A 27-year-old Shetland mare presented for assessment of a chronic, non-productive cough, increased RR, and mildly reduced exercise tolerance of a few months' duration. These signs had not responded to medical treatment and environmental management for recurrent airway obstruction or to a course of oral antibiotics. On examination, the mare was in good condition with normal HR and RR, but had a mildly increased respiratory effort and occasional coughing. Auscultation of the chest revealed reduced respiratory noises over the left hemithorax and increased respiratory noises diffusely over the right hemithorax. The rest of the mare's parameters were within normal limits. Routine hematology and serum biochemistry were unremarkable, apart from an elevated plasma fibrinogen (644 mg/dl [6.44 g/l]), and a mild anemia (PCV = 28%). Thoracic radiographs revealed a discrete soft tissue opacity just dorsal to the heart base (**117.1**). Transthoracic ultrasonographic examination showed an area of consolidated lung in the ventral left hemithorax, in the 6th–8th intercostal spaces. The pleural fluid appeared normal in volume and echogenicity. Endoscopic examination revealed a smooth, pink mass protruding into and obstructing the left mainstem bronchus (**117.2**). No other morphologic changes were observed in the upper or lower respiratory tracts.

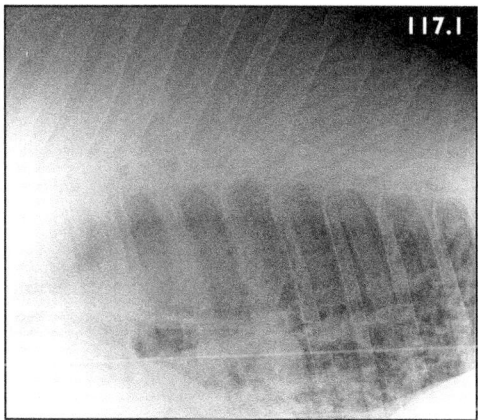 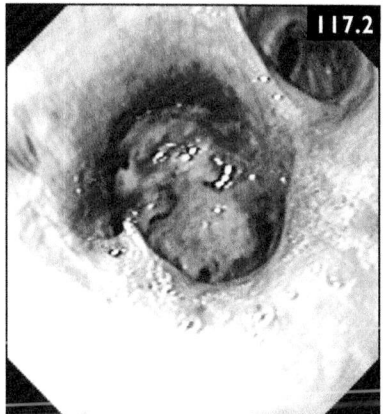

1 What is the presumptive diagnosis?
2 How would this suspicion be confirmed?
3 What treatment options should be considered?

CASE 118 A 5-year-old native pony mare was examined for an acute onset of respiratory distress. The pony had been fit and well previously, and lived out permanently within a herd. At the time of the initial examination, the pony was exhibiting severe inspiratory and expiratory distress and had a diffuse swelling of the neck. A moderate improvement was seen following treatment with IV dexamethasone, phenylbutazone, furosemide, and clenbuterol. On presentation the pony was in severe respiratory distress, with inspiratory and expiratory dyspnea and an RR of 60 bpm. Marked subcutaneous emphysema of the head and neck was present but there were no signs of external trauma. Auscultation of the thorax revealed harsh bronchial sounds and referred upper airway sounds. Cervical and cranial thoracic radiography revealed tracheal collapse in the mid-to-lower cervical trachea associated with intraluminal soft tissue thickening of the tracheal walls, severe peritracheal and subcutaneous emphysema, and pneumomediastinum (**118.1, 118.2**). The tracheal cartilage rings had increased radiopacity. The only significant abnormality detected on a routine hematology and biochemistry was hyperfibrinogenemia (plasma fibrinogen = 630 mg/dl [6.3 g/l]). Thoracic ultrasonography revealed mild roughening of the pleural surfaces over bilateral

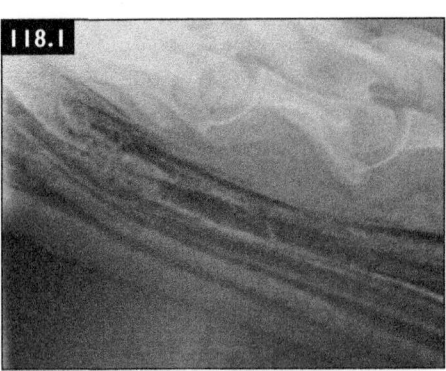

cranioventral lung fields. Endoscopic examination revealed a cobblestone appearance to the mucosal surface with over 100 small nodules along the ventral and lateral surfaces of the trachea, extending from the larynx to the primary bronchi (**118.3**). Tracheal aspiration was performed via the endoscope; cytologic examination revealed a normal population of cells apart from an increase in the number

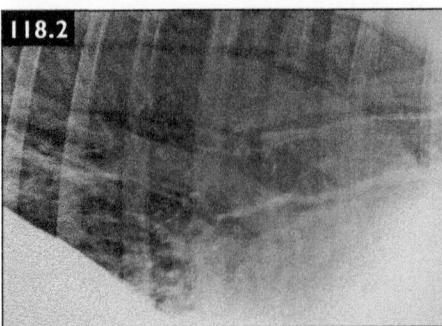

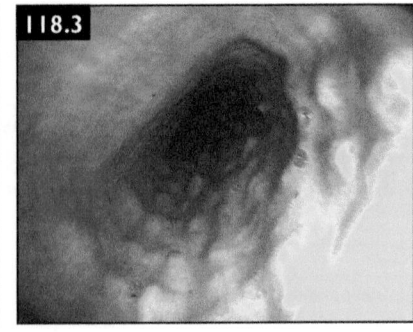

of mast cells (approximately 5%). Routine bacterial culture of the aspirate yielded no growth.

1 What is the likely cause of the swelling of the neck and the radiological findings?
2 What is the likely cause of the cobblestone appearance of the tracheobronchial mucosa?

CASE 119 A 6-year-old Warmblood mare presented with a bilateral 'crusty' nasal discharge (**119.1**) and pyrexia (rectal temperature 102.9°F [39.4°C]). The mare had recently traveled from continental Europe to the UK, and had been held in a quarantine facility for 72 hours prior to examination. The mare had been diagnosed with suspected bacterial bronchopneumonia at the quarantine facility and been treated with a combination of gentamicin and penicillin. On examination, the mare was dull but alert, and had mildly labored breathing with an abdominal expiratory effort. RR was 28 bpm and there was an intermittent harsh cough. Lung auscultation revealed bilateral adventitious lung sounds, including wheezes and crackles. Hematology and serum biochemistry were normal apart from hyperfibrinogenemia (plasma fibrinogen 550 mg/dl [5.5 g/l]). Thoracic ultrasound revealed no abnormalities apart from mild bilateral pleural roughening ventrally. Endoscopic examination revealed a moderate quantity of exudate within the trachea and mainstem bronchi (**119.2**). A tracheal aspirate revealed increased numbers of degenerate neutrophils and moderate numbers of gram-positive cocci. Microbial culture of the aspirate yielded a moderate growth of *Streptococcus equi* subsp. *zooepidemicus*.

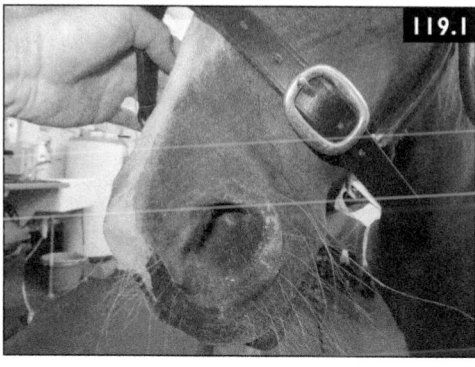 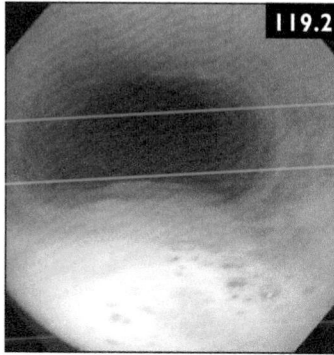

1 What other examinations should be performed?
2 What is the diagnosis?
3 How should this case be managed?

CASE 120 An 11-year-old gray Thoroughbred cross gelding presented with a 3-week history of dysphagia and quidding. The owner noticed that the horse appeared to be able to chew on the right side of the mouth but not on the left. Coincidental with the onset of dysphagia, a firm subcutaneous mass appeared below the base of the left ear (**120.1**) (no other dermal or subcutaneous masses are present). A routine dental examination had been undertaken 7 weeks previously, at which time no abnormalities were detected. On examination, the horse was quiet but alert. HR, RR, rectal temperature, and mucous membranes were all within normal limits. Examination of the mouth revealed displacement of the mandible to the right (**120.2**) and reduced lateral excursion of the mandible to the left. No other dental abnormalities were identified. The swelling at the base of the left ear was a firm, multinodular, non-painful soft tissue mass. Ultrasonography showed an ill-defined soft tissue mass with heterogeneous echogenicity that replaced the normal appearance of the temporomandibular joint (**120.3**). Oblique radiographs of the

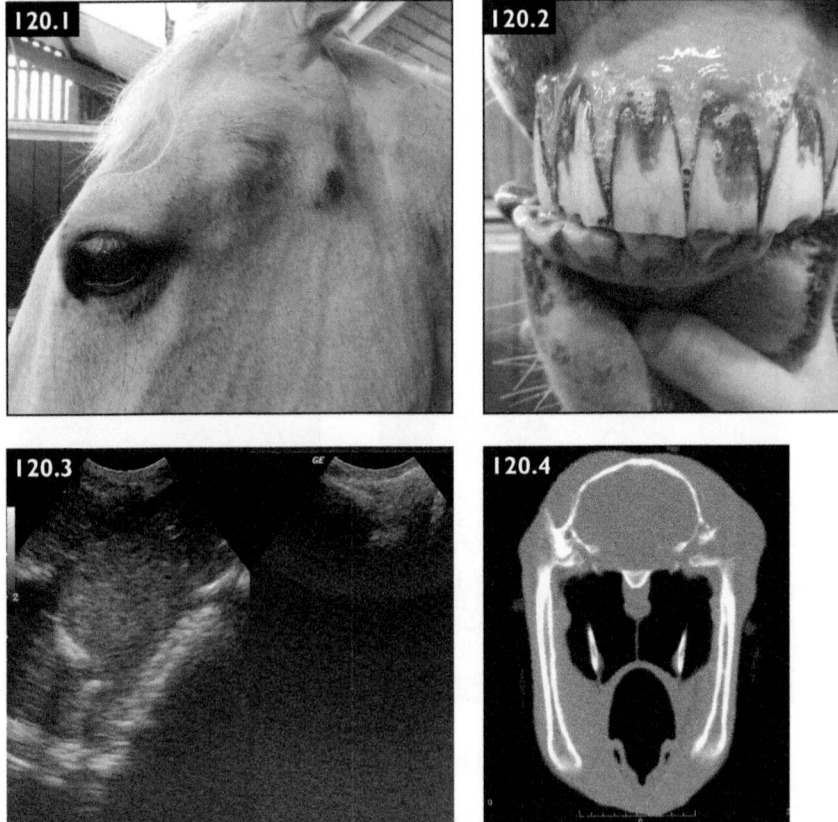

lateral aspects of the temporomandibular joints appeared normal. A standing CT scan was performed (**120.4**).

1 What abnormalities can be seen in the CT scan?
2 What further diagnostic procedures should be undertaken?

CASE 121 A 7-year-old Thoroughbred mare was examined because of recurrent colic of 1 year's duration and failure to maintain body condition during the past 2 months. Twelve months earlier the mare had been diagnosed with lymphocytic–plasmacytic enteritis and had a good response to dexamethasone treatment. Corticosteroid treatment had been tapered to every third day but during the prior 3 months the mare had lost 150 lb (67.5 kg). Anthelmintic treatments and husbandry were considered excellent.

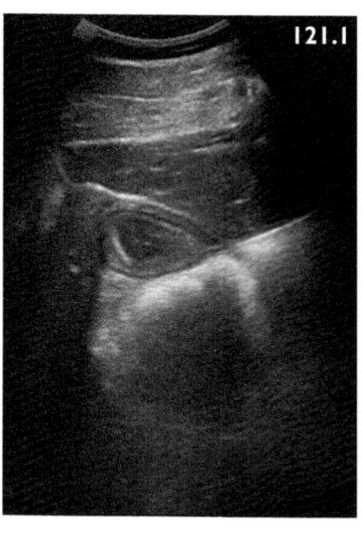

Clinical examination revealed very mild colic signs but unusual lethargy for this mare. Temperature, mucous membranes, RR, HR, and abdominal palpation per rectum were all normal. Abdominal ultrasound examination was normal except for thickening of the duodenum and jejunum (**121.1**). The only abnormalities noted on a CBC and biochemistry profile were marginally low plasma TP and albumin (5.6 g/dl [56 g/l] and 2.9 g/dl [29 g/l], respectively). Gastroscopy was normal. Fecal flotation for parasites and culture for *Salmonella* were negative. Rectal and duodenal biopsy both revealed a lymphocytic–plasmacytic enteritis/colitis similar to the findings from 1 year previously.

1 Other than continuing corticosteroid or other immunosuppressive therapy such as azathioprine, what additional treatments could be offered for the inflammatory bowel disorder (IBD)?

CASE 122

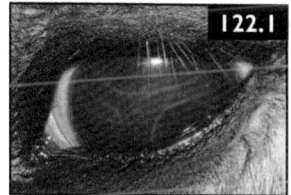

1 Identify the linear corneal lesions present in this picture (**122.1**). What are the possible etiologies?

CASE 123 A 10-month-old Thoroughbred Arab cross filly reared up and fell over backwards, hitting her poll on concrete 10 days ago. At the time of the accident she struggled for several minutes before being able to get up. When she did manage to rise to her feet she immediately showed severe ataxia in both her fore- and hindlimbs and demonstrated drooping of the right ear. The owner kept her stabled and treated her with phenylbutazone. She showed a slight improvement over the 10-day period with mild improvement in her coordination and partial return of ear movement. She continued to eat and drink normally. At the time of examination, the filly was moderately ataxic in all four limbs (grade 3/5). The right ear appeared paretic and there was mild ptosis of the right eye. There was a persistent head tilt (poll deviated to right) (**123.1**). The filly was bright and alert and appeared to have normal mentation. No other significant abnormalities were found on a routine neurologic examination. A standing lateral radiograph of the head was obtained (**123.2**).

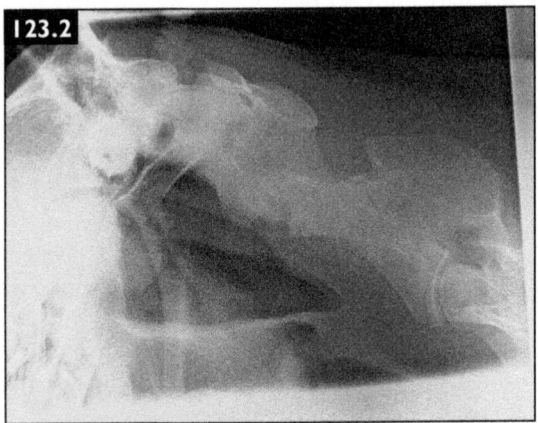

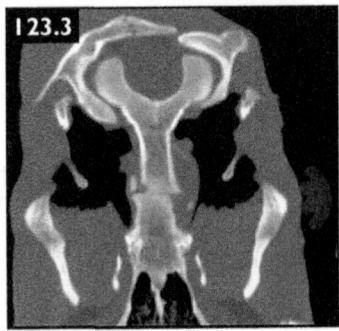

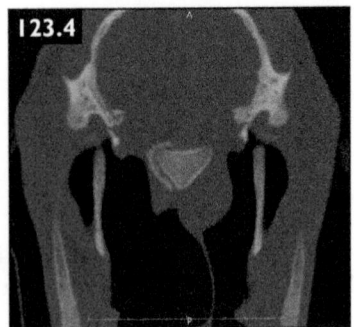

Endoscopic examination of the upper respiratory tract, including the guttural pouches, revealed no significant abnormalities apart from mild submucosal hemorrhages of the medial wall of the right guttural pouch; the left guttural pouch appeared normal. A standing CT examination of the head was performed (**123.3, 123.4**).

1 Can any abnormalities be identified in the skull radiograph?
2 What is the significance of the submucosal hemorrhages in the wall of the right guttural pouch?
3 What abnormalities can be seen in the CT images?

CASE 124 A 2-year-old Warm-blood cross gelding presented with a 3-week history of rapid weight loss, ventral edema, and inappetence (**124.1**). Routine hematology and serum biochemistry revealed leukocytosis with neutrophilia (10,200 cells/µl [10.2 × 10^9/l]), mild hyperfibrinogenemia (500 mg/dl [5.0 g/l]), hypoproteinemia, and hypoalbuminemia (2 g/dl [20 g/l]). As a part of the investigation of this horse's

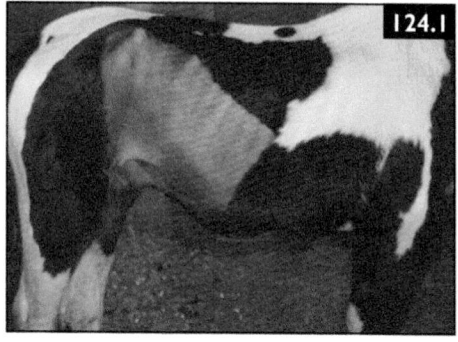

condition, standing diagnostic laparosopy was performed via a flank laparotomy to obtain some full-thickness intestinal biopsies.

1 What abnormalities can be seen in this laparoscopic image (**124.2**)?
2 What abnormalities can be seen in the large intestinal biopsy (**124.3**)?

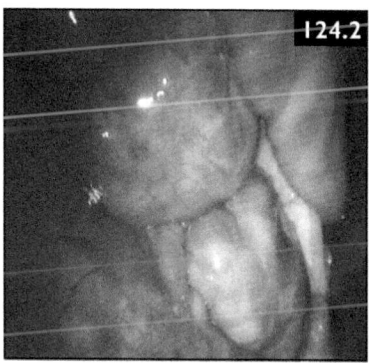

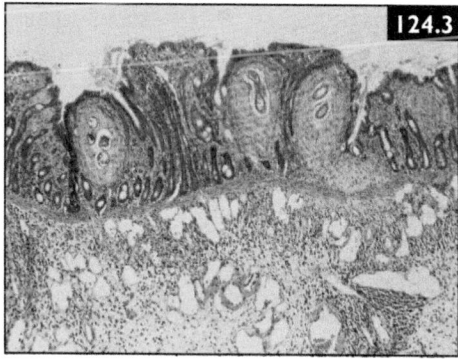

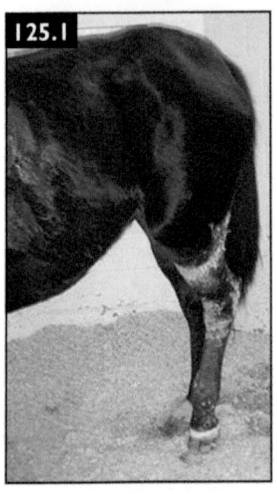

CASE 125 A 15-year-old Thoroughbred broodmare (7 months pregnant) was examined because of a non-painful swelling of the left hindlimb and a non-painful enlargement of the left half of the udder. A 'sweat' wrap had been applied, which successfully reduced the edema in the hindlimb (**125.1**). The mare had lost 150 lb (67.5 kg) over the past month, her appetite was decreased, and she had experienced a brief episode of bilateral uveitis 3 weeks earlier. Since the udder on the affected side was enlarged, it was assumed that the non-painful swelling in the leg was a result of impairment of venous return due to a primary mammary disease.

1 What are the differentials for an aberrantly enlarged mammary gland?

Ultrasound examination revealed prominent vessels in the left udder parenchyma and hypoechoic lymph nodes (4.3 × 2.3 cm) where the udder was continuous within the body wall. The mammary secretions were red and cytology showed many aged and degenerate RBCs and a substantial number of macrophages with abundant lipid material. Although this type of fluid is characteristic of a mammary tumor, no neoplastic cells were observed. The only abnormality on CBC and biochemistry was hypercalcemia (14.4 mg/dl [3.6 mmol/l]).

2 What does the hypercalcemia suggest?
3 What other blood test could be performed to help confirm that the hypercalcemia was associated with, or caused by, a neoplasm?
4 What would be the next diagnostic procedure(s)?

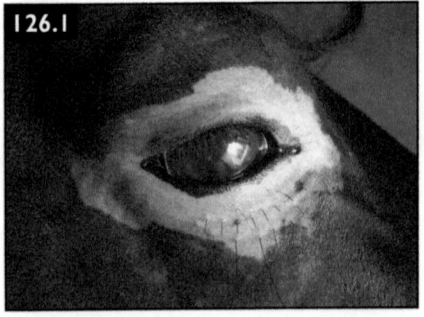

CASE 126 A 6-year-old crossbred gelding presented with bilateral extensive depigmentation around the eyes and lips that had been present for several months (**126.1**). The horse was well in all other respects and there were no other skin lesions or pruritus.

1 What is this condition?
2 What is its cause?
3 What treatments are available?

CASE 127 A 16-year-old Shetland pony gelding was referred after being found in the field stuck under a fence. When removed from the fence he was found to be ataxic and depressed. On examination the pony was profoundly depressed (**127.1**). He was severely ataxic (grade 4/5 in fore- and hindlimbs) and unable to walk without assistance. HR and RR were elevated

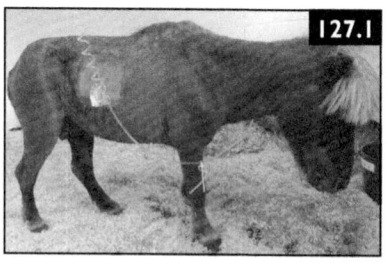

(58 bpm and 26 bpm, respectively) and rectal temperature was normal. He had proprioceptive deficits in all four limbs with the hindlimbs being more severely affected than the forelimbs. He had urinary incontinence and a maggot infestation of the dock. His menace response was absent bilaterally, a reduced pupillary light reflex was present, and facial sensation and tongue tone were reduced. The pony was noted to circle and head press. Hematology showed an elevated WBC count (13,000/µl [13 × 10^9/l) with a neutrophilia (12,200/µl [12.2×10^9/l]) and lymphopenia (700/µl [0.7 × 10^9/l]). Biochemistry showed a markedly elevated GGT (232 IU/l), markedly elevated bile acids (66 mmo/l), moderately elevated LDH (1,304 IU/l,), moderately elevated fibrinogen (711 mg/dl [7.11 g/l]), and mildly elevated AST (589 IU/l). Ultrasound examination of the liver showed areas of hyperechogenicity. A liver biopsy was taken and submitted for histopathology. In view of the neurologic signs, a blood sample was also submitted for equine herpesvirus serology, which yielded a low titer; the horse had not been previously vaccinated against EHV. Histopathologic examination of the liver biopsy resulted in a diagnosis of hemochromatosis.

1 What is hemochromatosis?
2 What is the prognosis for this pony?
3 What treatments should be recommended?

CASE 128 A 20-year-old Thoroughbred cross gelding presented with a 3-week history of an enlarging ulcerated gingival mass adjacent to teeth 202 and 203 (**128.1**). The submandibular lymph nodes were also enlarged. The horse appeared well in himself and was eating and drinking normally.

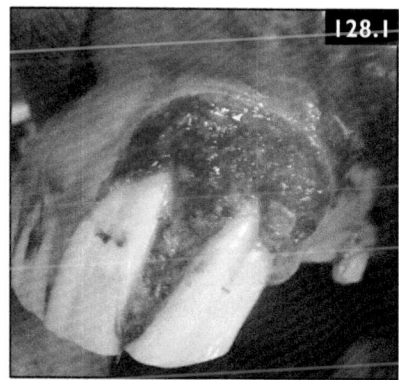

1 What conditions are suspected?
2 How could the diagnosis be confirmed?
3 What treatment options should be considered?

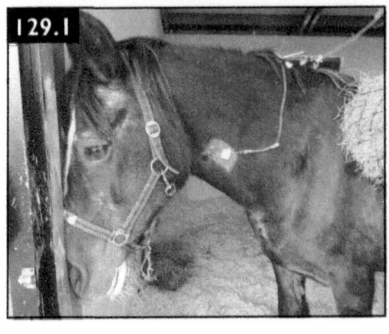

CASE 129 A 24-year-old Cob mare was examined for an acute onset of severe colic of several hours' duration. On examination the mare was tachycardic (HR = 60 bpm) and had pale, tacky mucous membranes with a CRT of 3 seconds. There was abdominal distension and diarrheic feces were present in the rectum. Severe 'pipe-stream' diarrhea subsequently developed. A diagnosis of acute colitis was made and the mare was treated intensively with IV isotonic fluids, plasma, 'low-dose' flunixin meglumine, pentoxyfylline, polymyxin B, and di-tri-octahedral smectite. After 48 hours of treatment, the cardiovascular parameters stabilized and the diarrhea started to abate. However, the mare suddenly started to show behavioral abnormalities, with circling, apparent blindness, and head-pressing (**129.1**).

1 What condition might be causing the behavioral abnormalities?
2 How can this be confirmed?
3 What treatments should be administered if the diagnosis is confirmed?

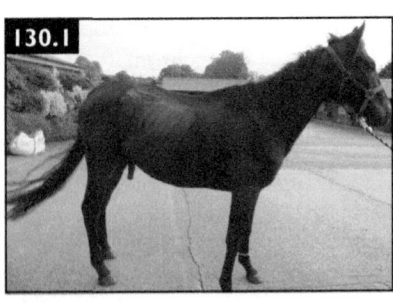

CASE 130 A 7-year-old Thoroughbred gelding had been progressively losing weight over the past 6 weeks. The horse had been in the current owner's possession for 4 years and there was no previous relevant medical history. The horse's appetite had been good up until the last few days before presentation, when he became dull and inappetent. A full physical examination revealed no significant abnormalities apart from poor body condition (**130.1**); no dental abnormalities were identified, and there was no diarrhea. Laboratory evaluations revealed hypoalbuminemia (albumin = 1.8 g/dl [18 g/l]) and mild anemia (PCV = 28%). A fecal worm egg count was zero. An oral glucose tolerance test revealed a malabsorption state (basal plasma glucose only increased by 10% at 120 minutes post dosing).

1 What are the common causes of malabsorption in the adult horse?
2 What further diagnostic tests could be performed to elucidate the precise cause of this horse's intestinal malabsorption?

CASE 131 An 8-year-old Thorough-bred gelding (**131.1**) used as a high-level dressage horse was 'resting' in the off-season with turnout and was believed to be completely healthy. Associated with the end of the work season, he was administered an oral herbal 'liver flush'. Two days later, the horse began to paw and stretch out indicating colic signs and was given a

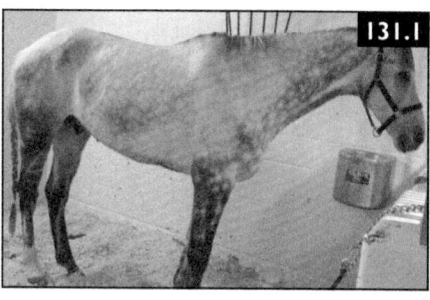

500 mg dose of flunixin meglumine that day and the following day. The colic signs persisted, and the horse became completely anorexic and developed soft manure. Temperature and HR remained normal. On referral examination, the horse was bright and alert, HR was 44 bpm, and rectal temperature was 101°F (38.3°C). Auscultation of the chest and abdomen were normal and GI borborygmi ausculted normally. Abdominal palpation per rectum did not reveal any anatomic abnormalities, but the horse appeared uncomfortable when the right quadrant (cecal area) was palpated. CBC was unremarkable except for a mild neutrophilia (8,500 cells/µl [8.5 × 10⁹/l]) and mild toxic changes. Plasma protein (6.6 g/dl [66 g/l]) and fibrinogen (200 mg/dl [2.0 g/l]) were normal.

1 What diagnostic procedures should be performed next?
2 Interpret the result of the procedure performed.

CASE 132 A 19-year-old Cob gelding presented with a multinodular mass of tissue on the plantar aspect of the left hind pastern (**132.1**). This lesion had been getting progressively larger over a period of 4 years. The horse was affected by chronic pastern dermatitis in the other three legs, with evidence of lichenification, hyperkeratosis, and fissuring. The left hindlimb was chronically swollen with edema. There was a chronic purulent discharge from the lesion and the horse was mildly lame.

1 What is the diagnosis?
2 What is the cause of this lesion?
3 What treatments should be recommended?

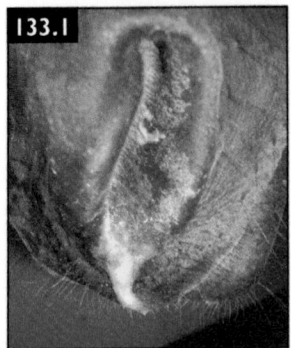

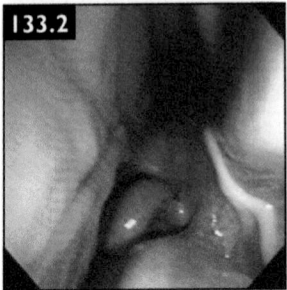

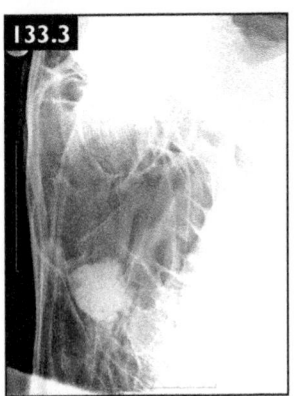

CASE 133 A 20-year-old Cob gelding presented because of a persistent left unilateral purulent nasal discharge. This horse, as well as several other horses on the same premises, was affected by strangles (*Streptococcus equi* subsp. *equi* infection) prior to the onset of the unilateral nasal discharge. All of the horses except this one recovered spontaneously. This horse had received several courses of antimicrobial therapy since the onset of the problem; this resulted in improvement of the clinical signs while the horse was on treatment, followed by worsening of signs when treatment finished. There was no cough or dysphagia and appetite was normal. On examination, the horse appeared bright and alert with normal HR, RR, and rectal temperature. Auscultation of the heart and lungs was unremarkable. There was enlargement of the left submandibular lymph nodes. There was a persistent left-sided purulent nasal discharge (not malodorous) (**133.1**). A swab from the left nasal passage yielded a profuse growth of *Streptococcus equi* subsp. *equi*. Endoscopic examination revealed purulent material draining into the caudal left middle meatus (**133.2**), presumably from the nasomaxillary opening; the nasopharynx, larynx, trachea, and guttural pouches appeared normal. A lateral-lateral radiograph of the frontomaxillary area was obtained (**133.3**).

1 What is the diagnosis?
2 What is the significance of the profuse growth of S. *equi*?
3 How should this horse be treated?

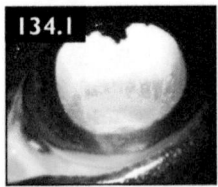

CASE 134 A 13-year-old Quarter Horse mare presents with a 3-week history of blepharospasm and epiphora. The horse has not responded to topical antibiotic therapy and the cornea is fluorescein negative (**134.1**).

1 What is the most likely diagnosis?

CASE 135 A 6-month-old Cob filly presented with a history of depression, inappetence, weight loss, and ventral subcutaneous edema of 2 weeks' duration (**135.1**). The filly had been normal prior to this history and had been weaned 1 month earlier. Fecal production was reduced and the feces were hard and dry. Rectal temperature was 102.2°F (39.0°C), HR = 80 bpm, and RR = 22 bpm. Palpation of the neck revealed hard and painful tissue

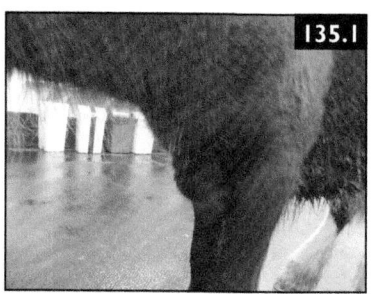

surrounding the nuchal ligament; similar tissue was palpated in the inguinal regions. Hematology and serum biochemistry revealed anemia (PCV = 24%; hemoglobin = 5.9 g/dl [59 g/l]), leukocytosis (WBCs = 14,100/µl [14.1 × 10⁹/µl]), hypoproteinemia and hypoalbuminemia (TP and albumin concentrations 5 g/dl [50 g/l] and 1.9 g/dl [19 g/l], respectively), increased AST (1,218 IU/l) and increased CK (594 IU/l). A thoracic radiograph was obtained (**135.2**) and abdominal ultrasonography (**135.3**) was performed.

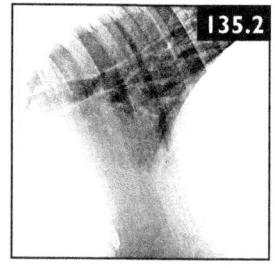

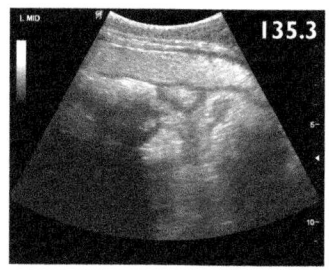

1 What abnormalities can be identified in the radiograph and ultrasonogram?
2 What is the provisional diagnosis?
3 What gross pathologic findings would be expected on necropsy with this disease?

CASE 136 A 3-year-old Thoroughbred colt presented with an abnormal hindlimb gait characterized by intermittent hyperflexion of either hock (**136.1**). The signs were more frequently seen when the horse was turning. The condition suddenly appeared 4 weeks prior to presentation and the signs were becoming more frequent. The horse appeared well in all other respects and did not show the signs when at rest.

1 What is this condition?
2 What is its cause?
3 How should this horse be treated?

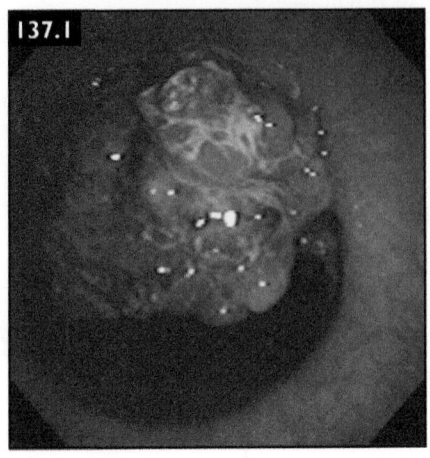

CASE 137 A 12-year-old Belgian Warmblood gelding presented with a 24-hour history of stranguria and hematuria. The owner reported that the horse had not had any previous problems with urination. On examination the horse was bright, alert, and responsive. All clinical parameters were within normal limits. Abdominal palpation per rectum revealed a partially full bladder with no palpable calculi. At the bladder neck an area of thickened soft tissue was palpable. Transrectal ultrasonography showed hyperechoic urine within the bladder and an ill-defined circular structure on the dorsal bladder wall on the left-hand side. The left ureter appeared thickened. Urinalysis revealed hematuria, proteinuria, and pyuria. Moderate amounts of epithelial cells, erythrocytes, and bacteria were seen on urine cytology. Bacterial culture of the urine yielded no growth. Hematology and serum biochemistry profiles were within normal limits. Cystoscopy revealed an intraluminal multinodular soft tissue mass originating from the left dorsal bladder neck (**137.1**). The left ureteral opening was obscured by the mass. The right ureteral opening appeared grossly normal.

1 What is the provisional diagnosis?
2 How could the diagnosis be confirmed?
3 What treatment options should be considered?

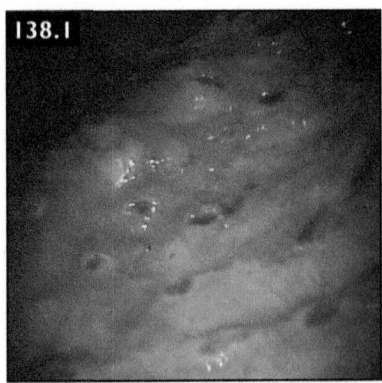

CASE 138 A routine gastroscopic examination (**138.1**) was performed in a 12-year-old Thoroughbred cross event horse gelding. The owner requested the examination because she felt that the horse's performance and exercise tolerance had declined slightly over the past few weeks.

1 What abnormalities can be seen on the gastroscopic image?
2 What clinical signs can be associated with this disease?
3 What treatment should be recommended?

CASE 139 A 6-year-old Warmblood mare presented with a 6-week history of progressive swelling medial and rostral to the left eye (**139.1, 139.2**). The mare had some nose bleeds 6 weeks prior to presentation, but these spontaneously resolved. The swelling was getting progressively larger and had changed from a soft swelling to a hard bony lump. There was also a smaller swelling in a similar position on the right side. Both eyes were demonstrating epiphora. The swellings were hard and painless. There was bilaterally symmetrical airflow through the nostrils and no nasal discharge was present. There was no palpable lymphadenopathy of the head. Endoscopic examination of the upper airways via both nasal passages was unremarkable. A lateral-lateral radiograph was obtained (**139.3**) and transcutaneous ultrasonography was performed (**139.4**).

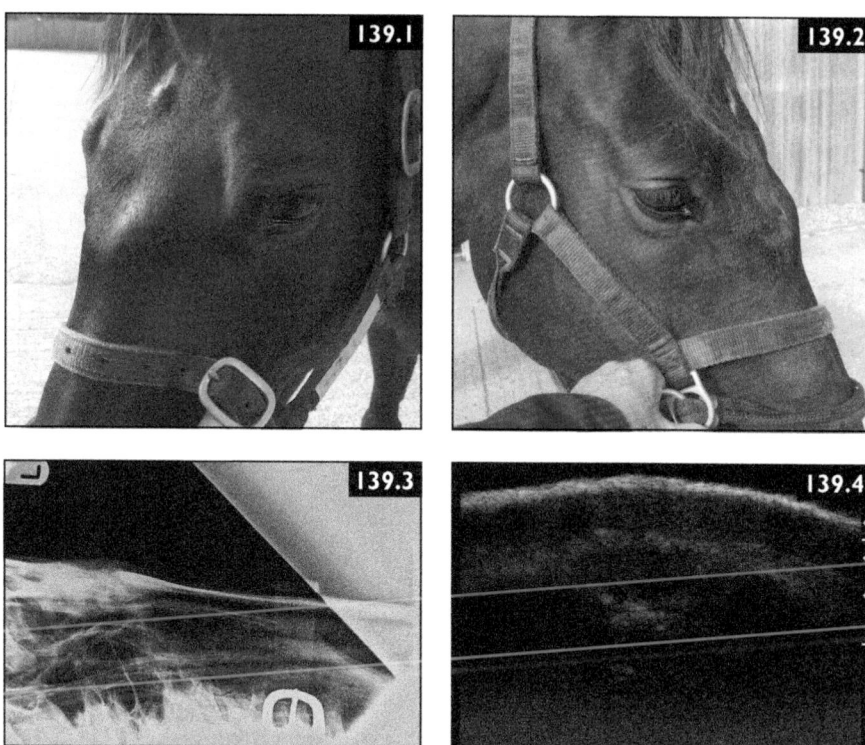

1 What is the diagnosis?
2 What is the likely cause of this condition?
3 How should this case be managed?

CASE 140 A 21-year-old Cob gelding presented with a 6-week history of progressively worsening dyspnea and inspiratory stridor with an occasional cough. The horse was reported to be fit and well otherwise. On examination, the horse demonstrated moderate inspiratory dyspnea, which worsened when he was excited or stressed. A harsh inspiratory noise was present. Ausculation of the chest was normal other than the presence of referred upper airway noise. A temporary tracheotomy was performed, which immediately alleviated the respiratory distress and stridor. An endoscopic examination per nasum (**140.1**) and via the tracheotomy (**140.2**) was performed.

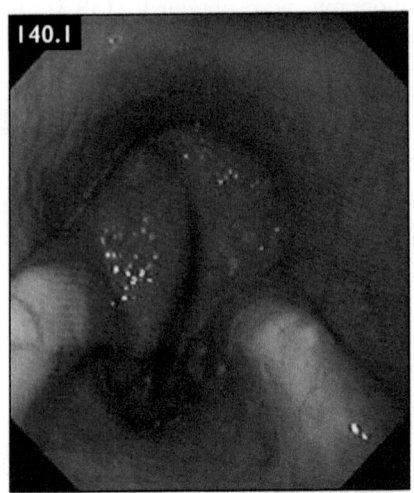 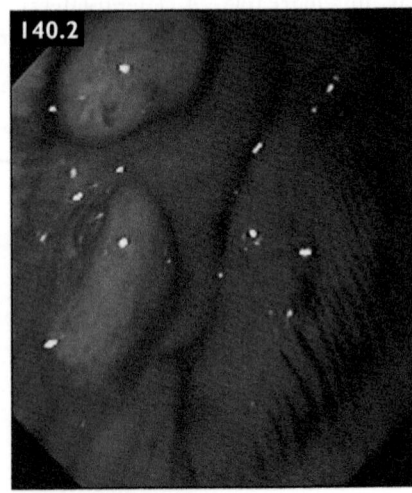

1 What is the likely diagnosis?
2 What is the cause of this condition, and what are the typical diagnostic features?
3 What treatment options are available?

CASE 141 A 5-year-old Irish Sport Horse gelding presented with multiple firm dermal nodules, 5–15 mm in diameter, on both sides of the neck and along the back (**141.1**). The nodules were painless and appeared to be organized in chains in the neck region.

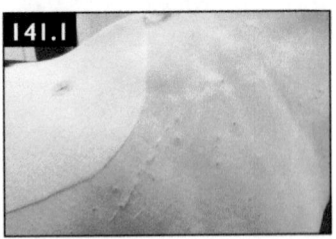

1 What is the most likely diagnosis?
2 How can the diagnosis be confirmed?
3 What is the etiology of these lesions?
4 How should this disease be treated?

CASE 142 A 20-year-old pony mare presented in acute respiratory distress. The pony was retired and lived out at grass; she had a history of mild chronic coughing over several years, which generally required no specific treatment. The owner reported that the pony suffered a similar episode of respiratory distress 2 years before, but that it resolved over a few days when treated with oral clenbuterol. The owner had already administered clenbuterol to the pony on this occasion, but it did not appear to have helped. On examination, the pony showed marked tachypnea (RR = 32 bpm) with flared nostrils (**142.1**) and abdominal effort on expiration (**142.2**). There was a mild bilateral mucopurulent nasal discharge and a spontaneous cough. Auscultation of the lungs revealed tachycardia (HR = 56 bpm), widespread crackles, and wheezes. Routine hematology and biochemistry were unremarkable apart from a mild hyperfibrinogenemia (fibrinogen = 480 mg/dl [4.8 g/l]).

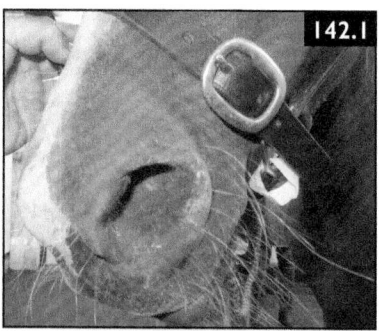
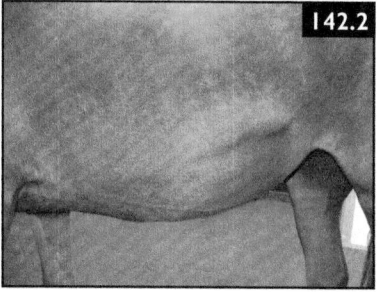

1 What condition is suspected, and what differentials should be considered?
2 What diagnostic tests can be performed to confirm the diagnosis?
3 How should this pony be treated?

CASE 143 A 20-year-old riding pony gelding presented with a swelling of the glans penis (**143.1**). The owner was unaware of the swelling until the horse was sedated for routine dental treatment, at which time the penis was extruded and the mass was visible.

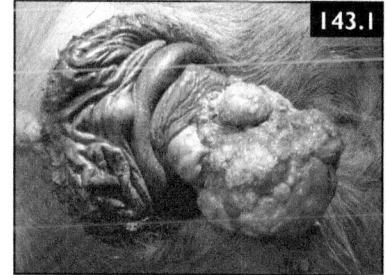

1 What is the most likely diagnosis?
2 What treatment options should be discussed with the owner?
3 What will likely happen if the owner decides not to treat this lesion?

103

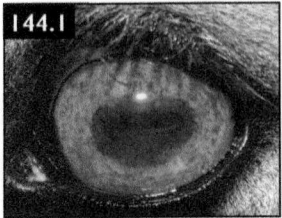

CASE 144 A 9-year-old Norwegian Fjord gelding presents for evaluation of anterior uveitis (**144.1**).

1 What clinical signs support this diagnosis?
2 What are the differential diagnoses?

CASE 145 A 5-year-old riding horse presented with a history of exercise intolerance with vibrant inspiratory and expiratory noises at exercise and coughing after eating. Endoscopic examination of the upper airways was performed (**145.1, 145.2**).

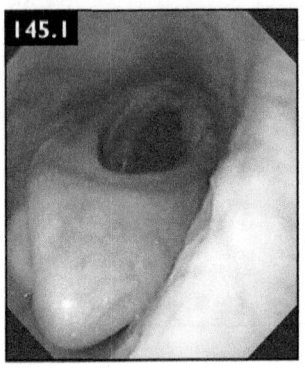

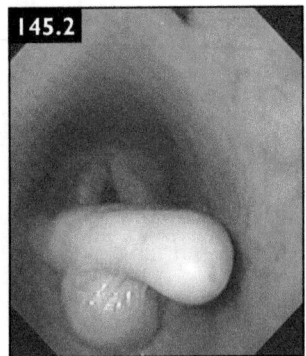

1 What is the diagnosis?
2 What is the etiology of this condition?
3 How can this condition be managed?

CASE 146 A 3-year-old Thoroughbred gelding presented with a single dermal nodule on the midline back (**146.1**). The nodule was 12 mm in diameter and non-painful. The nodule was present when the animal was a foal, but had gradually enlarged since then. The horse was being broken in and the owners were concerned that the nodule would become traumatized.

1 What is the likely diagnosis?
2 What is the cause of this lesion?
3 What treatment should be recommended?

CASE 147 An 18-year-old Irish Draft Thoroughbred cross gelding was successfully treated for severe colitis and diarrhea due to salmonellosis (**147.1**). Treatment had included IV fluids and plasma therapy. A few days after cessation of therapy and resolution of the diarrhea, the horse presented with a painful soft tissue swelling of the proximal neck centered on the site of IV catheterization of the right jugular vein (**147.2**). The horse was pyrexic (rectal temperature = 103.6°F [39.8°C]), depressed and inappetent. Routine hematology and serum biochemistry revealed leukocytosis, neutrophilia (10,200 cells/µl [10.2 × 10^9/l]) and hyperfibrinogenemia (760 mg/dl [7.6 g/l]). Ultrasound examination of the area was performed (**147.3**).

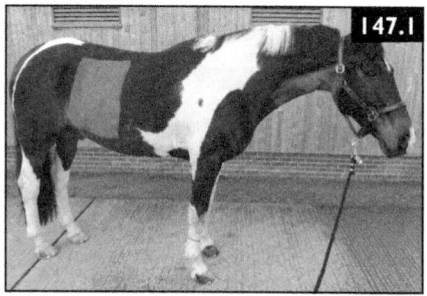

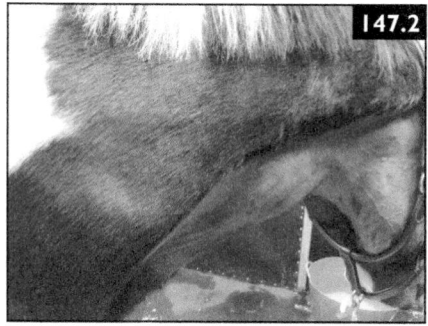

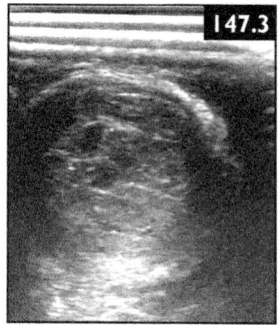

1 What is the diagnosis?
2 What further investigations should be performed?
3 What treatments should be considered?

CASE 148 A 4-year-old Cleveland Bay gelding, who had developed multiple raised depigmented papules on the medial surface of both pinnae 12 months earlier, presents because the lesions have now become scaling and flaking (**148.1**). The ears appear to be painful and the horse resents them being handled. There are no other skin lesions on the horse's body.

1 What is the diagnosis?
2 What is the likely etiology?
3 What treatment should be recommended?

CASE 149 A 7-year-old Sport Horse gelding presented with an acute onset of hemorrhagic diarrhea (**149.1**). Several hours prior to the onset of diarrhea, the horse had shown signs of moderate colic pain. On examination the horse was dull and had congested mucous membranes with a prolonged CRT (2.5 seconds). HR was 72 bpm and RR was 16 bpm. Routine hematology and biochemistry revealed: hypovolemia (PCV = 71%; TS = 5.1 g/dl [51 g/l]), azotemia (urea = 38.7 mg/dl [13.8 mmol/l]; creatinine = 3.96 mg/dl [350 µmol/l]), lactate concentration = 6.8 mmol/l. Ultrasonography showed diffuse thickening of the small intestinal wall (8 mm) and thickening of the large intestinal wall (10 mm). IV fluid therapy (compound sodium lactate solution), fresh plasma, flunixin meglumine, polymixin B, procaine penicillin, gentamicin, and metronidazole were administered. Di-trioctahedral smectite was given via nasogastric tube. Despite these treatments, the horse deteriorated rapidly and died. At postmortem examination, severe hemorrhagic enterocolitis was identified (**149.2**). The kidneys appeared enlarged and histologic examination revealed multifocal, proximal acute tubular cell necrosis and tubular cell degeneration. Abundant hard, brown, shell-like material measuring (0.2–0.6 mm) was dispersed within the gastric, intestinal, and colonic contents (**149.3**).

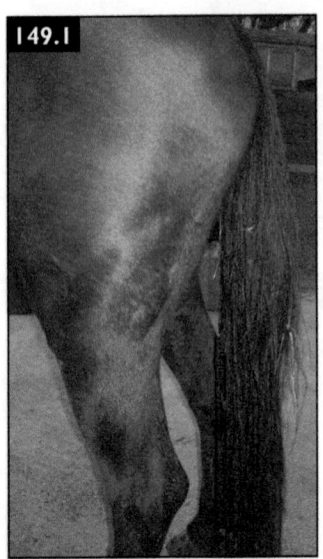

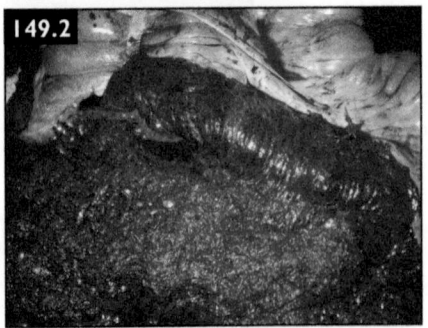

1 What condition is suspected?
2 What is the pathogenesis of this disease?

CASE 150 A 26-year-old Appendix gelding presents with a 2-month history of protrusion of the nictitans (**150.1**). The ophthalmic examination was otherwise unremarkable.

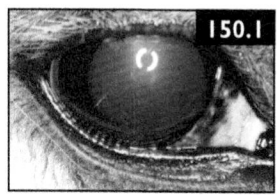

1 What are the differentials for this horse?

CASE 151 A 14-year-old Arabian cross gelding presented with a history of several bouts of bilateral forelimb lameness occurring over the previous 12 months. The horse was used for light hacking, kept permanently at pasture during the summer, and part-stabled during the winter. The bouts of lameness generally occurred during the spring/summer and resolved with time and phenylbutazone therapy (administered by the owner without veterinary advice). At the time of presentation, the horse had been bilaterally forelimb lame for 2 weeks without improvement on phenylbutazone treatment. On examination, he had a short-strided, laminitis-type gait with increased digital pulse amplitude in both front feet. There was a consistent pain response to the application of hoof testers to the sole at the toe in both front feet. Laminitis was diagnosed and radiographs obtained of both front feet (**151.1, 151.2**). The horse appeared well in all other respects apart from being moderately obese with patchy areas of subcutaneous fat deposition over the flanks (**151.3**), neck, and around the prepuce.

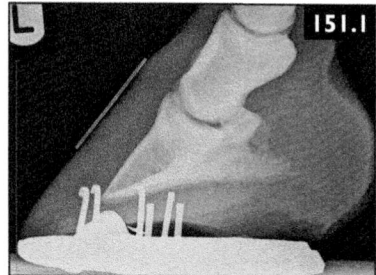

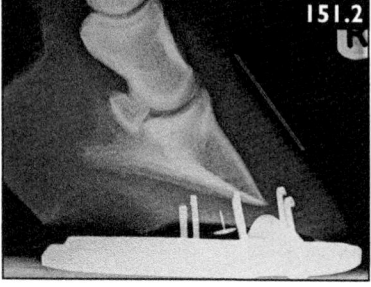

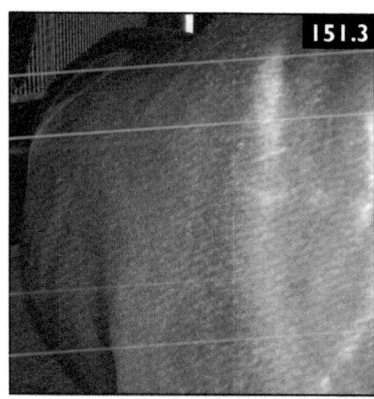

1 What abnormalities do the radiographs show?
2 What is the significance, if any, of the body condition score and fat deposits? What condition is suspected, and how does this relate to the lameness?
3 How can the suspicion be confirmed?
4 How should this case be managed?

CASE 152 A 10-year-old Warmblood cross gelding had a history of right-sided purulent aural discharge 2 months ago (**152.1**). At the time, the discharge appeared to resolve with a course of oral potentiated sulfonamides. Since then, the horse has demonstrated intermittent horizontal headshaking while eating and during exercise. The horse also has appeared mildly depressed and lethargic. Clinical and neurologic examinations were unremarkable. Routine hematology and serum biochemistry revealed mild hyperfibrinogenemia (470 mg/dl [4.7 g/l]). Endoscopic examination of the upper airways showed mild swelling of the proximal aspect of the right stylohyoid bone within the right guttural pouch. Routine radiographs of the head were unremarkable. A standing CT examination of the head was undertaken under sedation (**152.2**).

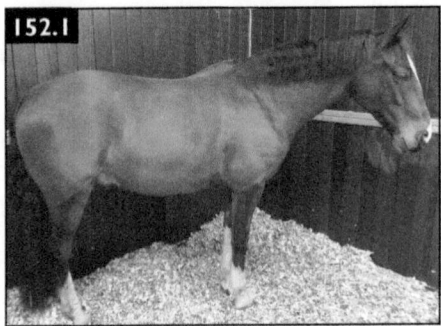 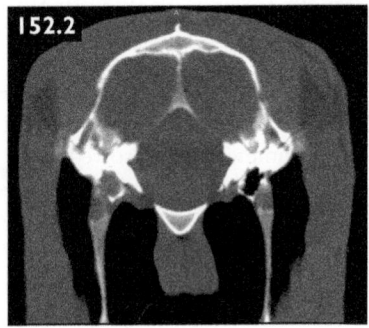

1 What abnormalities can be identified on the CT image, and what is the diagnosis?
2 What treatment options should be considered?

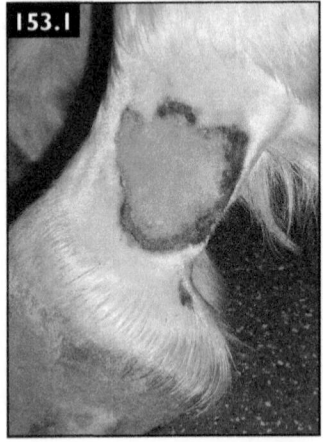

CASE 153 A 7-year-old Arabian mare presented with a 6-week history of dermatitis of the left hind lateral pastern region (**153.1**). The owner had been treating the lesion believing that it was a form of 'mud fever' (dermatophilosis); the horse had been kept stabled, and the affected area had been clipped, washed with mild chlorhexidine solution, and kept clean and dry. Despite these treatments the lesion was getting progressively larger.

1 What conditions are suspected?
2 What further diagnostic tests should be performed?
3 What treatments should be considered?

CASE 154 An 18-year-old Thoroughbred cross riding horse presented with a history of bilateral mucopurulent nasal discharge (**154.1**) and worsening dyspnea over a 6-week period. The horse had received a 7-day course of oral potentiated sulfonamides, which resulted in a temporary improvement in the nasal discharge, but it worsened again once the treatment stopped. The discharge recently became more profuse and malodorous. On examination, the horse was quiet with mild tachypnea (RR = 20 bpm) and inspiratory dyspnea at rest. Rectal temperature was normal and auscultation of the lungs was unremarkable. Endoscopic examination of the upper respiratory tract was performed (**154.2, 154.3**).

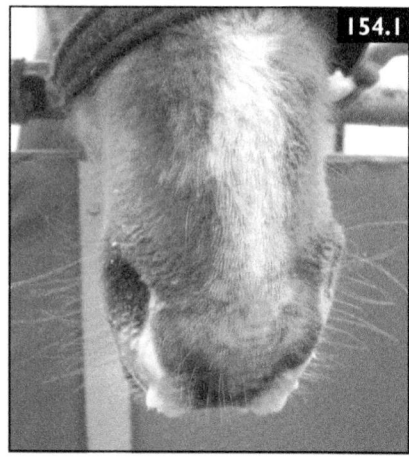

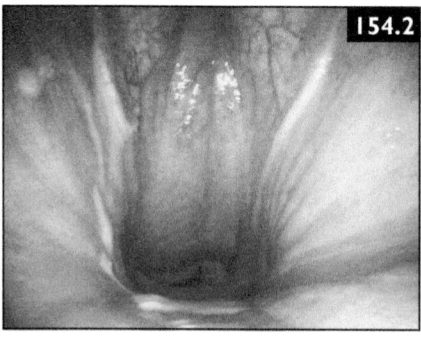

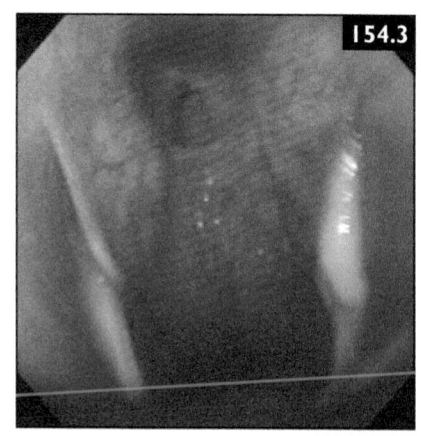

1 What is the diagnosis?
2 What is the etiology of this condition?
3 How should this disease be treated?

CASE 155 An 18-year-old American Saddle Horse gelding presents for evaluation of infected ulcerative keratitis (**155.1**).

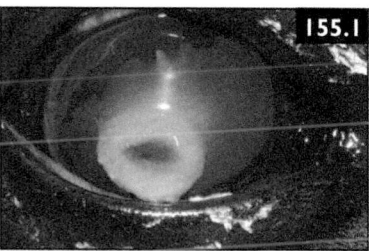

1 What diagnostic tests are indicated?
2 What are the most common etiologic agents causing melting ulcers in horses?

CASE 156 A 17-year-old Cob gelding presented with severe right forelimb lameness associated with soft tissue swelling of the caudal right cervical region and muscle atrophy over the right scapular area (**156.1, 156.2**). The swelling and muscle loss had been rapidly worsening over a period of 2 weeks. The lameness appeared suddenly 2 days prior to presentation. On examination, the horse was extremely lame on the right fore and reluctant to bear weight on this leg. Rectal temperature was normal, HR was 60 bpm, RR was 20 bpm. The swelling over the right caudal cervical region was painful to deep palpation. On palpation of the right scapular region, there was muscle loss over the supraspinatus and infraspinatus regions, and the spine of the scapula was more prominent than the contralateral side. Radiography of the caudal cervical and shoulder regions showed no identifiable osseous abnormalities. Ultrasonography of the bilateral scapular regions (**156.3**) and the right caudal cervical region (**156.4**) was performed.

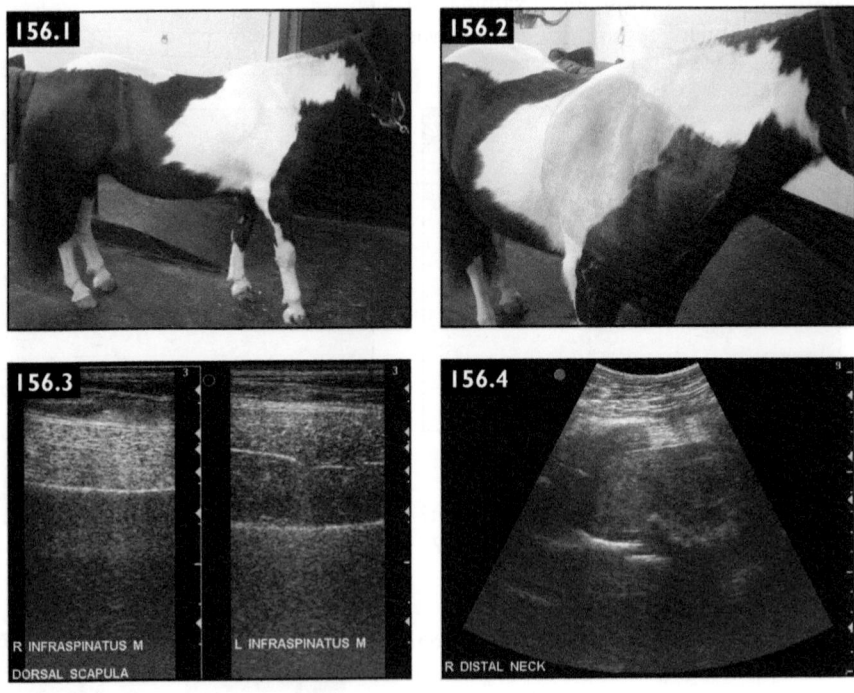

1 Describe the abnormalities seen in the ultrasound images.
2 What disease process is suspected, and how can the diagnosis be confirmed?

CASE 157 An 8-year-old Standardbred mare presents for a 2-week history of head shaking and vision impairment (**157.1**).

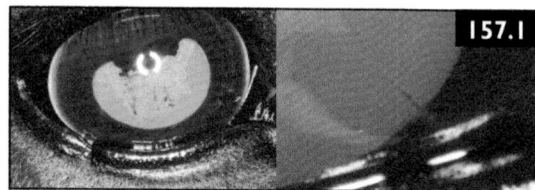

1 Describe the abnormalities and recommend treatment.

CASE 158 An 8-year-old Warmblood cross gelding presented with a 3-week history of bilateral mucopurulent nasal discharge (**158.1**). The horse was affected by a mild respiratory infection 8 weeks earlier that resolved after 10 days rest. Several other horses in the same yard were affected at the same time as this horse, but they all made complete recoveries with no complications. On clinical examination this horse was bright and alert, with normal RR, HR, and rectal temperature. The submandibular lymph nodes were mildly enlarged and the horse appeared to resent palpation of the parotid region. Endoscopic examination showed a discharge draining from the pharyngeal ostia of both guttural pouches (GPs). On passing the endoscope into the pouches, several white ovoid structures were observed in both pouches (**158.2**).

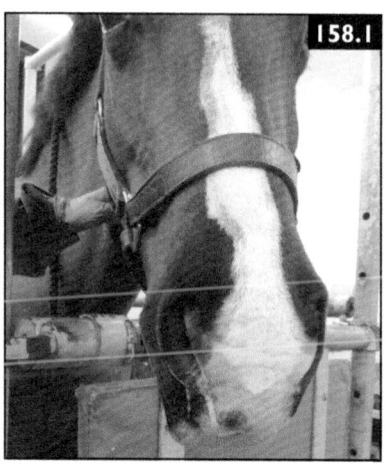 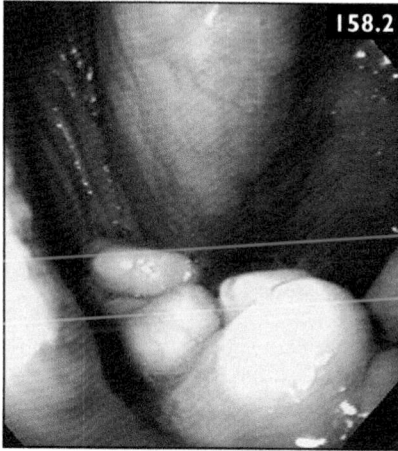

1 What are these structures?
2 What is the most likely cause?
3 How should this case be treated?

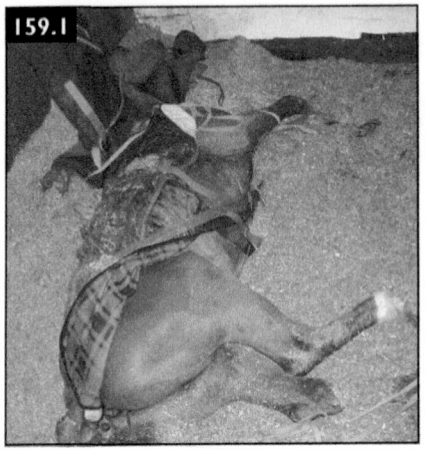

CASE 159 A 12-year-old Warmblood gelding had a 12-hour history of sudden onset of weakness and ataxia of the hindlimbs. By the time the horse was examined he had become recumbent (**159.1**) and unable to rise. The horse was kept in a large livery yard with 50 other horses; none of the other horses appeared to be affected, although the owner of the yard reported that several of them had been affected by a mild respiratory disease over the past few weeks, characterized by mild fever and mild bilateral mucopurulent nasal discharge. The horse was normothermic and mentally bright and alert. There appeared to be complete paralysis of the hindlimbs, tail, and anus. When encouraged to get up, the horse could adopt a dog-sitting position, but could not move his hindlimbs.

1 What condition should be suspected?
2 How can the suspicion be connfirmed?
3 What is the underlying pathogenesis of this disease?

CASE 160 A 20-year-old crossbred mare used for pleasure riding presents for examination of its left eye. The horse had developed a rapidly growing red plaque-like lesion of the cornea of the left eye (**160.1**); the lesion originated at the lateral

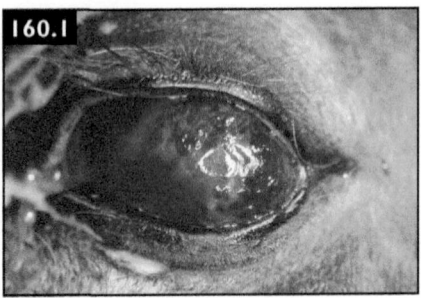

canthus and had been spreading over the cornea over the past 2 weeks. The owner thought the horse must have sustained a corneal ulcer and had been treating it symptomatically, but the lesion had been quickly getting worse. The mare had her right third eyelid removed surgically 18 months earlier for a focal squamous cell carcinoma (SCC) lesion of the nictitating membrane (confirmed histologically).

1 What is the likely diagnosis based on the gross appearance?
2 What treatments should be recommended?

CASE 161 A 36-hour-old pony filly presented with a history of depression, respiratory distress, and a 'rattling' respiratory noise. The birth was prolonged and the foal was slow to get up and suck from the dam. Since birth, the owner had needed to help the foal get to the teats and to supplement the feeding with a bottle. On examination the foal was depressed and tachypneic (RR = 60 bpm) (**161.1**). HR was 120 bpm and the rectal temperature was 101.1°F (38.4°C). An audible rattle was heard as the foal breathed. Relevant blood abnormalities included leukopenia (WBCs = 1,200/µl [1.2 × 10⁹/l]), elevated PCV (50%), elevated serum amyloid A (923 mg/l) and hyperfibrinogenemia (fibrinogen = 660 mg/dl [6.6 g/l]). Serum IgG concentration was normal (900 mg/dl). Auscultation of the lungs revealed marked wheezing and crackling ventrally on both sides of the chest. A thoracic radiograph was obtained (**161.2**).

 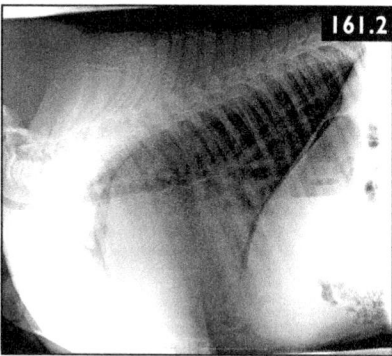

1 What abnormality can be seen in the thoracic radiograph, and what is the provisional diagnosis?
2 What other diagnostic imaging modality could be considered to help confirm the diagnosis?
3 What can cause this condition?

CASE 162 A 20-year-old Quarter Horse gelding presents for evaluation of a limbal mass noted 4 months previously (**162.1**).

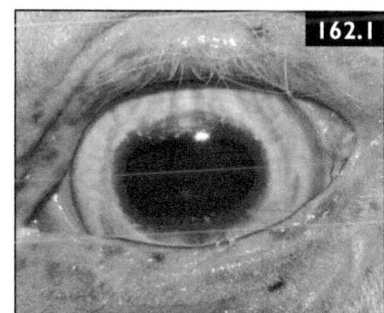

1 What is the most likely diagnosis?
2 What treatment is indicated?

CASE 163 A 13-year-old Irish Sport Horse mare had a history of an acute onset of foul-smelling, right-sided nasal discharge 8 weeks prior to examination. At the same time, the horse started showing dysphagia (quidding) and disorientated behavior. All the clinical signs improved after 10 days of a combination of oral and injectable antimicrobials. At the time of presentation, the mare had mild right-sided epistaxis. Radiographs of the paranasal sinuses obtained by the referring veterinarian revealed a radiodense, well-defined opacity just caudal and dorsal to the 111 tooth. A standing CT examination was performed. A transverse multiplanar reconstruction (MPR) at the level of the rostral aspect of the orbits (caudal to teeth 111 and 211) (**163.1**) and a transverse MRP at the level of 109/209 (**163.2**) are shown.

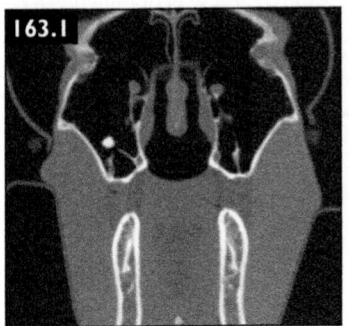 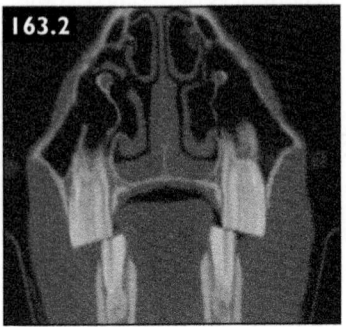

1 What abnormalities can be identified on the CT images, and how should they be interpreted?
2 What further examinations should be undertaken?

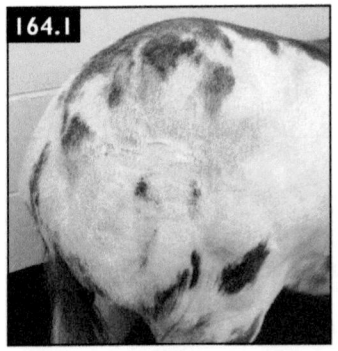

CASE 164 A 22-year-old female donkey (jenny) presented with a chronic draining wound over the gluteal area (**164.1**). The owners believed that there was a wound in this region 5 months previously, but it had never healed since then and constantly drained variable quantities of purulent material. The jenny objected to palpation of the area, but a large plaque of hard subcutaneous dermal and subcutaneous tissue, approximately 15 cm in diameter with several draining tracts at the adjacent skin surface, was noted.

1 What condition is suspected?
2 How could the diagnosis be confirmed?
3 What treatment options should be considered?

CASE 165 A 5-year-old Thoroughbred mare presented with a history of intermittent mild colic and weight loss over a period of 10 days. The mare was dull and inappetent, with mild abdominal distension. The bouts of colic had all responded to treatment with NSAIDs (phenylbutazone or flunixin meglumine). HR was 52 bpm, RR was 20 bpm, and rectal temperature was 101.1°F (38.4°C). Mucous membranes were pale but moist. Auscultation of the chest was unremarkable. No other significant abnormalities were detected on routine physical examination, but abdominal palpation per rectum revealed enlargement and caudal displacement of the spleen. Hematology and biochemistry revealed leukocytosis (WBCs = 12,800/µl [12.8 × 10^9/l]) and neutrophilia (PMNs = 11,500/µl [11.5 × 10^9/l]), hyperproteinemia (TP = 8.8 g/dl [88 g/l]) and hyperfibrinogenemia (fibrinogen = 688 mg/dl [6.88 g/l]). Abdominocentesis yielded an orange, slightly turbid sample of peritoneal fluid (**165.1**) with a TP concentration of 4.4 g/dl (44 g/l) and total nucleated cell count of 5,400/µl (5.4 × 10^9/l). Transcutaneous ultrasonography showed an enlarged spleen with areas of abnormal echogenicity (**165.2**).

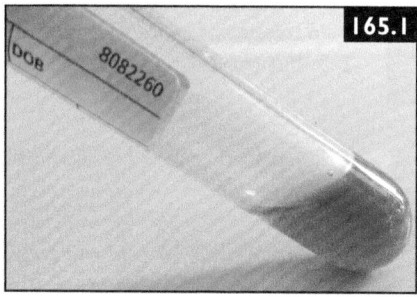 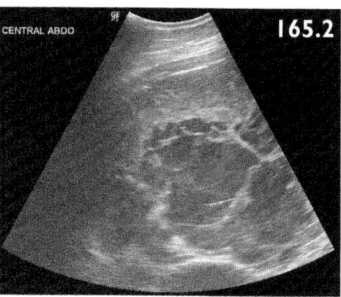

1 What is the most likely diagnosis?
2 What further diagnostic tests should be considered?

CASE 166 A 22-year-old Welsh pony gelding presents for a 3-month history of a slowly enlarging mass restricted to the nictitans (**166.1**).

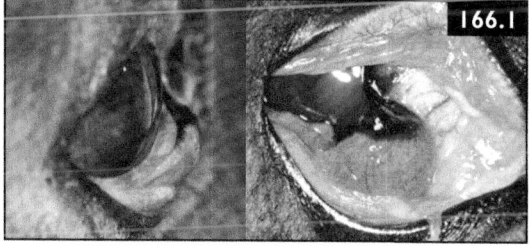

1 What are the differential diagnoses?

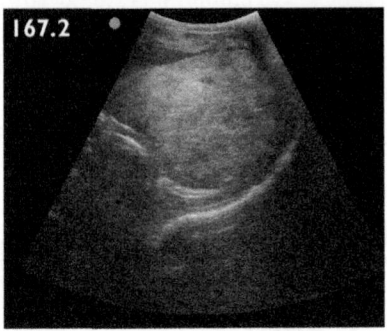

CASE 167 A 12-year-old Welsh Mountain Pony gelding presented with a swelling of the left proximal cervical area (**167.1**). The mass had been slowly enlarging over the past 3 months, and appeared to cause mild dysphagia. The pony was otherwise fit and well and had no previous relevant medical history. On examination, the pony's vital parameters were normal. The mass palpated as a firm soft tissue mass and was non-painful. Endoscopic examination of the nasopharynx and guttural pouches was normal. Routine hematology and serum biochemistry were normal. Ultrasound examination was performed (**167.2**).

1 What are the common causes of swellings in the proximal cervical region?
2 What should be suspected based on the physical examination and ultrasound findings?
3 What further tests are recommended?

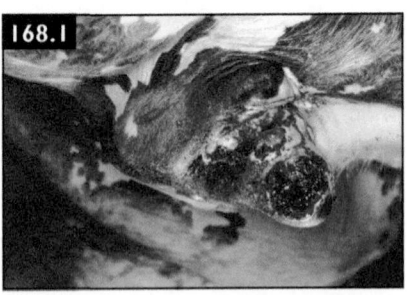

CASE 168 A 19-year-old crossbred mare presented with ulceration and swelling of the right mammary gland (**168.1**). The mare had delivered a foal 9 months earlier and developed swelling of the affected gland shortly after weaning the foal. The owner suspected mastitis, and the mare had received several courses of antibiotics with no improvement over the 6 weeks since the swelling was first noticed. The swelling had been getting progressively larger since that time and had recently become ulcerated. The mare appeared well in other respects and no abnormalities other than swelling of the mammary gland were found on a routine physical examination. The affected gland felt hard, hot, and painful.

1 What condition is suspected?
2 How would the diagnosis be confirmed?
3 What is the prognosis?

CASE 169 A 7-year-old pony gelding presented with a sudden onset of respiratory distress. Over the previous 6 weeks the pony had been inappetent and losing condition. On examination, the pony was dyspneic with inspiratory dyspnea, flared nostrils, and stridor (**169.1**). RR was 28 bpm and HR was 60 bpm. Rectal temperature was normal. Physical examination revealed poor body condition and the pony appeared depressed with periods of somnolence. In addition, he appeared to be blind and mildly ataxic. An endoscopic examination was performed (**169.2**).

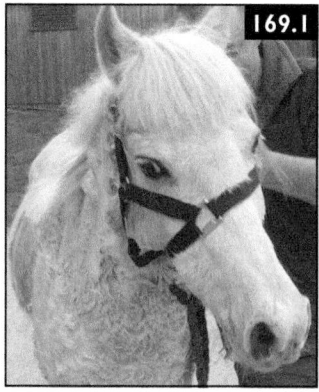 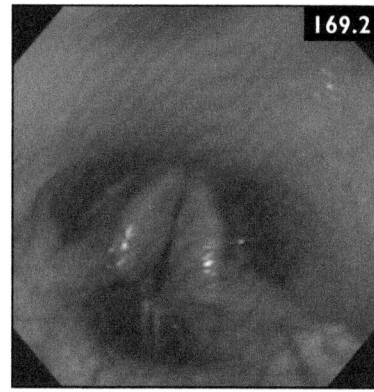

1 What abnormalities can be seen in the endoscopic image?
2 What disease process is suspected?
3 How could the suspicion be confirmed?

CASE 170 A 12-year-old Cob gelding presented with an 8-hour history of severe colic that had become unresponsive to treatment with analgesic drugs. On presentation the HR was 80 bpm, RR was 24 bpm; mucous membranes were pale and the CRT was prolonged (3 seconds). Transcutaneous ultrasound and abdominal palpation per rectum confirmed dilated immotile loops of small intestine. The horse underwent an exploratory celiotomy, during which entrapment of the mid-jejunum in a diaphragmatic hernia was diagnosed and treated. The horse made a good recovery from anesthesia but over the next 48 hours remained tachypneic (RR = 36 bpm). A lateral-lateral thoracic radiograph was obtained (**170.1**).

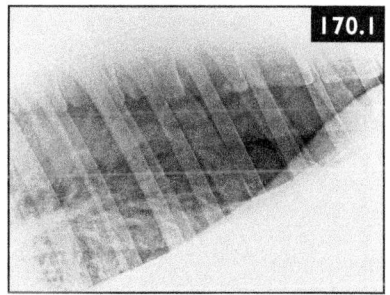

1 What abnormality can be identified in this radiograph?
2 What treatment should be recommended?

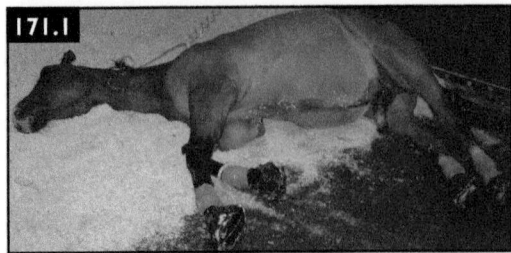

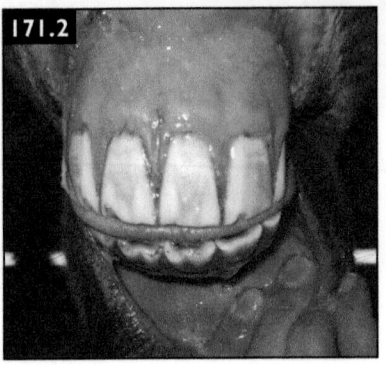

CASE 171 A 24-year-old Cob gelding presented after he broke into the feed room and ate a full 44 lb (20 kg) bag of barley. The horse was dull, depressed, and showing signs of mild abdominal pain (**171.1**). He was trembling, sweating, and had an elevated RR (28 bpm). HR was 60 bpm; mucous membranes were congested (**171.2**) with a CRT of 3 seconds. Abdominal borborygmi were reduced, but no specific abnormalities were detected on abdominal palpation per rectum. Hematology revealed leukopenia and neutropenia (WBCs = 3,800/µl [3.8 × 10^9/l]; neutrophils = 1,100/µl [1.1 × 10^9/l]).

1 What is the diagnosis?
2 What other complications are likely with this disease?
3 How should this case be managed?

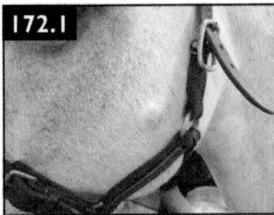

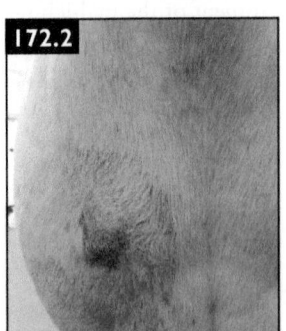

CASE 172 A 15-year-old Arabian cross pony mare presented with a 4-week history of a slowly growing cutaneous nodule on the left masseter region (**172.1**) and a deep subcutaneous mass over the right pectoral region (**172.2**). The pony appeared fit and well in all other respects, and no other abnormalities were detected on a full physical examination. Hematology and serum biochemistry profile were unremarkable. Fine needle aspirates from both masses had a similar cytologic appearance – eosinophils and mast cells.

1 What is the likely diagnosis?
2 What treatment should be recommended?
3 What is the prognosis following treatment?

CASE 173 A 26-year-old Cob Shire cross mare presented with a 48-hour history of inappetence and depression. Mucous membranes were pale pink with a CRT of 2 seconds. Abdominal sounds were reduced and the mare was reported to be passing slightly loose feces. There was mild dyspnea with a RR of 40 bpm; thoracic auscultation revealed muffled lung sounds ventrally. Temperature was 101.5°F (38.6°C) and HR was 56 bpm. The referring veterinarian had attempted abdominal paracentesis and had obtained blood on three separate attempts. Abdominal and thoracic ultrasound examinations were performed. An excess of swirling hyperechoic fluid was identified in both the peritoneal and thoracic cavities. Further evaluation of the abdomen revealed a heterogeneous echogenicity of the spleen with focal areas of increased echogenicity (**173.1**). Thoracocentesis yielded frank blood.

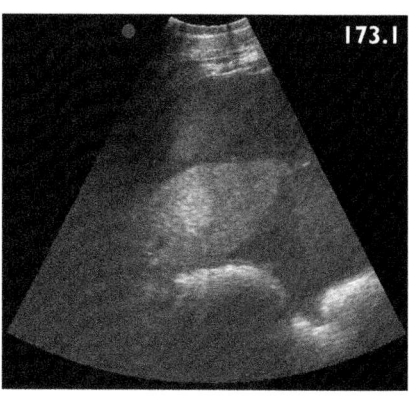

1 What condition is suspected?
2 How could the diagnosis be confirmed?

CASE 174 A recently acquired 5-year-old Cob mare presented for a mass protruding from the right eye (**174.1**). The previous history was unknown. The eye appeared uncomfortable with partial blepharospasm and a chronic mucopurulent ocular discharge. On close examination, the mass was noted to originate from the bulbar conjunctiva (**174.2**). The mass was excised and submitted for histopathology. The pathologist reported that the mass was composed of amyloid.

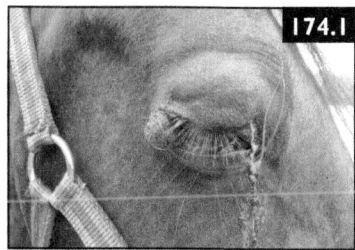
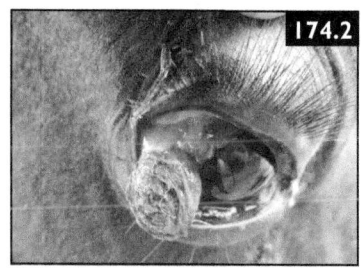

1 What is amyloid, and why does it occur?
2 What further evaluations should be recommended in this case?
3 What is the prognosis for this mare?

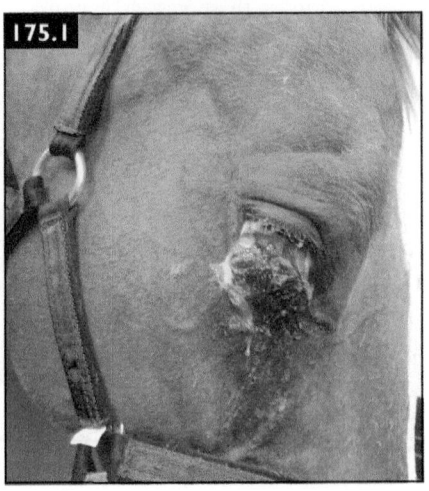

CASE 175 A 19-year-old Irish Draft Thoroughbred cross mare presented with a florid, ulcerated soft tissue mass protruding from the right eye (175.1). The mare was diagnosed with squamous cell carcinoma (SCC) of the third eyelid of the left eye 2 years previously. This had been treated by excision of the third eyelid followed by topical mitomycin therapy. The mass in the right eye appeared 6 weeks prior to this examination, and had enlarged rapidly since then. In addition, there were bilateral subcutaneous swellings of the parotid region and around the base of both ears. The mare had also recently started to make a mild inspiratory noise when exercising. The mare was otherwise well, eating and drinking normally, and not losing weight. The submandibular lymph nodes were mildly enlarged. Oral examination was unremarkable. Endoscopic examination of the upper respiratory tract revealed constriction of the nasopharynx (175.2). Endoscopic examination of the guttural pouches revealed multinodular masses at the floor of the medial compartment in both pouches (175.3).

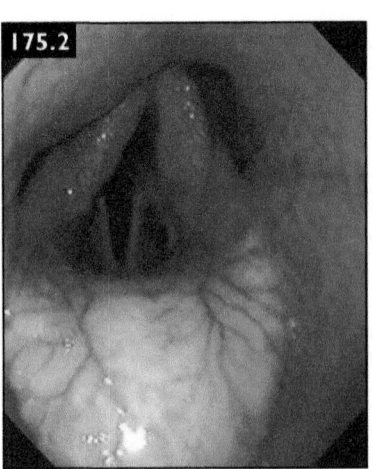
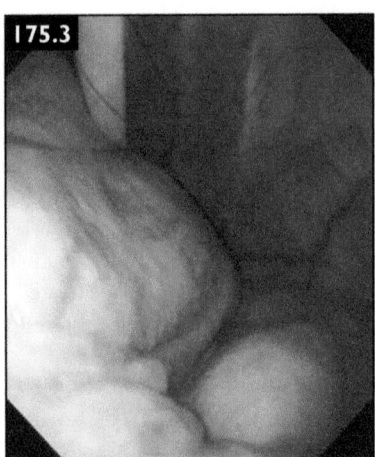

1 What is the most likely diagnosis?
2 How could the diagnosis be confirmed?
3 What treatment options are available?
4 What is the relevance of the history of SCC of the third eyelid of the left eye?

CASE 176 A 21-year-old Thoroughbred cross gelding presented with a 3-week history of intermittent mild right unilateral epistaxis (**176.1**). The horse appeared to be fit and well in all other respects and routine physical examination revealed no abnormalities other than epistaxis. Endoscopic examination revealed an abnormal mass in the region of the ethmoturbinates (**176.2**).

1 What condition is suspected?
2 What is the etiology of this disease?
3 What treatment options should be considered?

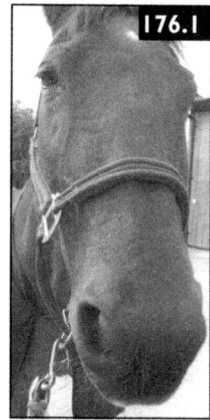
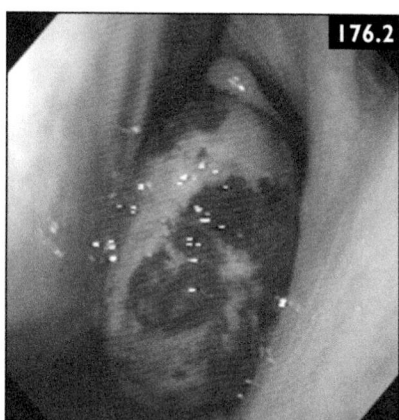

CASE 177 A 9-year-old Cob gelding presented with a 12-hour history of sudden onset of dysphagia, labored breathing, and stridor. The previous evening (after the horse came in from the field), he was noted to be 'squealing' when eating. On examination by the referring veterinarian, the horse appeared to be severely dysphagic and was drooling saliva intermittently. Rectal temperature and HR were normal, but the RR was elevated (36 bpm) and respirations were labored with an inspiratory stridor. A temporary tracheotomy was performed (**177.1**), which immediately alleviated the dyspnea and stridor. Endoscopic examination of the nasal passages, nasopharynx, and larynx identified swelling and inflammation of the larynx, more severe on the left side (**177.2**).

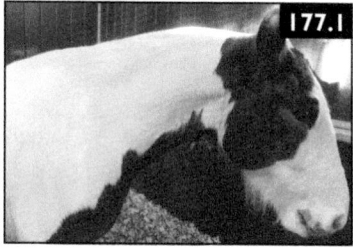
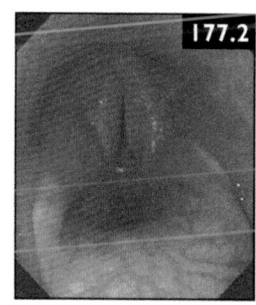

1 What conditions commonly cause sudden-onset dysphagia, dyspnea, and stridor?
2 What conditions should be suspected in this case?
3 What further investigations should be undertaken?

CASE 178 A 15-year-old Connemara mare presented with a 10-day history of mild depression, inappetence, weight loss, and recurrent mild bouts of colic (**178.1**). An oral glucose tolerance test was reported to be normal. On examination, HR was 48 bpm, RR was 14 bpm, and rectal temperature was 100.2°F (37.9°C). Mucous membranes were pale. No other significant abnormalities were detected on routine physical examination, but a mobile mass on the right side of the abdomen was identified on abdominal palpation per rectum. Hematology and biochemistry revealed leukocytosis (WBCs = 12,200/µl [12.2 × 10⁹/l]) and neutrophilia (PMNs = 10,600/µl [10.6 × 10⁹/l]), hyperproteinemia (TP = 8.4 g/dl [84 g/l]) and hyperfibrinogenemia (fibrinogen = 870 mg/dl [8.7 g/l]). Abdominocentesis yielded a slightly turbid sample of peritoneal fluid with a TP concentration of 2.8 g/dl (28 g/l) and total nucleated cell count of 2,800/µl (2.8 × 10⁹/l); cytologic examination was unremarkable. Transcutaneous ultrasonography showed a mass on the right-hand side of the abdomen (**178.2**).

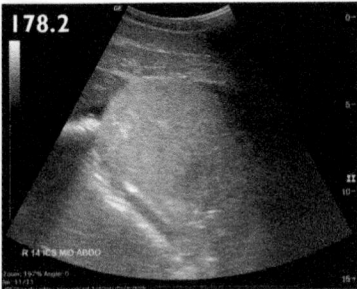

1 What abnormalities can be seen in the ultrasound image?
2 What further diagnostic procedures should be considered?

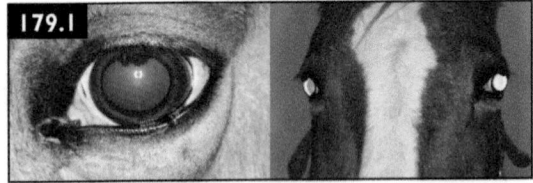

CASE 179 A 15-year-old Tennessee Walking Horse gelding presents for a 3-month history of progressive exophthalmos (**179.1**).

1 What are the differential diagnoses?

122

CASE 180 A 10-year-old Thoroughbred gelding presented with a history of left unilateral epistaxis. Ten days previously the horse had an episode of minor left unilateral epistaxis, which stopped spontaneously after a few hours. He then demonstrated more severe epistaxis the night prior to presentation. At the time of examination there was blood present at the left nostril (**180.1**) but no active bleeding. Endoscopic examination identified blood draining from the left guttural pouch (GP) ostium (**180.2**).

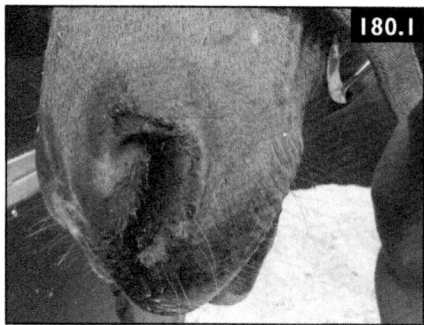

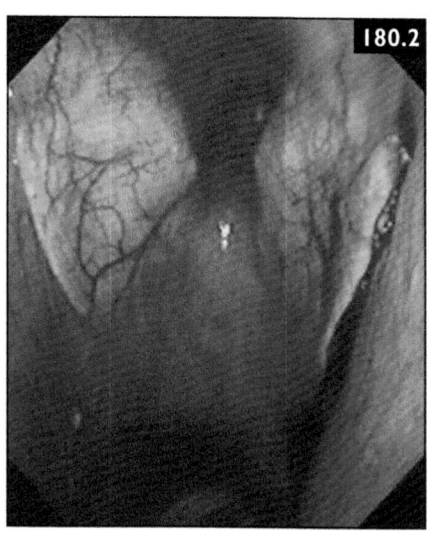

1 What are the possible causes of this epistaxis?
2 What could be done to obtain a definitive diagnosis of the cause?
3 How should this case be managed?

CASE 181 A 6-year-old Cob mare presented with a 3-week history of mild weight loss and a 5-day history of diarrhea (**181.1**). The feces had a semi-formed, 'cow-pat' consistency. Physical examination was unremarkable other than showing evidence of diarrhea. The mare was bright and alert but had a reduced appetite. Hematology and serum biochemistry revealed mild anemia (PCV = 24%) and hypoalbuminemia (albumin = 2 g/l [20 g/l]). Fecal examination revealed 1,700 strongyle epg. This horse and her pasture mates had been treated regularly (every 6–8 weeks) with either fendendazole or pyrantel pamoate; she received pyrantel pamoate 4 weeks ago.

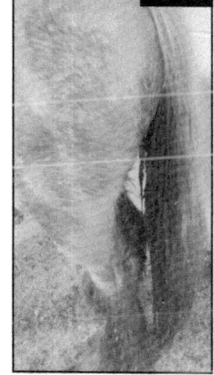

1 What parasite groups affecting adult horses would fenbendazole and pyrantel pamoate be expected to control?
2 How is the high fecal worm egg count explained in a horse that received a routine dose of pyrantel pamoate 4 weeks ago?
3 What further tests might be performed to evaluate this problem?

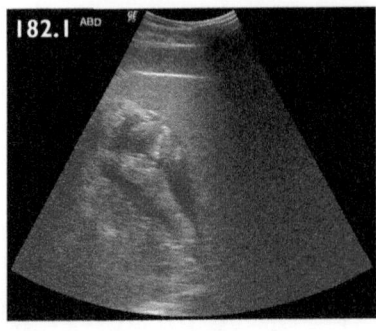

CASE 182 A 19-year-old Irish Sport Horse gelding presented with an 18-hour history of chronic, worsening colic signs, depression, and inappetence. On examination, the horse was showing signs of mild continuous colic. HR was 60 bpm, but the RR and rectal temperature were normal. Mucous membranes and CRT were normal. Intestinal sounds were absent in all areas. On abdominal palpation the rectum was empty and felt dry. No significant abnormalities were palpable on examination of the abdomen, but the horse showed a consistent pain response to palpation of the mid-abdomen. PCV was 53% and TS were 6.6 g/dl (66 g/l). Transcutaneous ultrasonography of the abdomen revealed moderately dilated loops of small intestine on both sides of the abdomen. The ultrasonographic appearance of the spleen is shown (**182.1**).

1 What abnormalities can be identified in the ultrasound image?
2 How should this case be managed?

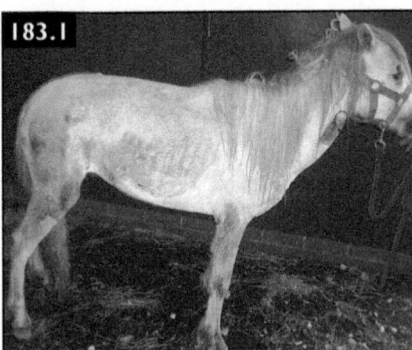

CASE 183 A 5-year-old pony mare presented with a 6-week history of weight loss and inappetence (**183.1**). In the past week, the pony had become severely depressed and developed dermatitis of the muzzle (**183.2**) and pasterns. Serum biochemistry revealed evidence of severe liver disease and liver failure (raised serum levels of GGT, AST, ALP, bilirubin, bile acids, and ammonia). The pony had been treated for the past 24 hours with IV polyionic fluids and dextrose, flunixin meglumine, B vitamins, pentoxifylline, and ceftiofur; the pony's general demeanor and appetite improved while on this treatment but she then started to show signs of abdominal pain and esophageal obstruction (excessive salivation, retching, coughing up saliva and food).

1 What is the cause of the dermatitis of the muzzle, and what is the underlying pathogenesis of this disease?
2 What is the likely cause of the colic and esophageal obstruction?

CASE 184 A 21-year-old New Forest Pony mare presented with a 3-week history of recurrent mild colic, depression, inappetence, and weight loss. On examination the pony appeared quiet and lethargic; HR was 40 bpm, RR was 16 bpm, rectal temperature was 100.0°F (37.8°C). Mucous membranes were pink with a CRT of 2 seconds. PCV was 38% and TP was 7.9 g/dl (79 g/l) with a plasma fibrinogen concentration of 880 mg/dl (8.8 g/l). The mare was not showing signs of

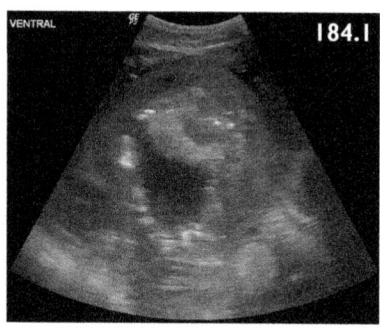

colic at the time of examination, but had recently received a dose of flunixin meglumine administered by the referring veterinarian. Abdominal palpation per rectum revealed a dry, tacky rectum with small amounts of dry feces. Multiple loops of distended small intestine were palpable (confirmed by transcutaneous ultrasonography). Abdominal palpation per rectum also revealed a firm, mobile, soft tissue mass in the right ventral abdomen. Transcutaneous ultrasonography of this area was performed (**184.1**). Repeated abdominocentesis attempts failed to yield a sample of peritoneal fluid.

1 What abnormalities can be identified in the abdominal ultrasonogram?
2 Based on the history and clinical findings, what disease process should be considered?
3 What further diagnostic tests should be recommended?

CASE 185 A client recently purchased an unbroken 4-year-old Warmblood gelding with a view to bringing the horse on as an eventer. Soon after purchase, the client noticed that the horse had a chronic cough, and when turned out to grass, a green-colored nasal discharge was apparent. Physical examination was largely unremarkable and the HR and RR were within normal limits. Auscultation of the lungs was normal, but the horse made audible inspiratory and expiratory adventitious noises when exercised. An endoscopic examination was performed (**185.1, 185.2**).

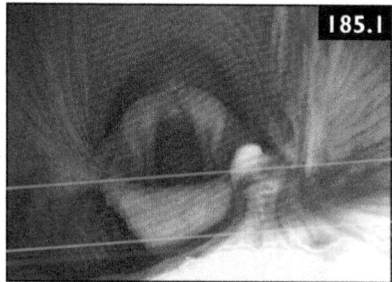

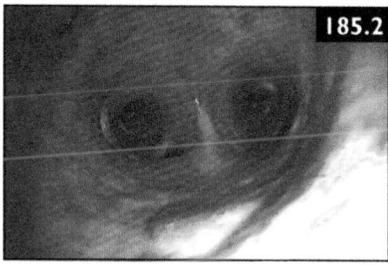

1 What abnormalities can be identified?
2 What advice should be offered to the client?

CASE 186 A 13-year-old Welsh Section C riding pony presented with a 6-week history of dyspnea, an exercise-related respiratory noise, cough, dysphagia, and nasal return of food (**186.1**). The referring veterinary surgeon had examined the pony endoscopically and made a provisional diagnosis of epiglottal entrapment. A lateral-lateral radiograph of the pharynx and larynx was obtained (**186.2**). Endoscopic examination *per nasum* (**186.3**) and *per os* (under GA) (**186.4**) was performed.

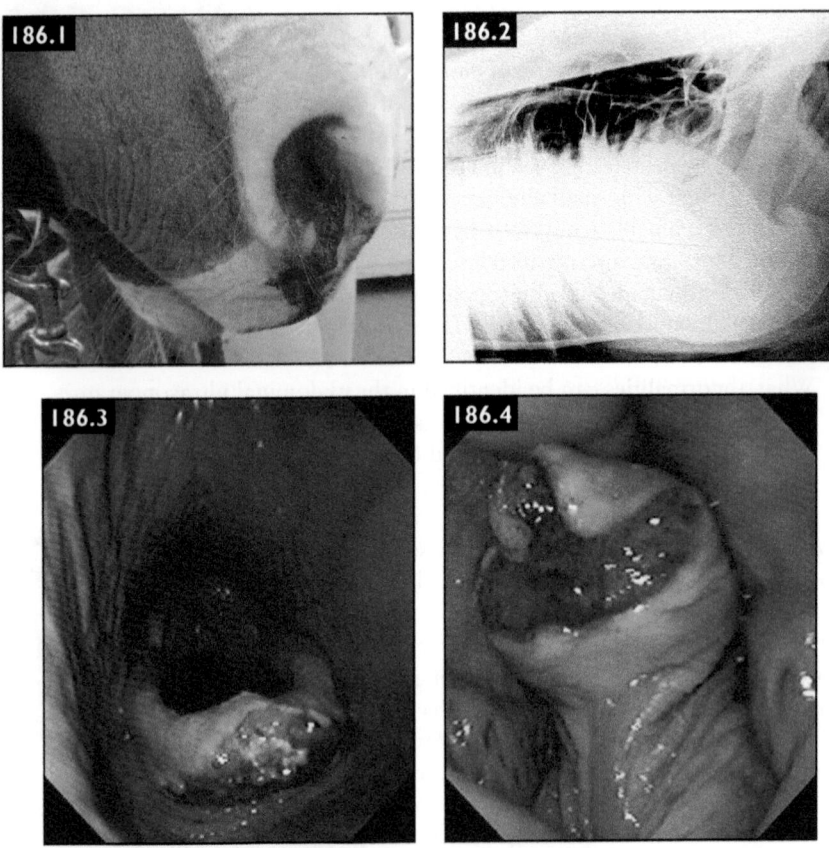

1 What abnormalities can be seen in the radiograph and the endoscopic images?
2 What are the differential diagnoses?
3 How should this case be further investigated?

CASE 187 A 5-year-old crossbred gelding was found in the field showing neurologic signs. There was no history of previous problems with the horse. The horse appeared to have been stuck in the fence because there was evidence of disruption of the fence posts and disturbance of the vegetation in one area (**187.1**). On examination the horse appeared distressed and showed marked ataxia (grade 4/5) of both the fore- and hindlimbs. There were no obvious CN signs. He was weak and there was an urticarial reaction confined to the left side of his body (**187.2**).

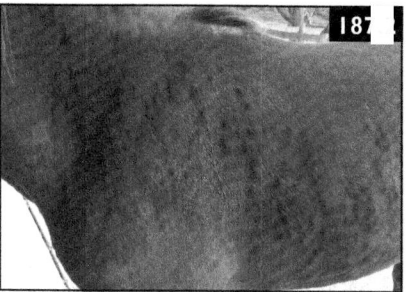

1 What is suspected to have caused the skin condition?
2 What treatment(s) should be recommended?

CASE 188 A routine endoscopic examination of the upper respiratory tract is performed as part of a pre-purchase examination of a 7-year-old gray Thoroughbred cross 3-day event horse. No abnormalities have been found during the physical examination or during strenuous exercise. Endosopy of both guttural pouches revealed these lesions (**188.1, 188.2**).

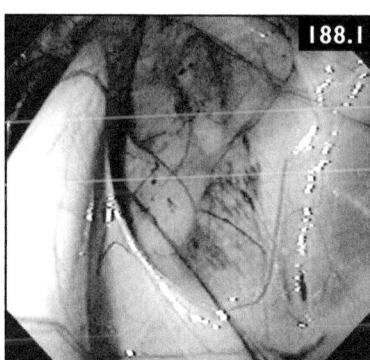

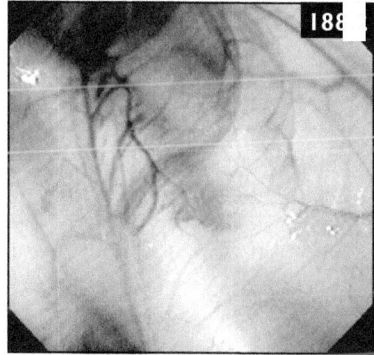

1 What are these lesions?
2 What advice should be given to the potential purchaser about the relevance of these lesions?

CASE 189 An 18-year-old Thoroughbred cross mare underwent a surgical procedure (removal of a sequestrum from the right hind cannon region) under GA. The horse was put into a padded recovery box following completion of the surgery, but failed to get up (**189.1**). After several hours, the horse could sit up in a dog-sitting position (**189.2**) but was unable to rise.

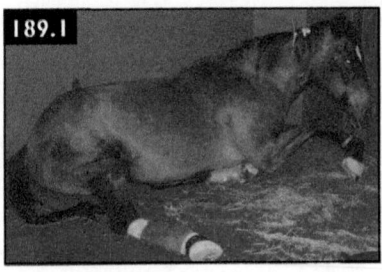 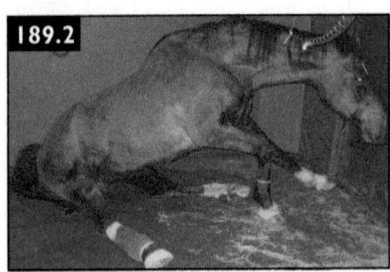

1 What abnormality is suspected?
2 What is the cause of this syndrome?
3 What can be done to reduce the risk of this disease?

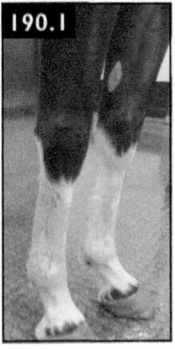 

CASE 190 A 7-year-old Thoroughbred cross colored (piebald) mare presented with skin lesions of the non-pigmented areas of both forelimbs and hindlimbs (predominantly in the cannon region) (**190.1, 190.2**). Other non-pigmented areas of skin were not affected. The lesions were more severe in the right forelimb, and more severe laterally. They were characterized by well-demarcated irregular areas of alopecia, erythema, oozing, and crusting. There was no pruritus, and the affected areas were painful to the touch. The lesions had been present for several weeks; the owner had been treating the affected areas with topical antiseptic shampoos and topical emollients believing that the condition was 'mud fever' (demaphilosis), but they had shown no improvement. Skin biopsy was undertaken; histologic examination revealed evidence of leukocytoclastic vasculitis.

1 What differential diagnoses should be considered?
2 What further investigation should be recommended?
3 How should this condition be treated?

CASE 191 An 18-year-old Thoroughbred gelding presented with a 6-week history of progressive facial swelling of the left-hand side (**191.1**), mild epiphora of the left eye (**191.2**), and a mild left unilateral mucoid nasal discharge. The horse was well otherwise, but had started to make a 'snuffling' respiratory noise when exercised. Airflow through the left nostril was reduced compared with the right. Submandibular lymph nodes were not enlarged. Ophthalmic examination was unremarkable. Endoscopic examination of the upper respiratory tract showed that the left nasal meati were narrowed, preventing passage of the endoscope on that side, but examination via the right ventral meatus was unremarkable. A routine lateral-lateral radiograph of the head showed diffusely increased radiopacity of the maxillary sinuses (**191.3**).

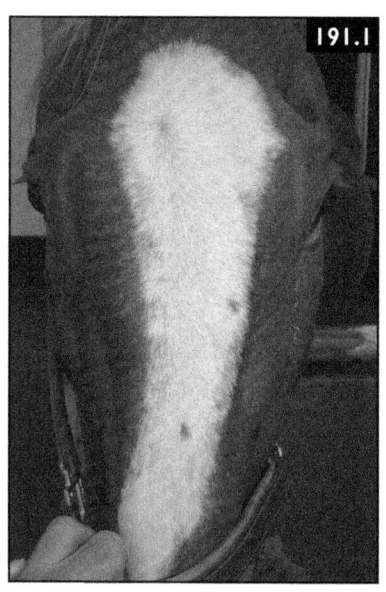

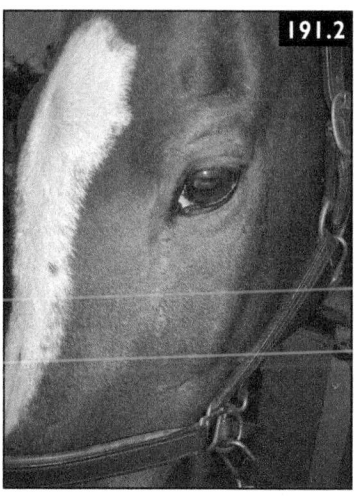

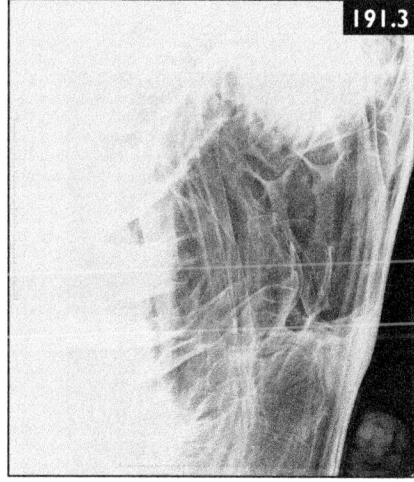

1 What are the differential diagnoses for unilateral facial swelling?
2 What is the most likely cause of the facial swelling in this case?
3 What further diagnostic tests should be considered?

CASE 192 An 8-year-old crossbred mare had a sudden onset of dysphagia, drooling saliva (192.1), and stridor. On physical examination there was pain on palpation of the parotid region, and the horse stood with the head and neck extended. Endoscopic examination revealed laryngeal inflammation and edema (192.2). Food material and saliva were present in the nasal cavities and nasopharynx.

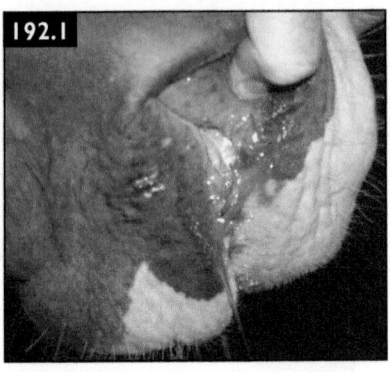 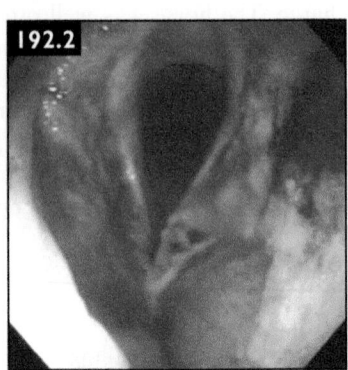

1 What conditions are suspected?
2 What other investigations should be considered?
3 How should this case be managed?

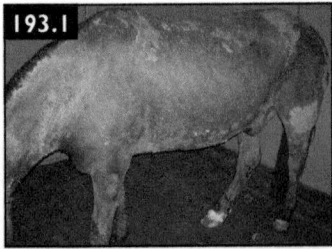

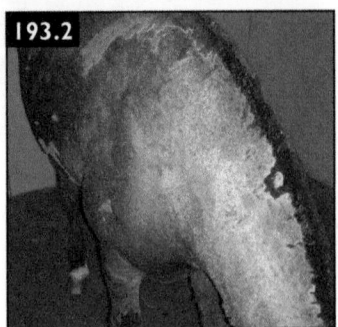

CASE 193 A 14-year-old crossbred gelding presented with a 10-day history of severe, widespread crusting and scaling skin lesions that progressed to extensive alopecia (193.1, 193.2). The skin felt hot and thickened. The lesions started on the face and limbs and then rapidly spread over the whole body. There was mild pruritus with patchy areas of excoriation. The horse was slightly depressed, but eating and drinking normally, and was not pyrexic.

1 What condition is suspected?
2 How could the diagnosis be confirmed?
3 What treatment should be recommended, and what is the prognosis?

CASE 194 A 19-year-old Welsh Mountain Pony cross mare presents with a history of suspected intermittent, variably severe pain around the base of both ears for 3.5 years. The mare frequently rubs both ears, and the owners report a "fluid sound from the ears" when ridden. At the time of presentation, the signs are thought to be predominantly right sided. The mare also has multiple dental abnormalities (displaced teeth, diastemata, periodontal disease, and overgrowths). She has been treated symptomatically with phenylbutazone and dental care. A mild mucopurulent discharge is identified from both external auditory canals. The mare resents palpation of the external ears and the parotid regions. Oral examination confirms multiple dental abnormalities as described previously. CT was performed (**194.1, 194.2**).

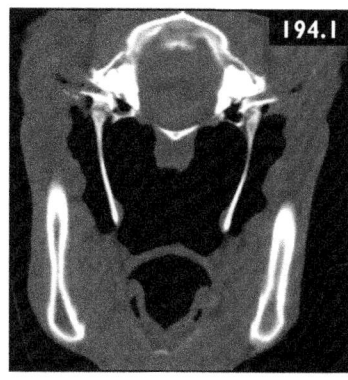

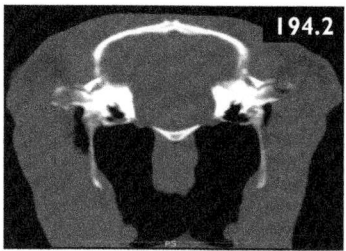

1 What further diagnostic techniques should be considered?
2 What abnormalities can be identified on the CT scans?

CASE 195 A 3-month-old Warmblood filly presented with a 3-day history of dullness, inappetence, and worsening dyspnea. She had been intermittently pyrexic (rectal temperature up to 103.6°F [39.8°C]). On examination the foal was showing inspiratory and expiratory dyspnea, with a RR of 56 bpm, HR

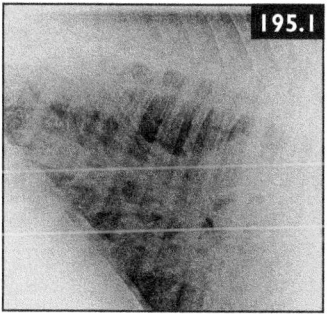

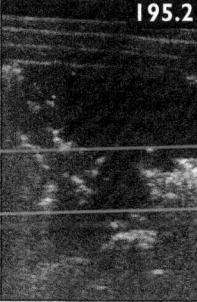

of 60 bpm, and rectal temperature of 101.1°F (38.4°C). Auscultation of the lungs revealed areas of harsh bronchial sounds and crackles with areas of dullness. Thoracic radiography (**195.1**) and ultrasonography (**195.2**) were performed.

1 What abnormalities can be seen in the radiograph and the ultrasonogram?
2 What is the most likely diagnosis?
3 How can the diagnosis be confirmed?

CASE 196 A yearling pony gelding presented with signs of generalized stiffness that had been getting progressively worse over the prior 2 days. The pony walked with a very stiff spastic gait, with flared nostrils, erect ears, and a raised tail head (**196.1, 196.2**). The pony was dysphagic and appeared unable to open his mouth. Widespread muscle tremors were observed when the pony was excited or attempted to move. When menaced, there was protrusion of the third eyelid (**196.3**).

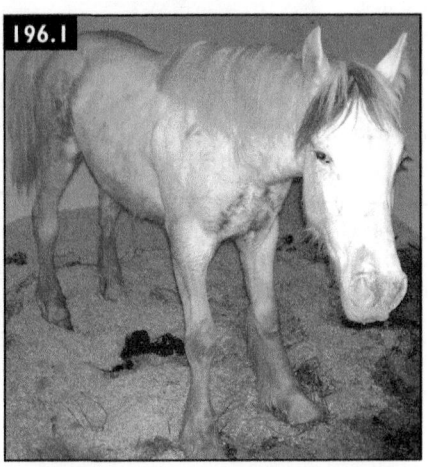
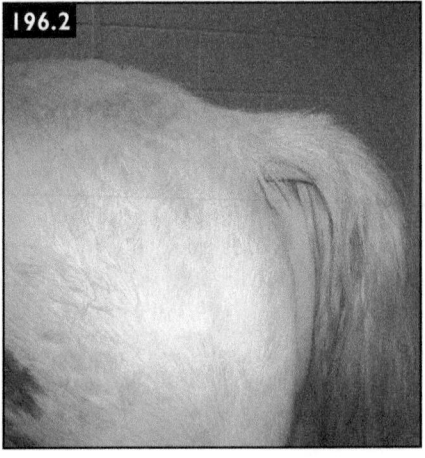

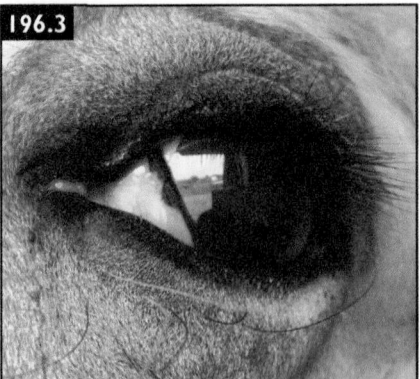

1 What is the diagnosis?
2 How should this horse be treated?
3 What is the prognosis?

CASE 197 A 5-year-old Warmblood mare presented with a history of a mild right unilateral malodorous nasal discharge (**197.1**). Two cheek teeth had been removed by intraoral extraction (the 109 and 209 teeth) 2 months previously because of sagittal fractures and periapical infection of both teeth. At the time of the extractions, the horse had a bilateral maxillary sinusitis that was treated by sinus trephination and lavage and a 3-week course of oral doxycycline. The horse responded well to treatment, but in the past few days the malodorous nasal discharge had appeared. Radiography of the horse's head did not identify any sinusitis/fluid lines or abnormalities at the extraction sites. The horse was referred for CT evaluation (performed in the standing, sedated animal). A transverse multiplanar reconstruction CT image through the alveoli of 109 and 209 cheek teeth following extraction is shown (**197.2**). The left side of the horse is on the right side of the image.

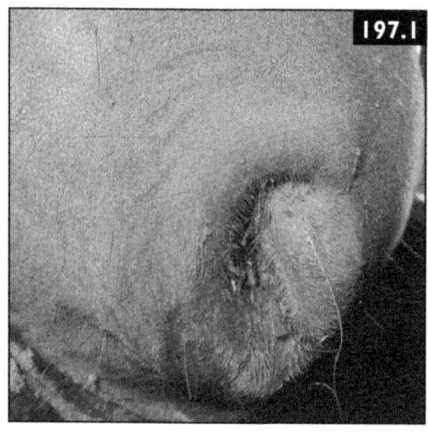
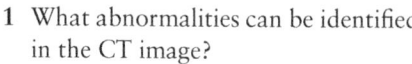
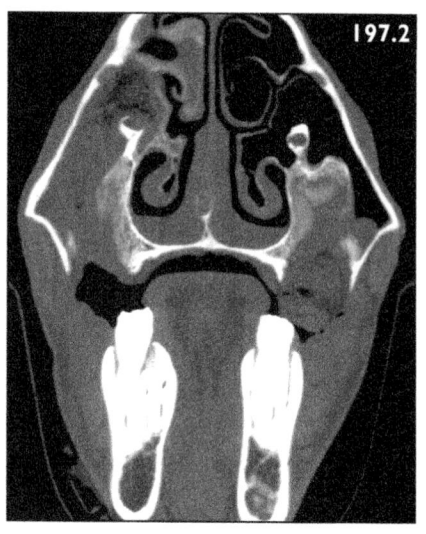

1 What abnormalities can be identified in the CT image?

CASE 198 A 24-year-old Thoroughbred mare presents with a 2-month history of blepharospasm, epiphora, and progressive corneal opacification (**198.1**).

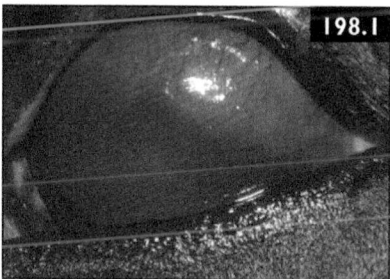

1 What are the diagnosis and treatment recommendations?

133

CASE 199 A 19-year-old Thoroughbred cross mare was examined for a malodorous, purulent, unilateral nasal discharge (**199.1**). A periapical infection of a maxillary cheek tooth (tooth 109) was suspected but could not be confirmed by oral examination or standard radiography. A standing CT examination was performed, which confirmed a periapical infection of tooth 109. The scans also identified an abnormality in the brain (**199.2, 199.3**).

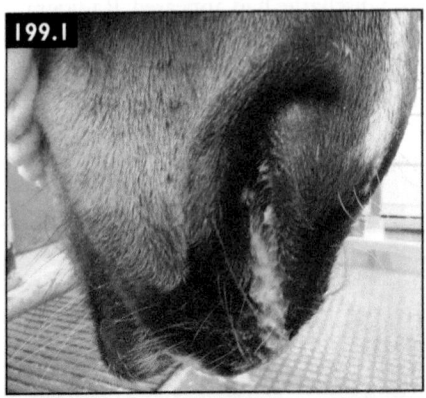

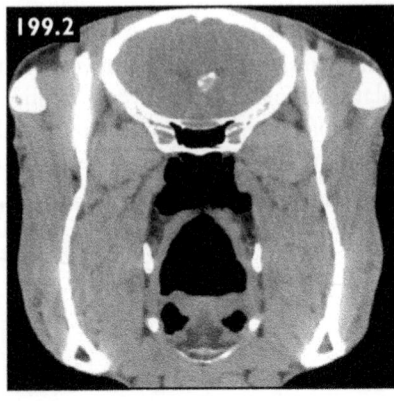

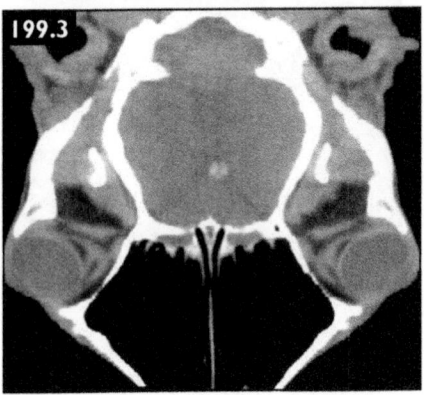

1 What abnormality can be detected in the brain on the CT scans?
2 What is the likely diagnosis?
3 What advice should be offered to the owners of this horse regarding this finding?

CASE 200 A 12-year-old Thoroughbred mare had a history of a left, unilateral, malodorous purulent nasal discharge that was occasionally blood tinged (200.1). The discharge had become more profuse over a period of 3 weeks. The horse was bright and alert and showed no other clinical signs. Endoscopic examination was performed via the left and right nostrils. Lesions were identified in both nasal cavities, more extensive on the left side (200.2, 200.3).

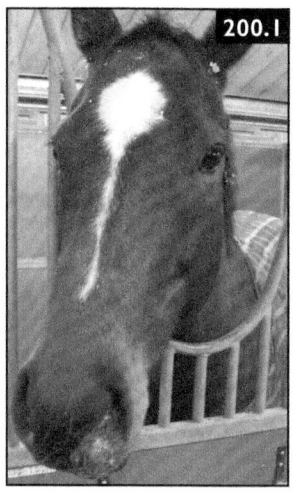

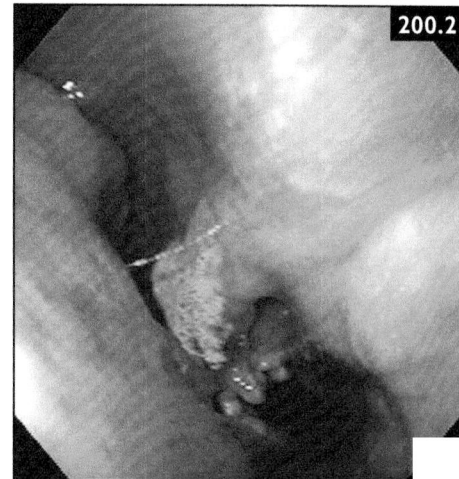

1 What condition is suspected?
2 How can this suspicion be confirmed?
3 How can this disease be treated?

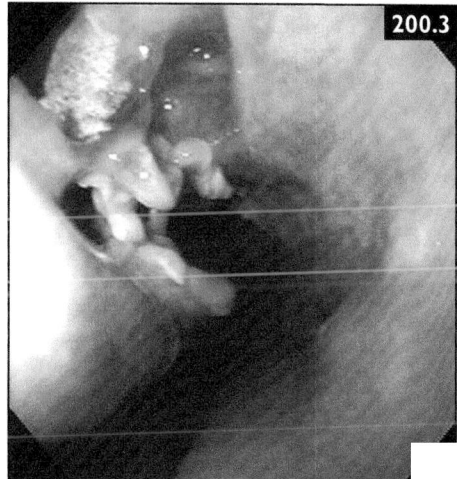

CASE 201 A newborn Cob colt presented collapsed and weak due to neonatal maladjustment syndrome (a.k.a. dummy foal, wanderer, barker, hypoxic ischemic encephalopathy, peripartum asphyxia syndrome, neonatal maladaptation syndrome) and hypoglycemia (**201.1**). The foal also had failure of passive transfer of immunity. The foal was fed via a nasogastric tube and had been given an IV plasma transfusion, and had been making good progress. At one week of age a diffuse white plaque over the dorsal surface of the tongue was noted (**201.2**).

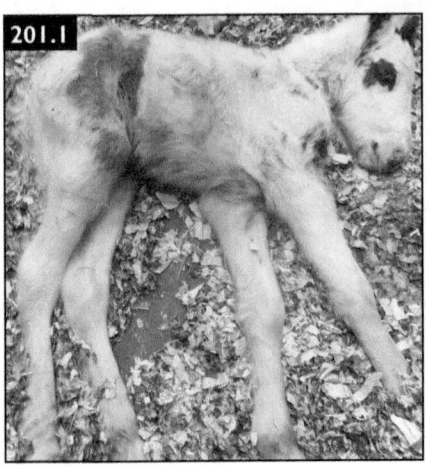
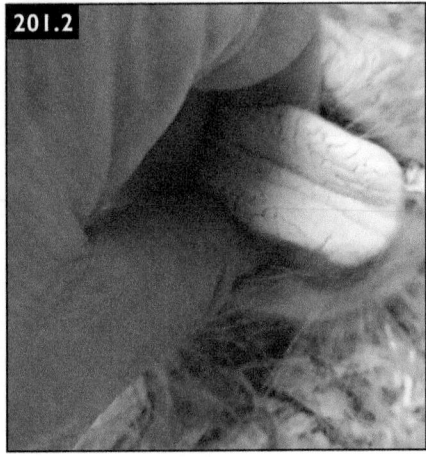

1 What is this lesion?
2 Why has this foal developed this disease?

CASE 1

1 Based on the age and acute onset of clinical signs, what would be the top differential diagnoses? Trauma, temporohyoid osteoarthropathy (THO), and EPM (in North America).

2 How can it be determined if the vestibular dysfunction is central or peripheral? Both peripheral and central vestibular disease cause loss of balance, asymmetric ataxia, and head tilt toward the side of the lesion (with the exception of paradoxical vestibular disease, which causes a head tilt contralateral to the side of the lesion). In some cases, central vestibular disease may also cause upper motor neuron proprioceptive deficits and weakness in addition to depression. Nystagmus can be horizontal or rotary and should have the fast phase of movement opposite to the side of the lesion, except for some cases of central vestibular disease in which the direction may change with different head positions or the fast phase may be toward the side of the lesion. (**Note:** Remember that C1 spinal cord disease may also cause signs of central vestibular disease via damage to the secondary spinovestibular tracts).

3 What diagnostics should be performed? Endoscopy of the guttural pouch: normal. Collection of CSF: normal (EPM titer was a weak positive on Western blot). CBC: normal except for mild leukocytosis. Radiographs and/or CT scan: declined by owners.

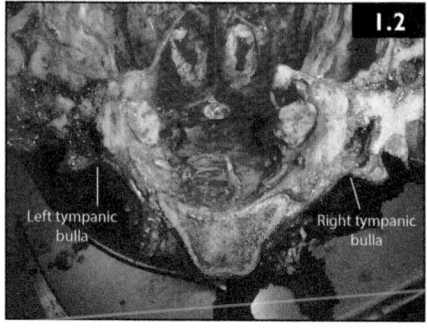

Follow up/discussion

NSAIDs and DMSO were administered IV and oral trimethoprim sulfa initiated. In addition, ponazuril for EPM was given for 9 days but with no improvement. At that time the owners elected euthanasia.

Postmortem diagnosis: severe *Aspergillus* infection of the left tympanic bulla (**1.2, 1.3,** courtesy Dr. Cameron Knight). Infectious otitis media/interna is a rare primary cause of peripheral vestibular disease in the horse. Although frequently blamed as a cause of THO, infection is rarely documented. *Staphylococcus aureus* is the most common bacteria isolated. Fungal infections of the tympanic bullae are exceedingly rare in the equine. One case has been

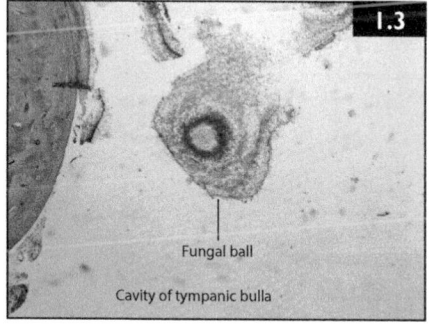

reported in an adult horse and a single case of yeast otitis has been reported in a neonatal foal in the intensive care unit. Systemic treatment for *Aspergillus* could be provided with intravenous amphotericin B or oral itraconazole, fluconazole, or voriconazole. Amphotericin B therapy is limited because of its potential to cause nephrotoxicity and/or thrombophlebitis and some *Aspergillus* strains are resistant. Itraconazole (solution) and fluconazole are well absorbed when given orally to horses and may be used as treatment for *Aspergillus*, but *in-vitro* susceptibility varies among *Aspergillus* strains. Voriconazole has the best bioavailability of the drug family, low toxicity, and best efficacy against several strains of fungi but is expensive.

References
Colitz CM, Latimer FG, Cheng H *et al.* (2007) Pharmacokinetics of voriconazole following intravenous and oral administration and body fluid concentrations of voriconazole following repeated oral administration in horses. *Am J Vet Res* **68**(10):1115–1121.
Pearce JW, Giuliano EA, Moore CP (2009) In vitro susceptibility patterns of *Aspergillus* and *Fusarium* species isolated from equine ulcerative keratomycosis cases in the midwestern and southern United States with inclusion of the new antifungal agent voriconazole. *Vet Ophthalmol* **12**(5):318–324.
Stewart AJ, Welles EG, Salazar T (2008) Pulmonary and systemic fungal infections. *Compend Equine* **3**:260–272.

CASE 2
1 Based on the history in this case, and without availability of ultrasound in the field, what diagnostic procedures should be performed? CBC and peritoneal tap.

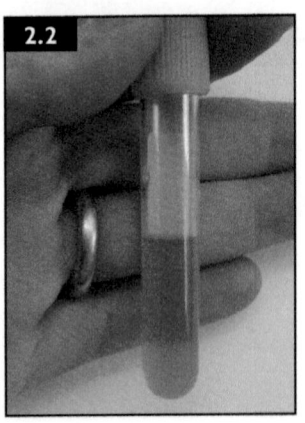

2.2

Follow up/discussion
CBC revealed a mild neutrophilia (7,300 cells/µl [7.3 x 10^9/l]) without a left shift. Fibrinogen was 300 mg/dl (3.0 g/l). The peritoneal fluid was dark orange (**2.2**) with a TP of 4.7 g/dl (47 g/l) and 220,00 nucleated cells/µl (22.0 x 10^9/l) (predominantly mildly degenerated neutrophils). No infectious agents were seen. Culture revealed *Actinobacillus equuli* subspecies *haemolyticus*. The organism was sensitive to ampicillin, aminoglycosides, cephalosporins, fluoroquinolones, and tetracyclines, and resistant to trimethoprim sulfa and penicillin.

Actinobacillus peritonitis is not a rare condition in otherwise healthy horses. It usually occurs spontaneously in adult horses, although foals as young as 9 months of age have been affected. The predominant clinical signs are lethargy, signs of depression, mild colic, anorexia, and fever. Approximately one-half of the cases have peritoneal fluid with suppurative inflammation and gram-negative rods. The organism is generally sensitive to most antibiotics, including trimethoprim sulfa and penicillin. This was not true in this case, which emphasizes the need to perform culture and sensitivity on the peritoneal fluid. It is unknown how the organism gains entrance to the peritoneal cavity, but migrating parasites or foreign bodies have been implicated. Pleuritis, pericarditis, and/or meningitis with the same organism may be seen separately or in combination with peritonitis. Interestingly, an increased number of cases of *A. equuli* pericarditis were seen in both sexes and all ages of horses in Kentucky in the same year as the mare reproductive loss syndrome outbreak.

References

Matthews S, Dart AJ, Dowling BA *et al.* (2001) Peritonitis associated with *Actinobacillus equuli* in 51 cases. *Aust Vet J* 79(8):563–569.

Sebastian MM, Bernard WV, Riddle TW *et al.* (2008) Mare reproductive loss syndrome. *Vet Pathol* 45(5):710–722.

CASE 3

1 What question should be asked regarding the swelling of the left hindlimb and neck? "Have there been any injections given in these areas?"

2 Are the swellings most likely septic or non-septic? An injection could cause a septic myositis or local tissue reaction, both of which could potentially cause thrombocytopenia and epistaxis. In this case, no injections had been given and the lack of moderate to severe pain on palpation rule out a septic cause of the swellings.

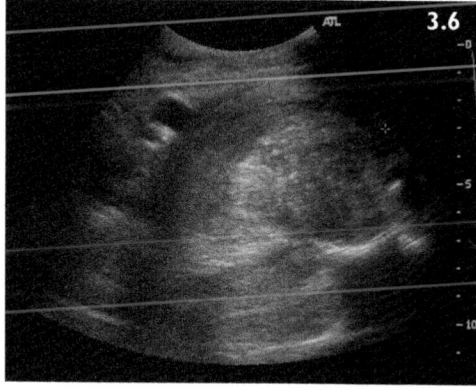

3 List some differential diagnoses for intermittent bilateral epistaxis. Include guttural pouch mycosis (almost always unilateral clinical signs), thrombocytopenia, and ethmoid hematoma.

4 What diagnostic test should be chosen next? Ultrasound of the neck (**3.6**), leg, abdomen, and thorax.

5 Given the severe thrombocytopenia, yet no evidence of infectious disease or recent drug administration, what general type of disease process would be the top differential? Neoplasia.

6 What diagnostic test should be chosen next? Needle aspirate of the mass. This revealed a soft tissue sarcoma.

7 What further diagnostics would help evaluate the thrombocytopenia more specifically? This should focus on determining if the thrombocytopenia was due to excessive consumption, lack of production, or increased destruction. Could include meticulous microscopic examination of the platelets for reticulation (indicative of increased messenger RNA) or for large platelets, either suggestive of increased platelet release from the marrow caused by excessive loss or immune-mediated destruction of mature platelets. Ultrasound or rectal examination to try to determine the size of the spleen would be subjective and rectal examination could result in rectal mucosal bleeding. Submission of blood to a laboratory for flow cytometry to determine the percentage of platelets bound with antibody would be the best test for immune-mediated thrombocytopenia. Bone marrow examination is necessary to determine if there was suppression of thrombocyte production.

8 Radiography of the lungs was performed to look for any metastases (3.5). What is the interpretation? Although metastasis had occurred to the lungs, it was not obvious on radiography. Some of the circles of increased radiodensity might have been neoplasia, but on first reading were thought to be 'end on' vessels. Lung radiographs are not a sensitive test for tumor metastasis in the horse due to the horse's size and subsequent loss of detail.

9 What is the most common neoplasm to invade muscle and also result in thrombocytopenia? Hemangiosarcoma.

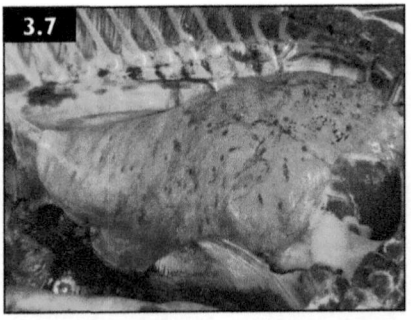

3.7

Follow up/discussion

Necropsy revealed hemangiosarcoma of the neck (in the muscle) with metastasis to regional lymph nodes, lungs (3.7), and kidney. The swelling of the hindlimb was believed to be due to hemorrhage, as no tumor was found in that location.

There are generally two forms of hemangiosarcoma in the horse: (1) the cutaneous or locally invasive form (most often in the eye), which may have a fair prognosis with proper treatment, and (2) the disseminated form (most often in muscle, lung, pleura, spleen, heart, and kidney), which has a grave prognosis. The cutaneous or ocular form can be confused with hemangioma, which is a local tumor. Hemangiosarcoma is most common in middle-aged horses. Systemic chemotherapy, such as doxorubicin, is unlikely to be curative for the disseminated form. Treatment with local cisplatin and/or surgical removal can be curative for cutaneous or local forms.

References
Johns I, Stephen JO, Del Piero F *et al.* (2005) Hemangiosarcoma in 11 young horses. *J Vet Intern Med* **19**(4):564–570.
Warren AL, Summers BA (2007) Epithelioid variant of hemangioma and hemangiosarcoma in the dog, horse, and cow. *Vet Pathol* **44**(1):15–24.

CASE 4
1 What is the primary differential diagnosis? The clinical signs, age, and physical examination findings are characteristic of foal pneumonia.
2 What are the most common etiologic causes for these signs? Bacterial causes (e.g. *Streptococcus zooepidemicus*, *Actinobacillus* spp., *Bordetella* spp., *Pasturella* spp., *Rhodococcus equi*). *R. equi* might be less likely in this case due to the cranioventral location of the disease seen on ultrasound, but it could not be ruled out based on that distinction. Viral infection may have predisposed the foal to bacterial pneumonia.
3 What diagnostic procedure should be performed next to gain further information? CBC and fibrinogen should be performed, but with any of the above differentials, a neutrophilic leukocytosis and hyperfibrinogenemia would be expected. Radiography could be performed, but ultrasound is nearly as sensitive – if not more so – in detecting foal pneumonia. A tracheal wash was performed and the sample submitted for cytology, culture, and sensitivity.
4 What treatment should be recommended based on these results? Cytologic evaluation identified *Pneumocystis carinii* (**4.1**). *Actinobacillus* spp. was grown from the bacterial culture. The foal was treated with trimethoprim-sulfa for the bacterial and fungal infections and with an immune

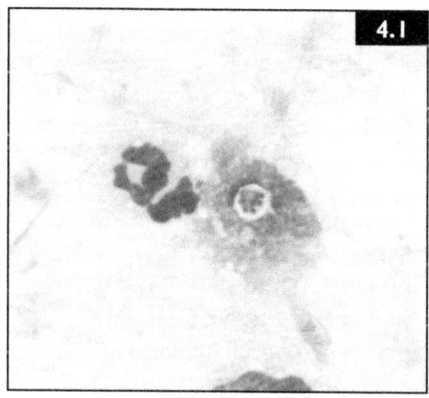

4.1

modulator because an abnormally low number of CD4+ T cells were measured. Additional immunologic testing, including IgG, IgM, and IgA concentrations, phytohemagglutinin response, and electrophoresis, was all normal. The foal responded well to treatment and was clinically normal in 2 weeks. *P. carinii* is a yeast-like fungal organism that is an uncommon pathogen in foal pneumonia but can be seen concurrently with *R. equi* pneumonia and immunodeficiency syndromes.

Reference

Clark-Price SC, Cox JH, Bartoe JT *et al.* (2004) Use of dapsone in the treatment of *Pneumocystis carinii* pneumonia in a foal. *J Am Vet Med Assoc* **224**(3):407–10, 371.

CASE 5

1 What is the most unusual blood laboratory observation in this mature horse, and how might it be associated with the clinical complaint? Although the low chloride concentration and the high anion gap are of interest, they are likely related to increased lactate, sulfate, or phosphate and the azotemia. The white color of the plasma indicates hyperlipemia, which is very rare in horses. In this case the hyperlipemia is likely related to the weight loss, intermittent inappetence, and the primary disease.

2 What additional laboratory test(s) should be performed based on these findings? Measure lipids in the plasma. Triglycerides were 1,635 mg/dl (15.42 mmol/l), cholesterol was 368 mg/dl (9.53 mmol/l), and non-esterified free-fatty acids were 0.54 mEq/l (0.54 mmol/l).

3 What are the two major physiologic pathway abnormalities that might have caused the marked hyperlipemia? Abnormal lipolysis and abnormal clearance of lipids (mostly triglycerides).

4 With regard to the ultrasound findings, what is the most likely diagnosis? Neoplasia.

5 In light of the postmortem findings (5.3), what is the most likely mechanism for this horse's hyperlipemia? Decreased clearance of very low-density lipoprotein, likely due to the lymphoma.

Follow up/discussion

The neoplasm in this horse was confined to the spleen and surrounding area and was believed to be a B-cell lymphoma. There was marked mineralization of the pulmonary artery, left atrium, and renal tubules, likely related to the hypercalcemia and hyperphosphatemia. There was no microscopic evidence of hepatic lipidosis. Inhibition of lipoprotein lipase has been reported in humans with lymphoma. Hyperlipemia severe enough to cause white-colored plasma is rare in equines, other than in sick ponies or miniature equines, although it is occasionally seen in horses with equine Cushing's disease. However, all of these types of hyperlipemic animals have hepatic lipidosis, unlike the horse in this case.

Reference

Kulkarni K, Kaur S, Sibal A *et al.* (2010) Severe lactic acidosis, hypertriglyceridemia, and extensive axial skeleton involvement in a case of disseminated Burkitt's lymphoma. *Int J Hematol* 91(3):546–548.

CASE 6

1 What is the most likely diagnosis for the chronic, fetid-smelling, unilateral nasal discharge? Septic sinusitis, tooth root abscess.

2 How could this diagnosis be confirmed? Radiography or CT. In this case, CT revealed an orosinus fistula and a tooth root abscess of the left upper 3rd molar (tooth 211 in the Triadan system).

3 What treatment was probably provided? A flap sinusotomy. The affected tooth was repelled, and packing material was placed in the tooth socket.

4 What was the likely cause of the complication in the left forelimb? The hard swelling of the triceps muscle, extreme pain, and unwillingness to bear weight, yet ability to extend the limb, suggests a myopathy rather than neuropathy. It is possible that there was some radial nerve paresis secondary to the myopathy and swelling. Fracture of the olecranon must also be considered.

5 What serum chemistry tests would help confirm this diagnosis? Measurement of muscle enzymes (CK and/or AST). Dark urine (due to myoglobin) would have also been supportive of myopathy. This horse was not tested for polysaccharide storage myopathy, but this would be recommended.

6 What could cause the unusual sweating pattern? It was likely due to damage to the sympathetic nerve pathways within the spinal cord at the most cranial site of the sweating (approximately the 3rd thoracic vertebra).

Follow up/discussion

CK becomes elevated more quickly than AST. A blood sample taken 3 hours following recovery showed a CK of 2,947 U/l and a mildly elevated AST of 387 U/L. Maximum values of both enzymes occurred on day 3 following surgery, with a CK of 44,085 U/L and an AST of 1,941 U/L. On day 15 following surgery, the AST was still increased at 977 U/L, but the CK was normal at 105 U/L, demonstrating the more prolonged half-life of AST in comparison with CK.

   Interestingly, the affected limb was the 'up' limb during anesthesia. Myopathies may occur in either 'up' or 'down' limbs during anesthesia, but peripheral neuropathies are almost always in the 'down' limb. Mild hypochloremia and hyperglycemia (likely due to stress) were also found on the chemistry panel following recovery from surgery. The exact cause of this sympathetic dysfunction was not proven, but injury to the sympathetic nerve due to neuropathic inflammation and pain was considered. The sweating resolved in 3 days.

References

Franci P, Leece EA, Brearley JC (2006) Post anaesthetic myopathy/neuropathy in horses undergoing magnetic resonance imaging compared to horses undergoing surgery. *Equine Vet J* **38**(6):497–501.

Wagner AE (2008) Complications in equine anesthesia. *Vet Clin North Am Equine Pract* **24**(3):735–752.

CASE 7

1 What would be a reasonable clinical workup for this case? Should include urinalysis, abdominal palpation per rectum, and ultrasound of the bladder. Urinalysis revealed a urine SG of 1.032 with 3+ blood and 2+ protein, no leukocytes. No neoplastic cells were seen on cytology. On rectal examination, a large firm mass was palpated as soon as the hand entered the rectum. This mass extended 6–10 cm beyond the brim of the pelvis and within the mass a smaller, possibly fluid-filled structure (perhaps the urinary bladder?) could be felt. A soft tubular structure was palpated on the right and was believed to be the ureter.

2 An ultrasound image is shown (7.1). What is the interpretation? A soft tissue mass can be seen within or next to a smaller, fluid-filled structure (likely the ureter, not the urethra as marked on the image).

3 What procedure could be performed to better visualize the urinary bladder? Cystoscopic examination with a sterilized 1 or 2 meter endoscope would allow

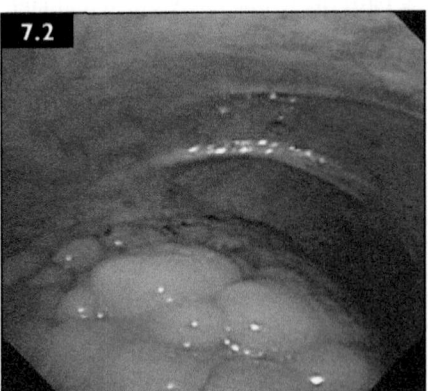

improved visualization of the lumen of the urinary bladder, ureteral orifices, and urethra (7.2).

4 What is the most likely diagnosis based on all the information? The age of the horse, clinical findings, urinalysis, ultrasound, and cystoscopic examination together would suggest bladder neoplasia.

5 What are the two main bladder neoplasms in the horse? Squamous cell carcinoma and transitional cell carcinoma.

6 **Which one is most likely in this horse, and why?** The horse's age could be compatible with either type of neoplasia, but the pale, rather homogeneous appearance of the mass and the absence of gross hematuria would be most supportive of a transitional cell carcinoma. Sarcomas involving the bladder muscle may also occur infrequently, often in younger horses.

7 **How could the diagnosis be confirmed?** A biopsy could be taken during cystoscopic examination. Alternatively, the tumor could be 'probed' with the scope and a urinalysis then collected for cytology; that is how the diagnosis of carcinoma was made in this case. Necropsy examination confirmed transitional cell carcinoma with partial obstruction of the right ureter.

Reference
Fischer AT Jr., Spier S, Carlson GP *et al.* (1985) Neoplasia of the equine urinary bladder as a cause of hematuria. *J Am Vet Med Assoc* **186(12):1294–1296**.

CASE 8

1 **How should the rectal prolapse be treated?** Epidural anesthesia should be given because of the severe mucosal damage to the rectum.

2 **What is the most likely cause of the rectal prolapse in this colt?** Severe parasitic infection, such as with the third-stage larvae of *Gasterophilus intestinalis*, which were passed in large numbers during the first 3 days of hospitalization.

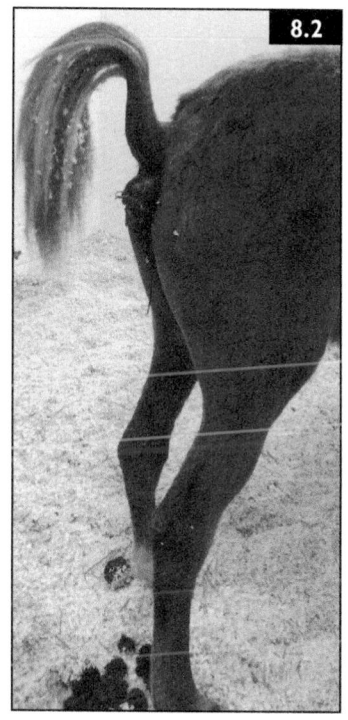

Follow up/discussion
Following placement of the epidural, the prolapsed tissue was soaked and cleaned with hypertonic saline and the mucosa was coated with a commercial hemorrhoid cream preparation. The prolapse was then manually reduced and a purse-string suture (**8.2**) was placed to prevent further prolapse. This was loosened 4–6 times daily to allow defecation, after which the rectum was treated as described above and replaced. The colt was also given mineral oil by nasogastric tube and had mineral oil mixed in its concentrate feed

twice daily to soften the manure. Prolapse no longer occurred with defecation within 4 days and the colt fully recovered.

Any condition causing prolonged tenesmus, irritation, injury, and edema of the rectal mucosa may predispose to the development of rectal prolapse. Rectal prolapse is occasionally seen in foaling mares and in horses with diarrhea and colitis. Of the horses with diarrhea, it may be most common in those cases caused by NSAID toxicity. Gasterophilosis was thought to be the inciting cause in this poorly managed colt and either *G. intestinalis* or *G. nasalis* may be associated with the condition. Prompt treatment for both the rectal prolapse and the initiating cause is necessary to prevent progressive damage to the rectal mucosa. If severe injury to the rectum has occurred, submucosal resection may be required.

References
Getachew AM, Innocent G, Trawford AF *et al.* (2012) Gasterophilosis: a major cause of rectal prolapse in working donkeys in Ethiopia. *Trop Anim Health Prod* 44(4):757–762.
Turner TA, Fessler JF (1980) Rectal prolapse in the horse. *J Am Vet Med Assoc* 177(10):1028–1032.

CASE 9

1 What is the most likely diagnosis in this mare? Middle uterine artery rupture. The uterine artery and its cranial and caudal branches are supported within the broad ligament (mesometrium) and it is a branch of the external iliac artery, which in turn is a branch of the aorta. The uterine artery is hypertrophied in pregnancy and rupture anywhere along it most often occurs periparturiently, resulting in rapid hemorrhage. A blood clot will form within the broad ligament, but blood can still leak into the abdomen. If the initial blood clot is dislodged or ruptures, massive, fatal intra-abdominal hemorrhage often ensues.
2 What factors have been associated with an increased occurrence of this condition? Rupture of the uterine artery is more common in multiparous, older mares and in assisted birth when the foal is pulled manually. In addition, dystocia can be a source of trauma. There is a predilection for right-sided uterine artery rupture, possibly because of increased tension on the right broad ligament due to relative displacement of the gravid uterus to the left by the cecum. There is also an association between low copper levels and increased vessel fragility, which would predispose to rupture.
3 How can the blood volume of a horse be calculated, and what percentage of this can be lost before a drop in blood pressure occurs? Blood volume is equal to 8–10% of body weight. Therefore, blood volume is 40–50 liters for a 500 kg mare. A horse can lose 20–30% of the intravascular blood volume (8–12 liters in this

500 kg mare) without a drop in blood pressure due to increased cardiac output and vasopressor, endocrine, and renal compensatory mechanisms.

4 What treatments should be employed in a case like this and what should be avoided (9.3)? What other clinical signs and bloodwork could help determine the mare's perfusion status? Uterine artery rupture can result in rapid death no matter what treatment measures are taken; however, survival is possible, especially in cases in which the majority of the hemorrhage occurs in the broad ligament. It is extremely important to keep a mare with suspected uterine artery rupture quiet and calm to avoid dislodging any clot formation. Horses with hypotension may be agitated so a quiet environment and *light* sedation (with acepromazine) may be necessary. However, if blood loss is very great (indicated by HR >100 bpm, white mucous membranes, weakness, ataxia, and increased blood lactate levels), then any sedation will likely result in collapse. Sometimes, transport is necessary for appropriate management of the mare, but it is risky and should be avoided if possible, as it increases stress. An attempt should be made to maintain permissive hypotension with a controlled volume of isotonic fluids, +/– blood transfusion (systolic pressures between 70 and 90 mmHg) in order to minimize any ongoing hemorrhage and the risk of clot disruption. If abdominal bleeding is stable and not causing respiratory distress, the abdomen should not be drained for two reasons: (1) increased abdominal pressure can help reduce ongoing bleeding into the abdomen; (2) approximately two-thirds of intra-abdominal erythrocytes will autotransfuse and components of the remainder will be utilized (e.g. protein, iron). Aminocaproic acid (30 mg/kg IV in 1 liter of 0.9% NaCl) is an antifibrinolytic and can help maintain clot stability. Whole blood transfusion is indicated if the PCV falls to <20% in 12 hours following IV fluid administration, to <12% over 24–48 hours, if lactate is high or if the PvO_2 and/or SvO_2 are low.

5 Why are the PCV and TS normal in this case? In acute hemorrhage all components of the blood are lost equally (unlike in hemolysis) and unless crystalloid therapy has been given IV, at least 4–8 hours are required in order for aldosterone and vasopressin to have a dilutional effect on PCV and plasma protein via the enhanced renal absorption of electrolytes and fluids. Horses with acute hemorrhage may die from lack of circulating RBCs in the face of normal PCV and plasma protein concentration.

References

Dechant JE, Nieto JE, LeJeune SS (2006) Hemoperitoneum in horses: 67 cases (1989–2004). *J Am Vet Med Assoc* **229(2)**:253–258.

Ueno T, Nambo Y, Tajima Y *et al.* (2010) Pathology of lethal peripartum broad ligament haematoma in 31 Thoroughbred mares. *Equine Vet J* **42(6)**:529–533.

CASE 10

1 What does the abnormal serum color mean, and what biochemical test would provide more information concerning this abnormality? Is this condition important, and what therapy should be used to treat it? It indicates hyperlipemia, which can be further assessed by measuring serum triglyceride levels – should be <66 mg/dl (0.73 mmol/l) in normal donkeys. Hyperlipemia occurs secondary to negative energy balance and the subsequent mobilization and utilization of fat for energy. It occurs commonly in donkeys with illness, stress, or any situation in which there is inadequate feed intake. Jennies and overweight donkeys are at higher risk of developing hyperlipidemia and hyperlipemia. Hyperlipemia is a very serious condition. Clinical signs include anorexia, depression, icterus, colic, weakness, incoordination and, if left untreated, death. Treatment should consist of force-feeding a mixture of glucose, electrolytes, and sodium bicarbonate and/ or proprietary 'critical care' meals. IV fluid therapy with B vitamins and/or partial parenteral nutrition should also be administered. Insulin therapy, broad-spectrum antibiotics, and flunixin meglumine may also be needed, depending on the severity of disease.

2 The donkey is dehydrated and may have been so for a prolonged period, since we know from the history that the illness has been progressing over 2 weeks. What

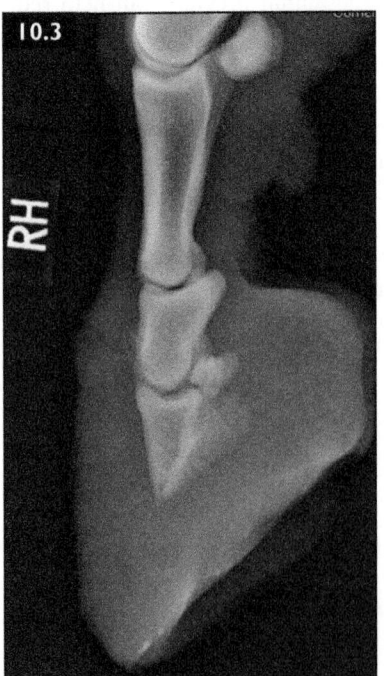

10.3

organ system is of immediate concern with respect to secondary effects, and how should this be investigated further? It is important to assess renal function in the face of marked (and in this case possibly chronic) dehydration. Further diagnostic modalities include serum biochemistry (especially creatinine, BUN, and electrolytes), urinalysis, and renal ultrasonography.

3 The donkey stopped shifting weight in the hindlimbs after abaxial sesamoid local nerve blocks were performed in both hindlimbs. What condition is most likely responsible for this clinical sign of weight-shifting and what could be done to further assess this and make the donkey more comfortable? The positive response to the abaxial nerve block indicates that the weight shifting was due to pain in the foot, most likely due to laminitis. Radiographs would help confirm this (10.3). Response to hoof testers is unreliable in donkeys. Laminitis is common in overweight donkeys and the clinical signs (lameness, abnormal stance, and reluctance to move) are often much more subtle than in horses due to their stoic nature. Many elderly donkeys have chronic laminitic changes and may suffer acute bouts that go unnoticed. The same is true for colic, which is why this donkey was not showing signs of abdominal pain, despite having an impaction. In this situation, the donkey initially developed a bout of acute laminitis, which caused her to have reduced feed and water intake due to a combination of pain and reluctance to move to food and water sources. Secondary hyperlipidemia then developed, compounding the anorexia and depression, and the impaction developed due to lack of water intake. Treatment involves housing in deep bedding, administration of NSAIDs for analgesia, and addressing the underlying cause. NSAIDs are well tolerated in donkeys and they appear to suffer GI and renal side-effects less frequently than horses. Sole padding can be used but the sole of a donkey's foot is naturally very thick and care should be taken not to elevate the heel, as this will result in more discomfort (especially if concurrent shoulder arthritis is present), unlike in a laminitic horse where heel elevation is desired.

4 What could be the cause of this, and how could it be remedied? Donkeys bond strongly with each other and other animals. Removal of a donkey from its home environment and herd mates can result in anorexia and depression in the absence of any clinical disease. In this case, a companion horse was shipped in to the hospital from the donkey's home farm and the donkey immediately became brighter and began to eat.

References

Burden FA, DuToi N, Hazell-Smith E *et al.* (2011) Hyperlipemia in a population of aged donkeys: description, prevalence, and potential risk factors. *J Vet Intern Med* **25(6)**:1420–1425.

Díez E, López I, Pérez C *et al.* (2012) Plasma leptin concentration in donkeys. *Vet Q* **32(1)**:13–16.

CASE 11

1 The horse presented to an equine referral hospital. Where should this patient be housed, and why? In isolation because of the history of fever and the leukopenia with neutropenia evident on the CBC.

2 What is the primary differential diagnosis in this horse based on the history, location, time of year, clinical signs, and hematology? The horse has hypoproteinemia, which can be due to protein loss from the GI tract, renal protein loss, or lack of protein production in the liver. The horse has a normal urinalysis and no azotemia, which rules out renal loss. There was no icterus or elevation of liver enzymes, and although liver enzymes can be low in end-stage liver disease, this filly was healthy and training well just a week previously, so did not have end-stage liver disease. Therefore, the protein was being lost through the GI tract, which was supported by the hypochloremia and hyponatremia, suggestive of reduced luminal absorption associated with some GI diseases. The hypermotile borborygmi also point to GI disease. The history of high fevers indicates an infectious or possibly neoplastic cause. Based on the location (NY) and the time of year (summer), together with the evidence of GI disease, Potomac horse fever (PHF) should be the primary differential diagnosis. The etiologic agent of PHF is *Neorickettsia risticii*, which is transmitted to horses by accidental ingestion of the metacercaria stage of trematodes (snails) present in larval and adult aquatic insects, such as caddisflies and mayflies. It should be noted that it is not uncommon to see normal manure in the early stages of disease and/or if the patient is dehydrated. Diarrhea will often ensue on rehydration, but some horses with PHF never develop diarrhea. Other differentials to consider in this case include right dorsal colitis secondary to the phenylbutazone administration, salmonellosis, *Clostridium difficile*, infiltrative bowel disease such as lymphosarcoma or granulomatous inflammatory bowel disease, and cyathostomosis.

3 What is the treatment of choice for this disease? Oxytetracycline (6.6 mg/kg IV q12h) for 5 days in addition to supportive care. Response to this treatment is usually good.

4 What important secondary complications can occur in these cases? Laminitis, which develops in 15–25% of cases and can often be a reason for eventual euthanasia.

Reference

Bertin FR, Reising A, Slovis NM *et al.* (2013) Clinical and clinicopathological factors associated with survival in 44 horses with equine neorickettsiosis (Potomac horse fever). *J Vet Intern Med* **27(6)**:1528–1534.

CASE 12

1 Interpret the echocardiographic findings. There are both clinical and echocardiographic findings supportive of right-heart dilatation and tricuspid insufficiency. The small-appearing left ventricle and atrium are likely due to volume contraction. The high velocity flow over the tricuspid valve (**12.3**) indicates obstruction of the pulmonary artery. If tricuspid insufficiency is the only problem (e.g. due to valvular degeneration), there would not be high velocity flow because there are usually only moderate pressure differences between the right ventricle and atrium. In this case, there must have been some right ventricular hypertrophy, likely related to pulmonary artery obstruction, causing the marked difference in pressures between the two chambers. The pulmonary artery (**12.4**) is larger than the aorta, suggesting dilatation of the pulmonary artery. There is no obstruction noted at the pulmonic valve.

2 What do the CBC abnormalities suggest? Normal arterial oxygenation with a modest elevation in lactate is suggestive of decreased perfusion. The elevated liver enzymes and ventral abdominal edema are likely the result of right-sided heart failure and congestion.

3 Which of the following is the most likely diagnosis: cor pulmonale due to pulmonic valvular stenosis; cor pulmonale due to diffuse pulmonary parenchymal disease; cor pulmonale due to obstruction of the pulmonary artery? Cor pulmonale due to obstruction of the pulmonary artery.

Follow up/disusion
On necropsy, a large firm mass was found in the pulmonary artery (plaque in the center of the opened pulmonary artery, **12.6**). Both the pulmonary artery and the aorta were 16 cm in diameter; the aorta should be significantly larger than the pulmonary artery. The right ventricular wall thickness was 2 cm and that of the left ventricle was 3 cm, which was not as large a difference as is typically seen. The mass in the pulmonary artery was composed of fibrous connective tissue, degenerate neutrophils, and a large number of fungal organisms, which were identified as *Aspergillus* spp.

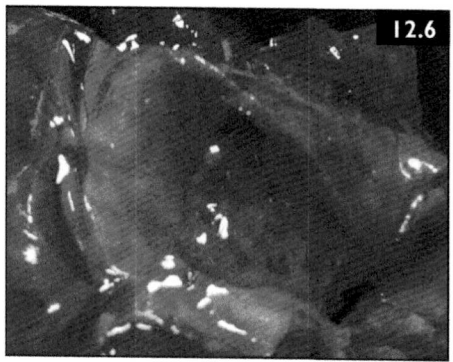

CASE 13

1 What are the rule outs for milk refluxing into the nasopharynx and nostrils in a newborn foal? Cleft palate, white muscle disease (selenium deficiency), persistent epligottic frenulum, hypoxic ischemic encephalopathy, megaesophagus, vascular ring anomalies, pharyngitis/stomatitis/esophagitis, esophageal choke, fourth branchial arch defect, and weak foals (e.g. because of systemic illness, electrolyte abnormalities, hypoglycemia).

2 What diagnostic procedures should be performed next to investigate the observation of milk refluxing from the nostrils? Examine the oral cavity to look for a defect in the hard palate (this was normal). Then endoscopy of the upper respiratory tract to look for abnormalities in the soft palate, distal pharynx, and larynx.

3 What abnormality is seen in this endoscopic image of the larynx (13.2), and how might it explain the clinical signs and future athletic use of the foal? The foal (unsedated) has rostral displacement of the caudal pillar of the soft palate over the right dorsal arytenoid cartilage and less abduction of the right arytenoid cartilage compared with the left arytenoid. External palpation of the larynx revealed a gap between the thyroid and cricoid cartilages, presumably due to an absence of the wing of the right thyroid cartilage. This defect was not felt on the left. This was highly suggestive of a right-sided fourth branchial arch defect, which is more common than a left-sided defect. A loss of muscle fibers at the proximal esophagus is also typically involved in the defect, which causes involuntary aerophagia (leading to chronic colic in some horses) and proximal esophageal dilation. In this case, the foal was believed to have been swallowing normally, but when the head was lowered after nursing there was reflux of residual milk from the esophagus. The foal did not have any noticeable milk reflux after a couple of days, which may have been related to further development of proximal esophageal muscle fibers and some minor improvement in proximal closure of the esophagus. Affected foals are unlikely to be successful racehorses because of instability of the larynx.

4 What routine test should be performed to make sure the foal had adequate absorption of maternal antibodies? How quickly after nursing can the test be accurately performed? The foal should be checked for adequate IgG between 14 and 24 hours after the first nursing. An on-site ELISA (Snap® Foal IgG Test, Idexx) is available to determine adequate passive transfer of colostral antibodies.

References
Holcombe SJ, Hurcomb SD, Barr BS et al. (2012) Dysphagia associated with presumed pharyngeal dysfunction in 16 neonatal foals. *Equine Vet J* 41(Suppl 41):105–108.
Menéndez IM, Mancha DAI, Fitch G (2011) Fourth branchial arch defects in full-siblings treated with a partial arytenoidectomy. *Equine Vet Educ* 23(12):626–629.

CASE 14

1 What is the neuroanatomic lesion in this horse? These clinical signs reflect a bilateral lesion to the mental branches of the mandibular branch of the trigeminal nerve (CN 5). The mental branch is sensory to the lower lip, muzzle, and gums. Peripherally, the mental nerve branch also carries some motor fibers from the buccal branch of the facial nerve (CN 7), accounting for the lower lip droop.

The weak closure of the lower jaw could be accounted for by a more proximal injury to the mandibular nerve before the mental nerve branches off.

2 How might this lesion have occurred? The focal and bilateral lesions suggest a local insult to the nerve branches. As the mental nerves emerge from the mental foramen in the mandible, they are relatively superficial and can be vulnerable to trauma. For example, a bilateral lesion could be caused if the horse had got his muzzle trapped in a narrow space and then panicked and pulled back with force or for a prolonged period. It is also possible that an injury from a bit might cause these signs, but there was no suspicion of that in this case. Finally, there is an idiopathic trigeminal neuropathy that occurs most commonly in dogs, but can also rarely occur in horses.

3 Although not affecting this horse, what important neurologic disease can produce these clinical signs as part of its constellation of possible symptoms? A dropped jaw and drooping lower lip can be part of the clinical signs of rabies, which can have a variety of neurologic presentations but is always a progressive, fatal disease.

Follow up/discussion

The etiology in this particular case was never determined. The horse was treated with dexamethasone (0.1 mg/kg IV q24h) for 2 days, thiamine (10 mg/kg IV q24h) for 5 days, and vitamin E (20 IU/kg PO q24h) for 14 days. The gelding improved gradually and, by 3 weeks after the initial onset, was described as normal by the owners. Dogs with idiopathic trigeminal neuropathy have been described to have a mean time to recovery without treatment of 22 days.

CASE 15

1 What are the differential diagnoses for diarrhea and fever in a 2-month-old foal? Include rotavirus, *Clostridium perfringens* A, *Clostridium difficile*, *Salmonella* spp., *Rhodoccocus equi*, acute strongyle infection, and cryptosporidiosis. It would be unusual to have an isolated case of salmonellosis or rotavirus, and infection with *C. perfringens* A is difficult to prove since the organism (and even its enterotoxin) can be found in the manure of normal foals. Cryptosporidiosis and acute strongyle infection are not common.

2 What diagnostics should be performed? An aerobic fecal culture, specifically looking for *Salmonella* spp. and *R. equi*; an ELISA or electron microscopy on the feces to explore the possibility of rotavirus; toxin testing for *C. difficile* and *C. perfringens* (in foals, toxigenic *C. difficile* may cause diarrhea without a history of antibiotics); a fecal smear or flotation to test for cryptosporidium. Fecal Gram stains are rarely clinically useful in foals.

3 Why might the hocks be swollen with little or no lameness? It might indicate an immune-mediated synovitis, which is most frequently found with *R. equi* infections.

4 What could be the cause of the dermatitis? The multifocal, raised, red nodules are not characteristic of contact dermatitis due to diarrhea, in which the dermatitis is typically more uniform throughout the diseased area. The fact that the nodules are raised suggests an infectious agent. Biopsy of one of the nodules revealed a granulomatous dermatitis that was PCR positive for *R. equi*. In this case, the large amount of *R. equi* in the feces apparently infected the perineal skin.

Reference

Slovis NM, Elam J, Estrada M *et al.* (2014) Infectious agents associated with diarrhoea in neonatal foals in central Kentucky: a comprehensive molecular study. *Equine Vet J* 46(3):311–316.

CASE 16

1 What is the most likely reason for the lameness? Septic arthritis secondary to septic thrombophlebitis and bacteremia.

2 What diagnostic procedures could be performed to determine an etiologic cause? The joint fluid should be aspirated from the carpus and evaluated by visual inspection for viscosity, followed by laboratory determination of cell count, TP, cytologic evaluation, and bacterial culture. The jugular vein abscess could also be aspirated following ultrasound examination. Blood culture(s) could be performed, but might require discontinuing antibiotics for 24 hours. It is uncommon to obtain positive cultures from joint fluid in foals with bacteremic-induced septic arthritis because the bacteria are found mostly in the synovial membrane. In this foal, the thrombosed jugular vein was aspirated and grew a *Staphylococcus* sp. (not MRSA) resistant to most antibiotics except enrofloxacin, gentamicin, and amikacin.

3 What are the treatment options? Antibiotics, ideally bactericidal, are indicated. Enrofloxacin was not chosen because of the age of the foal. Although the foal was azotemic at hospital admission, which in the face of isosthenuria indicated likely renal failure, a rapid resolution of the azotemia meant that gentamicin could be cautiously administered. Potassium penicillin was also given IV for gram-positive coverage. NSAIDs were administered IV. Amikacin (250 mg) was administered by regional IV perfusion to the affected limbs once daily for 3 days. The joint was lavaged once with isotonic sodium bicarbonate and amikacin. Omeprazole was administered PO. IV fluids were administered via a catheter placed in the lateral thoracic vein.

4 **What would be the benefits of therapeutic drug monitoring in this case?** Verifying that the gentamicin peak concentration (taken 30–60 minutes after administration) was 10× the MIC of the *Staphylococcus* grown would be important when treating a life-threatening infection. Also in this foal recovering from AKI to know that the gentamicin trough level was <1.0 µg/ml.

5 **What are the differentials and cause for this rapid swelling and new lameness?** Fracture (pelvis or femur) or a rupture of a tendinous insertion in the caudal thigh. In this case, no fracture was identified on radiographs and the foal was much improved the following day. If this was a rupture of the gastrocnemius insertion in the caudal thigh area, it must have been incomplete, as the reciprocal mechanism between the hock and stifle remained intact. A more likely cause was rupture of the gracilis muscle. This generally has a good prognosis for recovery, although fibrotic myopathy might occur later in life.

Reference

Smith LJ, Marr CM, Payne RJ *et al.* (2004) What is the likelihood that Thoroughbred foals treated for septic arthritis will race? *Equine Vet J* 36(5):452–456.

CASE 17

1 **What are the three most common causes of colic and depression in a neonatal foal?** (1) Meconium impaction is by far the most common cause of colic in a foal less than 3 days of age. This can usually be resolved medically with enemas (soapy water +/– acetylcysteine retention) +/– oral laxatives and analgesics. Many breeding operations administer a prophylactic enema shortly after birth to help prevent a meconium impaction. (2) Enterocolitis can cause colic, often beginning with depression and accompanied by pyrexia. The most common causes of enterocolitis in a very young foal are *Salmonella* spp., *Clostridium perfringens*, or secondary to generalized septicemia. (3) Uroperitoneum is a common cause of colic in foals, with colts more frequently represented than fillies. It is most often due to a ruptured bladder, but rupture of the ureters and urethras is also possible.

2 **Given the information provided above, what is the most likely etiology of the clinical signs in this 48-hour-old colt?** The signalment (48-hour-old colt), clinical findings (abdominal distension, mild colic, straining, and depression), ultrasound findings (increased anechoic free fluid) and, most importantly, the laboratory work abnormalities (hyponatremia, hyperkalemia, hypochloremia, and azotemia) all point to a diagnosis of uroperitoneum.

3 **What diagnostic test should be performed next to help confirm the diagnosis?** Abdominocentesis. The fluid should be clear to yellow, with a creatinine level double or more than that of the serum creatinine (i.e. indicating that the abdominal

fluid was urine). In addition, careful ultrasound examination can often identify a tear in the bladder or urethra.

4 What ECG abnormalities might be expected with hyperkalemia? Spiked/tented T waves, blunted or absent P waves, prolonged QRS duration, prolonged PR interval, and shortened QT interval can be seen with hyperkalemia. If the hyperkalemia is severe enough, life-threatening bradycardia can ensue.

5 What initial medical treatments should be administered to a foal with this condition? Definitive treatment for a ruptured bladder is usually surgical. However, the patient needs to be stabilized medically first, otherwise it will be a poor anesthetic candidate. IV fluids without potassium should be administered (0.9% NaCl or 1.3% [isotonic] sodium bicarbonate). IV dextrose (5% CRI in IV fluids or 50 ml of 50% dextrose) should also be given to help lower potassium by causing it to move intracellularly. If there are ECG abnormalities associated with the hyperkalemia, IV calcium borogluconate should be administered as a cardioprotectant. Slow, controlled peritoneal drainage will help lower potassium and creatinine, improve the foal's comfort by reducing pressure in the abdomen and on the thorax, and improve cardiopulmonary dynamics prior to inducing anesthesia.

Prophylactic broad-spectrum antimicrobials should be administered in the face of IV catheter and intraperitoneal drain placement followed by abdominal surgery. In addition, foals with ruptured bladders may also be septic (sepsis is another predisposing factor for bladder rupture). Finally, the IgG level should always be checked in a young foal to ensure adequate passive transfer. If this has not occurred, a plasma transfusion should be administered as soon as possible.

Reference
Butters A (2008) Medical and surgical management of uroperitoneum in a foal. *Can Vet J* **49**(4):401–403.

CASE 18

1 What is the most likely cause of the swelling? A *Streptococcus equi* subsp. *equi* abscess caused by the vaccinations. There are two possibilities for this occurrence: (1) when the modified-live intranasal vaccine was given, there was contamination of the neck or the West Nile virus vaccine needle with *S. equi*; (2) the intranasal vaccine resulted in bacteremia. Regardless, the IM vaccine likely caused enough hyperemia at the site such that the *S. equi* was able to colonize and form an abscess. This is why no additional vaccines should be administered when modified-live strangles vaccines are given.

2 What is the best treatment for this condition? The abscess should be drained and cultured to determine the antibiotic sensitivity of the organism (**18.2 shows**

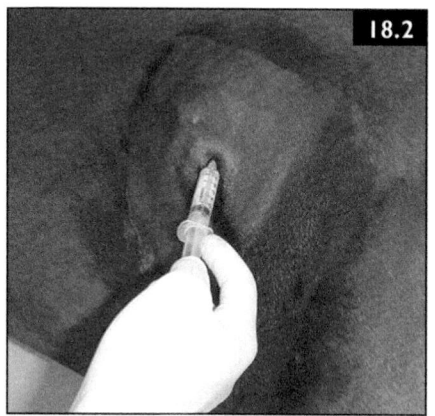
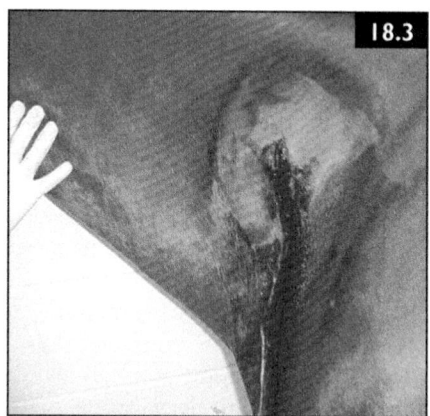

aspiration of the abscess; **18.3** shows the incisional drainage). A 1–2 week course of antibiotics based on known or expected antibiotic sensitivity would be reasonable.

3 Should this horse be isolated? Yes, while the abscess is draining. All areas in contact with the draining exudate should be disinfected properly. Horses infected with the vaccine strain of *S. equi* rarely transmit the infection to other horses but caution should still be used.

4 How long does the organism survive in the environment? Less than 24 hours in sunlight, but may last for several days in indoor environments.

5 Is there a way to determine if the horse is a chronic shedder? Culture and PCR of a guttural pouch wash will detect most carriers. A single nasopharyngeal flush detects fewer carriers and a nasal swab for culture or PCR would be negative in many carrier horses. Guttural pouch examination via endoscopy was performed in this case and saline washes of both guttural pouches were PCR negative for *S. equi*.

6 Is there a way to confirm that the organism cultured from the abscess is the vaccine strain organism? The bacterial colonies of the vaccine strain may look more mucoid than the natural strain. PCR testing can confirm that the abscess was due to the vaccine strain. A positive PCR does not necessarily mean that live organisms are present.

References

Kemp-Symonds J, Kemble T, Waller A (2007) Modified live *Streptococcus equi* ('strangles') vaccination followed by clinically adverse reactions associated with bacterial replication. *Equine Vet J* **39**(3):284–286.

Weese JS, Jarlot C, Morley PS (2009) Survival of *Streptococcus equi* on surfaces in an outdoor environment. *Can Vet J* **50**(9):968–970.

CASE 19

1 Botulism was the initial tentative diagnosis, but some of the clinical findings were not characteristic of botulism. Which ones? Most horses with botulism have diffuse neuromuscular weakness including tail tone, eyelid tone, and dysphagia. This horse's normal eyelid tone, tail tone, and ability to eat were not characteristic of botulism

2 List other differentials for the skeletal muscle weakness in this horse. Include equine motor neuron disease, magnesium toxicity, lead poisoning, organophosphate toxicity, hypocalcemia, hypokalemia, hyperkalemia, ionophore toxicity, tick paralysis, and primary myopathies.

Follow up/discussion

Further historical investigation of this horse revealed that the horse had been accidentally fed lasalocid (a calf coccidiostat) for at least 6 days. The horse was ultimately euthanized but no ionophore testing was performed. Postmortem histopathology revealed axonal swelling and corresponding nerve and muscle degeneration throughout the cervical to sacral area. There was no evidence of cardiac necrosis and cardiac troponin I testing was not available at the time. The tentative diagnosis was ionophore toxicity with peripheral neuropathy and myopathy.

The most common cause of ionophore toxicity in the horse is monensin, which generally causes acute colic and sometimes diarrhea immediately followed by signs of cardiac dysfunction. Ataxia and skeletal myopathy leading to recumbency may occur in some cases. Both lasalocid and salinomycin have also been reported to cause skeletal and cardiac myopathy in addition to neuropathy in horses. Salinomycin is reported to be more toxic than monensin on a mg per kg basis, but equine toxicity from this product is rare, presumably due to its less common use in livestock feeds in comparison with lasalocid and monensin. Lasalocid is reported to be considerably less toxic on a mg per kg basis than monensin. The prognosis for ionophore toxicity is guarded to poor, and poor if cardiac involvement is present.

References

Aleman M, Magdesian KG, Peterson TS *et al.* (2007) Salinomycin toxicosis in horses. *J Am Vet Med Assoc* **230**(12):1822–1826.

Decloedt A, Verheyen T, DeClercq D *et al.* (2012) Acute and long-term cardiomyopathy and delayed neurotoxicity after accidental lasalocid poisoning in horses. *J Vet Intern Med* **26**(4):1005–1011.

Divers TJ, Kraus MS, Jesty SA *et al.* (2009) Clinical findings and serum cardiac troponin I concentrations in horses after intragastric administration of sodium monensin. *J Vet Diagn Invest* **21**(3):338–343.

CASE 20

1 What is the interpretation of the foal at this point? The foal has failure of passive transfer, is tachycardic, and the laboratory changes are suggestive of sepsis in spite of being normothermic. The extremely low WBC count suggests either gram-negative sepsis or viral infection.

2 The placenta weighed 12% of the foal's body weight. What is the interpretation of this information? A normal equine placenta should not be >11% of the foal's body weight. Increases are likely associated with edema due to infectious causes (placentitis) or toxic insults.

3 What is the interpretation of the rapid deterioration of the foal and the gross pathology findings? The diffuse consolidated appearance of the lungs in a foal that had a fairly normal RR and effort for the first 24–36 hours of life suggests an acute interstitial pneumonia. The multifocal hemorrhagic lesions in the small intestine with normal mucosa suggest a vasculitis.

4 What are the most likely differentials at this time? Interstitial pneumonia due to EHV-1, EVA, or possibly adenovirus.

Follow up/discussion

The foal was positive for EHV-1 in the liver, lung, and lymphoid follicles of the small intestine. Analysis of the virus via PCR revealed that it was the 'neuropathic' strain of EHV-1. Approximately 70% of equine herpes myelopathy cases are due to a mutation that has commonly been termed the 'neuropathic strain'; 25% of EHV-1 abortions are a result of infection with this strain. One horse developed equine herpes myelopathy 7 days after likely having been in contact with the placenta from this case. Neonatal death due to EHV-1 is rare in foals but does occur. It is also extremely rare for neonatal EHV-1 to transmit infection and neurologic disease to adult horses. The contact with the heavily virus-laden placenta in this case was likely the cause of the infection in the adult horse. Interestingly, the infected adult horse had received steroids for another problem, which may have predisposed to the high EHV-1 viremia and subsequent neurologic signs. This foal did not transmit the infection to any other adult horses, although he had contact with several other horses during his viremia. This highlights the complexity of the development of neurologic disease, and even infection, due to EHV-1. The adult horse had a gradual improvement in the neurologic signs.

Reference

Pronost S, Cook RF, Fortier G *et al.* (2010) Relationship between equine herpesvirus-1 myeloencephalopathy and viral genotype. *Equine Vet J* 42(8):672–674.

CASE 21

1 What generalized type of disease does this horse have? Vasculitis.

2 What are the differential diagnoses for this type of disorder? Vasculitis is most often an immune-mediated inflammatory disease of blood vessels and may be associated with bacteria, drugs, viral agents, and other environmental or chemical factors such as sun exposure. Purpura hemorrhagica is one of the most common vasculitic disorders of the horse.

3 What tests could be run to confirm the diagnosis and identify initiating antigens? A skin biopsy, which revealed a leukocytoclastic vasculitis, edema, hemorrhage, and hyperkeratosis. Other tests include endoscopy of the guttural pouch and PCR for *Streptococcus equi* subsp. *equi*, serology for equine viral arteritis, and testing (serology or PCR) for anaplasmosis.

4 What treatments should be recommended? Once daily dexamethasone (20 mg IV) until an excellent clinical response is observed. The dose is then tapered over a 10-day period. Because of the crusting of the skin, systemic penicillin and a medicated shampoo containing ketoconazole and chlorhexidine could also be given.

5 What is the primary concern with the drug administered in this case? Corticosteroid-induced laminitis. Short-term treatment with moderate doses of dexamethasone rarely causes laminitis in horses unless they are predisposed due to an endocrine disorder (e.g. Cushing's disease or insulin dysfunction/metabolic syndrome). Morgan horses are a breed known to have a high incidence of equine metabolic syndrome. In this case, the horse did not have a phenotype (i.e. abnormal fat deposits) suggestive of equine metabolic syndrome.

6 This horse (21.3) had another immune-mediated disease that often causes a generalized dermatitis and can appear similar to the horse in this case. What is this disease? Pemphigus foliaceus.

Follow up/discussion

Cutaneous vasculitis in horses is a common problem and the inciting antigen/antibody is not confirmed in >50% of cases. Painful edema of the limbs and other areas of the body along with scales and crust, as seen in this case, are the most common clinical findings. Many cases of purpura hemorrhagica have petechiation of mucous membranes and occasional fever. Diagnosis is based mainly on clinical findings since skin biopsies are not highly sensitive in confirming the disease. Treatment consists mostly of immunosuppressive therapy with corticosteroids or azathioprine and addressing any inciting infection such as *Streptococcus equi*. Pentoxifylline, which inhibits cytokines, has also been used in treating vasculitis in the horse and other species. The dose of dexamethasone is variable and should be given to effect to control the signs. Doses as high as 80 mg/day have been used, although 30–40 mg daily is a standard starting dose. Dexamethasone appears to be a more effective treatment than oral prednisolone, even at equivalent doses.

This horse responded to treatment, but the signs recurred when prednisolone (400 mg PO daily) was substituted for dexamethasone. Reinstitution of dexamethasone therapy at 20 mg IV once daily followed by tapering over another 10 days was effective in resolving the disease.

Treatment for vasculitis and pemphigus foliaceus can be very similar; however, horses with vasculitis generally completely recover while pemphigus foliaceus can be controlled but not often cured.

References

Pusterla N, Watson JL, Affolter VK *et al.* (2003) Purpura haemorrhagica in 53 horses. *Vet Rec* **153(4):**118–121.

White SD, Affolter VK, Dewey J *et al.* (2009). Cutaneous vasculitis in equines: a retrospective study of 72 cases. *Vet Dermatol* **20(5–6):**600–606.

Zabel S, Mueller RS, Fieseler KV *et al.* (2005) Review of 15 cases of pemphigus foliaceus in horses and a survey of the literature. *Vet Rec* **157(17):**505–509.

CASE 22

1 What is the diagnosis for these clinical signs in this foal? Neonatal isoimmune thrombocytopenia. Historically, an immune reaction to navel dips was implicated in the pathogenesis of this disease, but that is likely erroneous.

2 What laboratory test should be recommended? CBC. In this case, platelets = 3,000/µl (3 × 10^9/l) and neutrophils = 2,800/µl (2.8 × 10^9/l). Although thrombocytopenia is the most dramatic abnormality on the CBC of foals with this disorder, many foals also have neutropenia.

3 How should the foal be treated, and what complications could occur? The foal should be confined to a stall or small paddock to limit exercise and the possibility of hemarthrosis. Administration of corticosteroids is recommended and causes a rapid rebound in the platelet count. Generally, 0.1 mg/kg of dexamethasone is given PO, IM, or IV, with a 1-week tapering dose. Since injections may result in hematoma formation, a small gauge needle (≥20 gauge) should be used and the vein tightly compressed after blood collection or injection. The platelet count should be repeated, as the adverse effects of maternal antibodies on platelets can last for 3 or more weeks. This foal developed a hematoma after the IM injection of dexamethasone, but it resolved without complication and the platelet count quickly rebounded. The foal was also treated with sucralfate, omeprazole, and trimethoprim sulfa.

4 What recommendations should be made to the owner regarding the management of future foals from this mare? Colostrum from this mare would be at risk of producing the same condition in future foals. Unlike neonatal isoerythrolysis, there is no proven test to determine the risk for neonatal isoimmune thrombocytopenia in foals.

5 **What does the mean platelet volume (MPV) indicate?** It is a measurement of mean platelet size and an increased value indicates that the bone marrow is responding appropriately to the thrombocytopenia, as the marrow releases immature platelets that are larger than normal (i.e. have increased RNA). An increased MPV therefore suggests that the thrombocytopenia is due to excessive consumption or destruction and not lack of production.

Reference
Buechner-Maxwell V, Scott MA, Godber L *et al.* (1997) Neonatal alloimmune thrombocytopenia in a quarter horse foal. *J Vet Intern Med* **11**(5):304–308.

CASE 23

1 **What are the major concerns as to the clinical conditions affecting this horse?** That the horse has dysphagia, probably due to a neurologic disease, and possibly aspiration pneumonia.

2 **What does the thoracic ultrasound suggest?** There is cranioventral pulmonary consolidation, most likely due to aspiration pneumonia. It is typical for aspiration pneumonia to be more severe on the right side of the lungs, as seen in this case.

3 **What organisms are commonly associated with this type of pneumonia?** Include *Streptococcus equi* subsp. *zooepidemicus*, *Actinobacllus* spp., *Enterobacter* spp., *E. coli*, *S. aureus*, and numerous anaerobic organisms.

4 **What is the most likely diagnosis, and how is it causing the clinical signs?** A fungal lesion, which likely damaged the medial pharyngeal branch of the vagus nerve and caused the dysphagia, leading to the subsequent aspiration pneumonia.

5 **What are the treatment options for the guttural pouch lesion?** Topical antifungal treatment would be preferred. Local treatment options are nystatin powder, natamycin, miconazole, clotrimazole, or 2% enilconazole solutions. Irritating products such as iodine or chlorhexidine products should never be placed in the GP because severe and often irreversible paralysis will occur. Additional therapeutic options include itraconazole PO and sodium iodide IV.

Follow up/discussion
Due to the location of the lesion in this case (not in close proximity to an artery), it seemed unlikely that arterial occlusion would be of benefit. However, a previous case had a deep fungal plaque over the mid-region of the stylohyoid bone, causing a brown nasal discharge as the only clinical sign. In that case, arterial ligation was successful in resolving the lesion after 2 weeks of topical clotrimazole had been unsuccessful. In the current case, enilconazole solution was infused into the pouch twice daily and the horse was cross-tied with the head in a neutral position for 30 minutes after each infusion. This has been shown to maintain the enilconazole

solution in the pouch and in contact with the lesion (**23.4**). The horse made a complete recovery following 3 weeks of enilconazole treatment in the GP and 6 weeks of IM procaine penicillin for the aspiration pneumonia. (Recheck endoscopy of the GP [**23.5**] shows resolution of the fungal plaque with only mild hyperemia in the area of the previous infection).

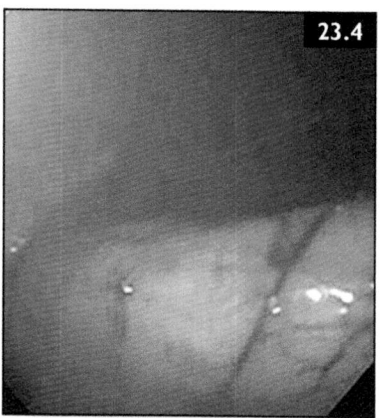

GP diseases are common in the adult horse. The most common disorders include empyema and chondroids, which are usually (but not always) caused by *Streptococcus equi* subsp. *equi* infections, and mycosis, which may cause either hemorrhage or dysphagia and, less commonly, chronic nasal discharge. Neoplasia (most commonly melanoma or sarcomas) may also occur in older horses. There are several CNs that course through the GP, and CN IX (glossopharyngeal), CN X (vagus), and the sympathetic nerves are most commonly diseased by fungal plaques. These are rarely damaged by even severe bacterial infections within the GP. Clinical signs associated with damage to CNs IX and X are dysphagia, laryngeal paralysis, and soft palate displacement. Prognosis is extremely grave if dysphagia is

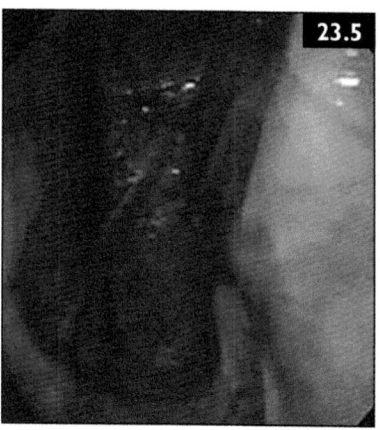

severe. In this case the dysphagia was mild, most likely because the fungal injury damaged only a branch of the vagus nerve.

Reference

Davis EW, Legendre AM (1994) Successful treatment of guttural pouch mycosis with itraconazole and topical enilconazole in a horse. *J Vet Intern Med* 8(4):304–305.

CASE 24

1 What would be the next diagnostic test? Radiography of the limbs (**24.2**).
2 What is the most likely diagnosis for the limb lesions? Hypertrophic osteopathy.

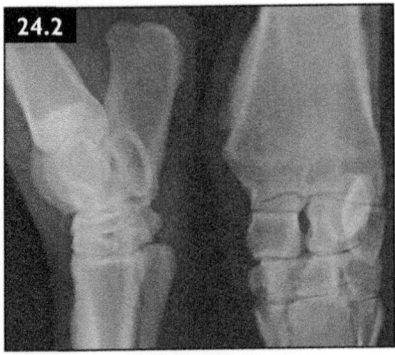

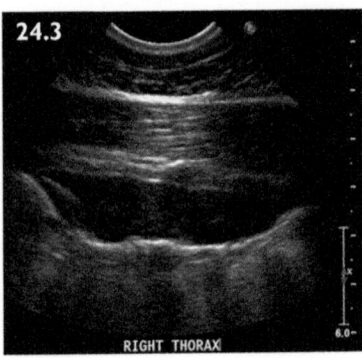

3 With this differential diagnosis and observation of an increased RR, what should be done next? Ultrasound examination of the chest (**24.3**).

4 What did this test show? Several areas of pulmonary parenchymal consolidation (i.e. lack of aeration) on both sides in the mid-chest.

5 Multiple lung lesions, fever, weight loss, lymphopenia (1,100 cells/µl [1.1 × 10^9/l), and elevated fibrinogen (600 mg/dl [6.0 g/l]) might be suggestive of what pulmonary disease? Equine multinodular pulmonary fibrosis or neoplasia. The multiple areas of consolidation in the mid-thorax would make pneumonia less likely but still possible (either bacterial or fungal).

6 How could this diagnosis be confirmed? Transtracheal wash or bronchoalveolar lavage. Also a lung biopsy if the previous tests were not diagnostic. The transtracheal wash revealed a predominantly histiocytic inflammation. Lung biopsy showed moderate, locally extensive fibrosis. The sample was negative for equine herpesvirus (EHV)-5 on PCR.

7 How could this case be treated? With corticosteroids (dexamethasone starting at 0.1 mg/kg every day with a gradually tapering dose over 40 days).

Follow up/discussion

The horse was not given valacyclovir due to the negative EHV-5 testing. He was also treated with topical cidofovir in both eyes. The fever quickly resolved following initiation of the corticosteroid treatment and the horse made a remarkable recovery with resolution of the ultrasound abnormalities in the lung. The bony lesions in the limbs did not resolve, but the horse has remained sound.

Reference

Tomlinson JE, Divers TJ, McDonough SP *et al.* (2011) Hypertrophic osteopathy secondary to nodular pulmonary fibrosis in a horse. *J Vet intern Med* **25**:153–157.

CASE 25

1 What does the radiograph demonstrate? There is incomplete ossification of the tarsal bones. The small tarsal bones and distal epiphysis of the tibia are slightly small and rounded.

2 What is shown in the new radiograph? It shows tarsal valgus (it was bilateral, with the right side being more severely affected). The tarsus/metatarsus is laterally deviated. The lateral tarsocrural joint space is abnormally wide and the proximal aspect of the lateral trochlear ridge is not distinct. A patchy decrease in bone opacity is seen in the distal tibial epiphysis and metaphysis and the physis was questionably wide. This probably was a result of asymmetrical growth in the distal tibia due to physitis/epiphysitis and/or asymmetrical ossification of the tarsal and tibial bones.

3 What treatment should be recommended? Bilateral sleeve casts should be applied from above the tarsus down to the fetlocks. The hooves should not be incorporated in the cast in hope of decreasing development of tendon laxity. The legs should be physically straightened as the cast is being applied. The casting should be performed with the foal anesthetized with a short-acting anesthetic.

4 Based on these radiographs, what is the most likely diagnosis, and what treatment should be recommended? The distal tibial metaphysis and epiphysis have an approximately normal size, shape, and radiographic opacity. The distal tibial physis is improved from the previous films, but is minimally wide and irregular medially. The lateral trochlear ridge is still shorter than the medial trochlear ridge, but improved from the previous films. There is no longer thickening of the soft tissues. Due to the improvement in symmetry of the trochlear ridges, lack of lameness noted, and owners who did not want to show or sell the foal, transphyseal bridging of the distal tibia was not recommended.

Reference
Dutton DM, Watkins JP, Honnas CM *et al.* (1999) Treatment response and athletic outcome of foals with tarsal valgus deformities: 39 cases (1988–1997). *J Am Vet Med Assoc* **215**(10):1481–1484.

CASE 26

1 What diagnostic tests should be performed to determine the cause of the urine discoloration? Urine discoloration occurs when abnormal pigments are present in the blood and are subsequently filtered by the kidney. The common causes of pigmenturia include hemoglobin, myoglobin, frank blood, and bilirubin. Assessment of the serum will determine if hemoglobinemia or hyperbilirubinemia is present (**26.2**). Urine dipstick analysis will indicate the presence of protein, blood, or bilirubin; however, it will not differentiate between hemoglobin and myoglobin.

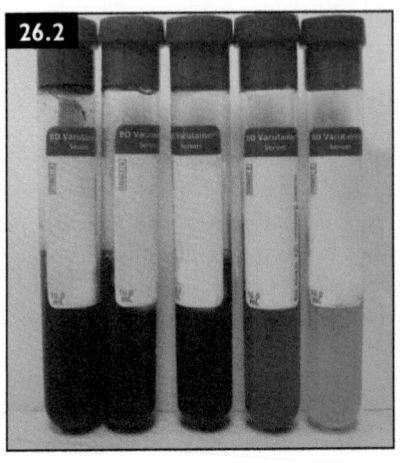

26.2

Centrifugation of the urine can determine if hemorrhage or pigmentation is present (RBCs spin down, pigments do not).

2 What is the likely cause of this problem? The serum was hemolyzed and severe anemia was present (PCV = 11%). Hemolytic anemia occurs secondary to oxidative, infectious, immune-mediated, and iatrogenic causes. After further discussion with the owners, the cause in this case was identified as exposure to fallen branches of a red maple tree and subsequent ingestion of the wilted leaves.

3 What secondary problems should be considered when treating this case? Oxidative damage caused by toxic compounds in the maple leaves can also cause methemoglobinemia, which may result in sudden death due to hypoxia. Acute tubular necrosis caused by the pigmenturia can result in AKI. Severe hemolytic anemia can result in disseminated intravascular coagulation due to activation of the coagulation cascade by inflammatory mediators. Other complications include laminitis and organ damage secondary to hypoxia.

Reference

Alward A, Corriher CA, Barton MH *et al.* (2006) Red maple (*Acer rubrum*) leaf toxicosis in horses: a retrospective study of 32 cases. *J Vet Intern Med* 20(5):1197–1201.

CASE 27

1 List the initial differentials for this horse. Include endocarditis, thoracic or abdominal abscess, purpura hemorrhagica, and neoplasia.

2 What additional diagnostic procedures should be performed? Ultrasonography of the chest and abdomen, which was unremarkable, as was abdominal palpation per rectum. CBC was normal except for a mature neutrophilia of 8,600 cells/μl (8.6 × 10^9/l). Plasma fibrinogen was normal (200 mg/dl [2.0 g/l]). Chemistry panel was normal except for a mild decrease in albumin (2.6 g/dl [26 g/l]), low iron (23 μg/dl [4.1 μmol/l]), and a marked increase in muscle enzymes (CK = 2,014 U/l; AST = 1,766 U/l). Transtracheal wash (TTW) and peritoneal tap were unremarkable, although *Aspergillus fumigates* was cultured from the TTW. Although there was no known exposure to strangles, purpura hemorrhagica remained one of the differentials and serum was submitted for *Streptococcus equi* subsp. *equi* titer, which was moderately positive (1:600), neither confirming nor ruling out purpura hemorrhagica. Twenty-four hours later, a blood culture was performed, which had no growth.

3 What is the abnormality on the ECG? Uniform ventricular tachycardia.

4 Does this arrhythmia need to be treated? If so, what treatments could be used? Treatment should be attempted because of the high HR and the signs of cardiac compromise, even though it is uniform and there is no R on T. The most common treatments are magnesium sulfate or lidocaine administered IV. The horse was given 20 g of magnesium sulfate IV over 20 minutes and an additional 20 mg of dexamethasone IV. The ventricular tachycardia persisted for an additional 3 hours before reverting to sinus tachycardia (HR = 64 bpm).

5 Interpret the echocardiogram. There is tachycardia, mild aortic valve thickening, and decreased fractional shortening of the left ventricle (20%). All diastolic chamber sizes are within normal limits.

Follow up/discussion
Daily dexamethasone (20 mg) was continued for 3 days and the horse remained afebrile, the pectoral edema resolved, the HR decreased to 44–52 bpm, the fractional shortening improved to 30%, and the muscle enzymes decreased to nearly one-half the hospital admission levels. The horse was discharged from the hospital on dexamethasone (12 mg PO q24h). Two days after returning to the farm, the attending veterinarian performed a physical examination and the horse's HR was 64 bpm. The dexamethasone dose was increased to 30 mg PO q24h and slowly tapered over 2 weeks, at which time the heart rate was 40 bpm. A recheck examination at the referral hospital 4 months later revealed a clinically normal horse with normal BCS, HR of 44 bpm, and no complaints from the owner regarding the health of the horse. The echocardiogram was normal with a fractional shortening of 44.25%. The left atrial and left ventricular diastolic diameters were nearly identical to the values recorded 4 months earlier, indicating no enlargement of the heart had occurred. Muscle enzymes (AST and CK) were then within normal range (349 and 204 U/l, respectively), as was serum cTnI (0.03 ng/ml). The horse was given no treatment at that time, but a recheck examination 12 months later was recommended prior to the horse being exercised. At that time, all echocardiogram findings were normal and a 24-hour Holter monitor revealed no arrhythmias and an average HR of 44 bpm. The horse has remained clinically normal for 1 year and is in full work.

Myocarditis is not a rare disease in the horse and should be considered in horses with fever, tachycardia, and elevations in muscle enzymes. Ventricular arrhythmias may develop in some cases and can be life threatening. Unfortunately, the cause of fever, myocarditis, and skeletal myopathy is usually unproven, but influenza, EHV-1, streptococcal myopathy, purpura-like syndrome, babesiosis,

African horse sickness, and anaplasmosis should all be considered. In this case, serologic testing could not confirm the cause of the disease.

Increases in blood cTnI are specific for myocardial injury. Horses with myocarditis have high levels of cTnI and measurement is important in both the diagnosis and management. Horses with ventricular arrhythmias and other primary cardiac diseases may have normal or abnormal cTnI levels. It might be helpful to measure cTnI in the clinical management of those cases, in order to determine the need for corticosteroid treatment or period of rest. Horses with acute severe hypoxia or systemic inflammation may have marked elevations in cTnI, often corresponding to the severity of illness. Changes in cTnI following treatment are usually more important in critically ill horses than the level at hospital admission.

Treatment of myocarditis relies mainly on anti-inflammatories and corticosteroids, which typically result in a rapid decrease of the fever, HR, and cTnI concentration. Furthermore, an improvement in function should be seen, evidenced by lowered HR and improved fractional shortening on echocardiography.

CASE 28

1 What is the interpretation of the peritoneal fluid analysis, and what other test should be performed? Mild, non-septic peritonitis with chyle (a milky fluid consisting of lymph and emulsified fat from either intestinal lymphatics or the pancreas). Amylase and lipase were each only slightly increased in the peritoneal fluid versus the plasma (47 U/l versus 31 for amylase and 323 U/l versus 134 for lipase), suggesting that the chylous ascites was possibly (but unlikely) a result of acute pancreatitis. Serum cholesterol was elevated at 184 mg/dl (4.77 mmol/l).

2 How would the effusion be treated? If the foal is stable with no signs of intestinal strangulation, treatment with IV fluids and restriction of fat in the diet (i.e. decrease milk intake) would be recommended. If the chylous ascites did not diminish but the foal remained stable with no evidence of intestinal strangulation, the use of TPN to completely decrease intestinal fat would be recommended.

3 What would be the most likely diagnosis for the lameness? Clostridial myositis associated with the flunixin meglumine injection given 24 hours earlier.

4 Since there was no rupture of the bladder, how could the electrolyte abnormalities be explained in light of the new clinical sign? Hyponatremia, hypochloremia, and hyperkalemia can occur with acute muscle disease, presumably associated with the movement of sodium and chloride into muscle cells and potassium out of the muscle cells.

5 What is the black hypoechoic structure in 28.1? It is most likely a distended urachus. In fact, the foal developed a patent urachus on the second day of hospitalization and was treated by dipping the umbilicus with 2% chlorhexidine q6h and maintenance of excellent stall hygiene. The urachus closed on day 4 of hospitalization.

6 How should the swollen leg be treated initially? Under short-acting injectable anesthesia and local anesthesia, make an incision into the leg for drainage. Infuse oxygen into the wound and flush it with saline. Suture a Penrose drain into the wound (**28.4**).

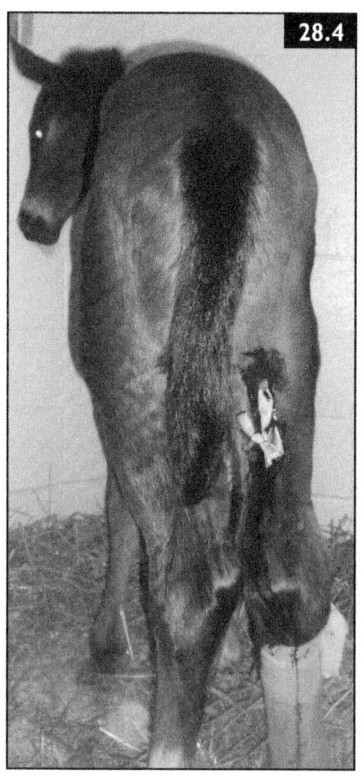

Follow up/discussion
Clostridium perfringens-like organisms were seen on cytology from a sample taken from the incision at the time of drainage. The filly was treated with IV penicillin and PO metronidazole and made a complete recovery. The ultrasonographic appearance of the peritoneal fluid returned to normal and a definitive cause of the chylous effusion was not determined.

Transient chylous effusion is sometimes seen in foals with colic and, in many cases, both the colic signs and chylous effusion resolve with supportive treatment. Congenital chylous ascites is a rare disease defined as the accumulation of chylomicron-rich lymphatic fluid within the peritoneal cavity, resulting from abnormal development of the intra-abdominal lymphatic system. Congenital intestinal lymphangiectasia was suspected to be the cause of the chylous peritoneal effusion and colic in this foal. Treatment as mentioned above is recommended to decrease the chylous effusion because it causes irritation to peritoneal membranes, resulting in mild colic. Although chylous effusion would be expected to contain a large number of lymphocytes, this may not be the case if the leakage has stopped and neutrophils then replace the lymphocytes due to the inflammation. In another foal with abdominal pain and chylous ascites, the only abnormality found on exploratory surgery was dilated lymphatics observed in the mesentery. That foal recovered following the exploratory surgery. Acute pancreatitis may also cause chylous effusion, but did not appear to be the cause in this foal. Chylous peritoneal effusion should be included in the differential diagnosis of young foals with history or signs of abdominal discomfort associated with increased free peritoneal fluid detected by ultrasonography. Abdominocentesis and fluid analysis are the confirmatory diagnostic tools. Based on human literature, the prognosis is generally good when it is caused by congenital intestinal lymphangiectasia.

Clostridial myositis is not common in foals but, similar to adult horses, may occur after IM injections of flunixin meglumine and other non-antibiotic products.

Reference

Cesar FB, Johnson C R, Pantaleon LG (2010) Suspected idiopathic intestinal lymphangiectasia in two foals with chylous peritoneal effusion. *Equine Vet Educ* 22(4):172–178.

CASE 29

1 What is the tentative diagnosis? Sabulous cystitis, which is a common condition in male horses.

2 What would the urinalysis and urine culture results likely show on the sample taken at the initial evaluation? Initial urinalysis typically reveals pyuria and crystalluria without bacteriuria.

3 What would the urine culture results likely show on all subsequent samples? After initial catheterization, bacteriuria will be present with a variety of both gram-positive and gram-negative bacteria.

4 What are the treatment options for this horse? Repeat flushing the bladder with saline and antibiotics based on the culture and sensitivity results.

5 What are possible predisposing factors? There are multiple predisposing causes of sabulous calculi and cystitis in adult male horses. The most common is back pain associated with skeletal abnormalities. This makes the horse unable to completely void the bladder, resulting in non-septic inflammation of the bladder due to ventral gravitation and accumulation of calcium carbonate sediment over several months. In other cases, sabulous cystitis may be associated with neurologic disorders such as equine herpes myelitis or polyneuritis. In these cases there are typically other overt neurologic signs associated with distal spinal cord disease. Mares may also have a similar condition associated with foaling injury and subsequent incontinence.

Follow up

The bladder was flushed with saline on several occasions, each resulting in improvement of clinical signs for 1–3 months. Antibiotics were administered based on culture and sensitivity of any organism found at >10,000/ml in the urine. Eventually, the horse had little improvement with bladder lavage and the cultured bacteria became highly resistant to antibiotics, causing the owners to request euthanasia. Thick sabulous debris could be seen on the ground from the dribbling of urine at that time. A full necropsy was performed and microscopic neurologic examination of the pelvic and hypogastric nerves was normal. The bladder muscle was severely inflamed. The horse had severe spondylitis of the lumbar vertebrae.

Reference

Rendle DI, Durham AE, Hughes KJ *et al.* (2008) Long-term management of sabulous cystitis in five horses. *Vet Rec* **162(24)**:783–787.

CASE 30

1 What are the most likely differential diagnoses for the diarrhea? *Clostridium perfringens* type C would be the leading differential. *Salmonella* spp. or other clostridial infections (*C. difficile* or *C. perfringens* type A) should also be considered. The age, hemorrhagic diarrhea, and isolated case on the farm make *C. perfringens* type C the most likely diagnosis.

2 What should be the immediate treatment plan? Treat the foal aggressively with IV fluids and plasma, gastroprotectants, and antibiotics that are effective against *Clostridium* spp. and any other bacteria that may translocate from the diseased bowel. *Enterococcus* spp. and *E. coli* are enteric organisms that frequently translocate to the blood in foals with enteritis.

3 After reviewing the laboratory findings, how can the results be interpreted, and what would be the treatment plan? The foal had numerous electrolyte abnormalities with the hypokalemia requiring immediate attention. Assuming that the foal was urinating, the most appropriate IV fluids would be 0.45% NaCl plus 2.5% dextrose, supplemented with 20–40 mEq/l of KCl. The foal should not be given bicarbonate for the acidemia at this point for fear of further lowering the serum potassium concentration and increasing the risk of fatal cardiac arrhythmia. The serum sodium should be corrected over 24 hours, as rapid correction of hyponatremia can cause irreversible brain disease.

4 What complications can be associated with neonatal foal diarrhea, and how can these be prevented? Gastric and duodenal ulcers are common complications in foals with enteritis, so gastro-protectants should always be part of the treatment plan. Electrolyte abnormalities and bacteremia are also common. Peritonitis may occur in more severe cases of *C. perfringens* type C or *Salmonella* spp. Early treatment with IV fluids, appropriate antibiotics, and gastroprotectants will help prevent peritonitis (30.3 shows a different

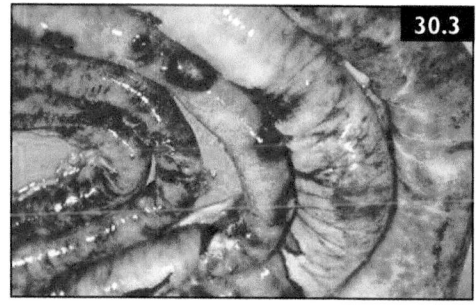

30.3

case with peritonitis). Another common problem is rapid cachexia, which can be prevented by parenteral nutrition. Contact irritation of the perineum and injury to the skin over pressure points (i.e. the hocks) are common complications and can sometimes be prevented by excellent nursing care. Renal failure may occur due to hypotension and administration of nephrotoxic drugs. Fluid therapy and proper use of potentially nephrotoxic drugs will prevent renal failure in most cases.

Discussion

Treatment of foal and weanling diarrhea includes both specific and non-specific therapy. Specific therapy is based on knowledge of the cause of the diarrhea, which in many neonatal cases is unproven. The infectious causes of neonatal diarrhea include neonatal septicemia syndrome including *in-utero* sepsis, *C. difficile*, possibly *C. perfringens* type A, *C. perfringens* type C as an isolated case in foals <5 days of age, rotavirus (the most common infectious cause), *Salmonella* spp., coronavirus, cryptosporosis, and very rarely EHV-1 and adenovirus. *C. perfringens* type C most commonly causes acute hemorrhagic diarrhea. It is mostly seen as a sporadic disease and affected foals usually have consumed colostrum, which is thought to inactivate trypsin and allow the organism, when present, to proliferate in the bowel.

The antibiotic of choice will depend on the etiologic agent. For clostridial diarrheas, metronidazole is given PO or IV in addition to broad-spectrum antibiotics to help protect against bacterial translocation. Antibiotic therapy for bacterial translocation is generally more important in foals with severe enteritis than in adult horses with colitis. Antibiotics should provide coverage against common gram-negative enteric bacteria, including *Enterococcus* (the most common bacteremic pathogen in foals with diarrhea), and *Clostridium* spp. A combination of IV penicillin and amikacin in addition to metronidazole given PO is the author's standard of care in foals that are moderately to severely ill. Amikacin should not be given to foals with diarrhea unless IV fluids are also being administered. Lactase is a specific treatment for foals with moderate to severe rotavirus diarrhea, but might be helpful in most neonatal foal enteritis cases regardless of the etiology. Bio-Sponge® paste is administered to most foals PO with moderate to severe diarrhea regardless of the etiology.

Non-specific treatments include IV fluid therapy (crystalloids and colloids), gastroprotectants including anti-ulcer medication, parenteral nutrition, and probiotics. Analgesics should be given if the foal shows any colic signs, which is a very common presenting complaint in foals with enteritis. Butorphanol and/or limited doses of ketoprofen or flunixin meglumine can be used to control the pain. Foals with enteritis should not be prevented from nursing unless the disease and nursing is causing colic and abdominal distension. For some very mild enteritis cases, such as mild rotavirus diarrhea, no treatment is required if the foal's condition is good.

It can be difficult to correct electrolyte abnormalities in neonatal foals and it must be done without causing adverse effects (e.g. hypokalemia, metabolic acidosis, hyponatremia, hypernatremia, or too rapid correction of sodium disorders). Nursing care of the neonatal foal with diarrhea is extremely important in order to prevent complications such as perineal dermatitis and thrombophlebitis. Lastly, proper biosecurity to prevent the spread of the infectious agents deserves immediate attention because some 'outbreaks' are hard to stop.

References

Frederick J, Giguère S, Sanchez LC (2009) Infectious agents detected in the feces of diarrheic foals: a retrospective study of 233 cases (2003–2008). *J Vet Intern Med* **11**:1–7. (Unfortunately this study did not include *R. equi.*)

Guy JS, Breslin JJ, Breuhaus B *et al.* (2000) Characterization of a coronavirus isolated from a diarrheic foal. *J Clin Microbiol* **38(12)**:4523–4526.

Hollis AR, Wilkins PA, Palmer JE *et al.* (2008) Bacteremia in equine neonatal diarrhea: a retrospective study (1990–2007). *J Vet Intern Med* **22(5)**:1203–1209.

Magdesian KG, Hirsh DC, Jang SS *et al.* (2002) Characterization of *Clostridium difficile* isolates from foals with diarrhea: 28 cases (1993–1997). *J Am Vet Med Assoc* **220(1)**:67–73.

Netherwood T, Wood JL, Townsend HG *et al.* (1996) Foal diarrhoea between 1991 and 1994 in the United Kingdom associated with *Clostridium perfringens*, rotavirus, *Strongyloides westeri*, and *Cryptosporidium* spp. *Epidemiol Infect* **117(2)**:375–383.

Pearson EG, Hedstrom OR, Sonn R *et al.* (1986) Hemorrhagic enteritis caused by *Clostridium perfringens* type C in a foal. *J Am Vet Med Assoc* **188(11)**:1309–1310.

CASE 31

1 What should be done next? Lavage the joint with sterile crystalloid solution. Change the systemic antibiotic treatment to IV penicillin and gentamicin. This treatment was continued for 2 more days with minimal improvement in the lameness.

2 How could the infection in this horse be explained? The horse may have been infected with MRSA spread by the owner or other hospital personnel who were caring for the other two patients with MRSA thrombophlebitis. It is also possible that the horse was a 'carrier' of MRSA at the time of its wound and had a 'self-infection' of the wound.

3 What would be the treatment recommendations? Change the antimicrobial treatment to enrofloxacin (7.5 mg/kg IV q24h) and continue flushing the joint. Regional limb perfusion with vancomycin (300 mg in 60 ml of 0.9% saline) twice daily could be performed. This would likely maintain joint fluid concentrations >4 µg/ml for 24 hours.

4 **The horse became non-weight bearing on the leg (31.3). What would be a concern regarding the opposite hindlimb?** Because of the severe and prolonged lameness unresponsive to phenylbutazone in the left hindlimb, there should be concern about support limb laminitis in the opposite limb.

5 **When considering how frequently vancomycin and enrofloxacin should be administered to this horse, what should be taken into account?** Vancomycin is a time-dependent antimicrobial, which means that the fluid concentrations should be >MIC of the offending organism during the entire treatment period. Enrofloxacin is considered a concentration-dependent antimicrobial, which means that higher peak levels would have higher efficacy (hence, 7.5 mg/kg q24h versus 5.0 mg/kg q12h).

Follow up/discussion

The horse was administered a morphine epidural via an epidural catheter, as needed, to improve weight bearing on the affected limb. The right foot was also kept in an ice boot. Although not performed in this case, removing weight bearing on the right limb by sling support could also have been useful.

Unfortunately, in spite of 6 additional days of treatment with enrofloxacin, increasing the vancomycin dose to a record 1 gram, and joint lavage every other day, the horse had minimal change in lameness and MRSA was still cultured in small numbers from the joint fluid on a repeat aspirate. The horse was euthanized due to the poor prognosis and lack of response to treatment.

MRSA should be considered highly contagious from horse to horse, human to horse, and vice versa. There are many different ribotypes of MRSA and determining where the infection came from (e.g. human, horse) can be challenging. Regardless, all individuals handling horses, and especially those inspecting wounds, catheters etc., should wear disposable gloves in facilities where MRSA cases have been documented. Chlorhexidine is the preferred skin disinfectant if MRSA is a concern. The antimicrobial sensitivity of MRSA can vary. Community-acquired infections can be susceptible to several antimicrobials, while hospital-acquired MRSA is typically resistant to most antimicrobials. However, hospital-acquired infections are often susceptible to fluoroquinolones and amikacin, and nearly always susceptible to vancomycin. Although equine veterinarians commonly harbor MRSA in their nasal cavities, there are no known studies that have demonstrated a higher risk of wound or surgical infection in equine veterinarians. In one situation, two horses developed MRSA thrombophlebitis on a single weekend associated with placement of a jugular catheter, and the veterinary technician placing the catheters was culture positive on nasal swab. Following nasal treatment and subsequent negative culture, the technician resumed hospital duties with no other suspect transmissions.

References

Orsini JA, Snooks-Parsons C, Stine L *et al.* (2005) Vancomycin for the treatment of methicillin-resistant staphylococcal and enterococcal infections in 15 horses. *Can J Vet Res* **69**(4):278–286.

Rubio-Martínez LM, López-Sanromán J, Cruz AM *et al.* (2006) Evaluation of safety and pharmacokinetics of vancomycin after intraosseous regional limb perfusion and comparison of results with those obtained after intravenous regional limb perfusion in horses. *Am J Vet Res* **67**(10):1701–1707.

Sieber S, Gerber V, Jandova V *et al.* (2011) Evolution of multidrug-resistant *Staphylococcus aureus* infections in horses and colonized personnel in an equine clinic between 2005 and 2010. *Microb Drug Resist* **17**(3):471–478.

Weese JS (2010) Methicillin-resistant *Staphylococcus aureus* in animals. *ILAR J* **51**(3):233–244.

CASE 32

1 What is the tentative diagnosis? The clinical signs and ultrasound findings are consistent with pulmonary edema of unknown cause. Echocardiography revealed significant mitral regurgitation with an enlarged left atrium (13.5 cm). This was supportive of cardiogenic pulmonary edema. A mitral valve leaflet was observed moving into the left atrium during systole.

2 What is the most likely cause of the acute left heart failure and cardiogenic pulmonary edema in this case? A necropsy specimen is shown (32.3). Chordae tendineae rupture. It is very unusual that no murmur was auscultable, even with multiple internists and cardiologists listening to the heart.

Discussion

Chordae tendineae are fibrous cords of tissue that secure the tricuspid and mitral valves to the walls of the right and left ventricles. Chordae tendineae prevent the tricuspid and mitral valves from being displaced by the pressure of blood flow through the chambers of the heart. Chordae tendineae are important in maintaining optimal left ventricular systolic function. Rupture of the mitral chordae tendineae is more common than rupture of the tricuspid ones, and would be expected to have more dramatic clinical signs because of the increased pressures in the left side of the heart. It is unusual for horses to present with predominant signs of left heart failure, but when this happens one of the main differentials would be acute or subacute rupture of the chordae tendineae. Horses with ruptured mitral valve chordae may have a history of sudden onset of acute distress with predominantly respiratory symptoms. On auscultation there will usually be a widespread pansystolic murmur with an extension of the area of cardiac auscultation. Two unusual features in this case were the absence of a murmur and apparent resolution of clinical signs for 4 days. Some horses with ruptured mitral chordae tendineae do not develop clinical signs of pulmonary edema and may live for several weeks, but are exercise intolerant.

References

Holmes JR, Miller PJ (1984) Three cases of ruptured mitral valve chordae in the horse. *Equine Vet J* **16**(2):125–135.

Marr CM, Love S, Pirie HM *et al.* (1990) Confirmation by Doppler echocardiography of valvular regurgitation in a horse with a ruptured chorda tendinea of the mitral valve. *Vet Rec* **127**(15):376–379.

CASE 33

1 What immediate treatment should this horse receive? IV fluids and, if urine is produced from the fluid treatment, acepromazine and analgesics such as NSAIDs. Some veterinarians also use dantrolene. Warm compresses should be applied to affected muscles if possible.

2 What clinical and laboratory diagnosis is most likely responsible for the stiffness, tachycardia, and distress? Explain the reason for the laboratory findings provided. The horse most likely had exertional rhabdomyolysis. The clinical signs, history, and marked elevations in muscle enzymes (CK, AST) were compatible with this diagnosis. The low sodium and chloride were likely due to sweat loss and muscle damage, which cause abnormal redistribution of these intracellular electrolytes. Likewise, potassium may have been high due to leakage from intracellular sites or from decreased glomerular filtration. The calcium might have been low due to sweat loss; chloride and calcium are lost in excess in equine sweat in comparison with blood composition. The low bicarbonate was due to the severity of the disease; horses with milder forms of myopathy without perfusion deficits are more likely to be alkalotic due to hypochloremia. With perfusion deficits and more severe myopathy, lactic acid accumulates and titrates the bicarbonate. Bicarbonate therapy is rarely indicated in horses with myopathy unless it is an attempt to decrease the toxic effect of myoglobin on renal tubular cells. The extremely high PCV was likely due to both dehydration and pain-induced splenic contraction. The marked elevation in creatinine, especially in comparison to BUN, suggested that the horse was in renal failure. The BUN is often more drastically elevated than the creatinine with prerenal azotemia.

3 Interpret the ultrasound findings. The kidney size and architecture were within normal limits, which is not unusual for acute kidney injury (AKI) in horses unless there is severe renal edema. The cortex remained the normal 1–2 cm in thickness and the kidney was approximately 9 × 15 cm. The renal pelvis did seem to have a prominent fluid pocket for an unexplained reason, since there was no other evidence of obstruction.

4 What is a major concern at this time, and what monitoring and treatments should be used? The horse has developed anuric AKI. This could have resulted from the combination of hypotension and myoglobin tubular toxicity. Assuming enough IV fluids were given to correct prerenal factors, then additional treatments are warranted in an attempt to change intrarenal hemodynamics and promote glomerular filtration and diuresis. A large number of drugs can be used in oliguric/anuric myoglobinuric nephropathy in an attempt to promote diuresis (e.g. dopamine, fenolypam, mannitiol, dobutamine, and furosemide). However, none are proven to work.

Follow up/discussion
In this case, dopamine (5 µg/kg/minute) was administered along with furosemide, and diuresis ensued. The urine was discolored (**33.3**). The dopamine infusion was stopped after 24 hours and anuria recurred, so the dopamine was reinstated for another 24 hours along with IV fluid therapy. Although diuresis occurred with

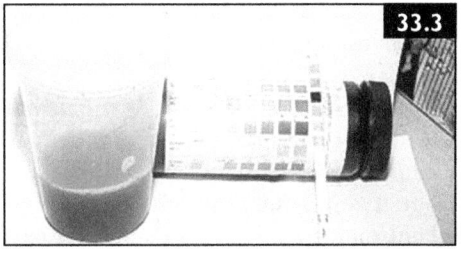

both dopamine treatments, the creatinine did not decrease until day 4 of treatment. The horse recovered and his creatinine was 1.8 mg/dl (137.3 µmol/l) by day 8. Anti-oxidant therapy such as IV or PO vitamin C can also be used for myoglobionuric renal disease since an oxidative component of the disease has been documented.

AKI can be caused by either intrinsic causes or obstruction/rupture of the urinary tract. Intrinsic causes of AKI include those disorders that cause significant (>50%) functional decreases in the glomerular filtration rate (GFR), mostly due to abnormal glomerular filtration pressures, morphologic nephron damage, or both. In many cases of AKI in large animals, there may be minimal morphologic changes seen on light microscopy of a kidney biopsy in spite of life-threatening functional changes. It is important to remember that effective glomerular filtration is determined by perfusion pressure at the glomerulus, which is a result of both afferent renal artery blood flow and pressure balances between afferent and efferent arterioles, and the filtration efficiency of the glomerulus. Ischemia is often blamed for AKI but hyperemia may exist in many cases due to dilation of the efferent arterioles. It is also important to remember that in many cases of intrinsic AKI, especially those with tubular necrosis, not every nephron is affected equally. Some may be totally non-functional, and, if the basement membrane is lost, they may remain non-functional. Some are dysfunctional but may recover, while others may be relatively unaffected and are responsible for providing most of the renal function during the episode of AKI.

The most common site for nephron damage in AKI in large animals is the proximal tubule (i.e. acute tubular necrosis [ATN]). Acute glomerulopathies are rare. Tubular damage may be due to necrosis or apoptosis and is affected by GFR, tubular casts, and peritubular blood flow. The most common predisposing systemic disorders that cause abnormalities in glomerular perfusion pressures resulting in ATN are septic shock conditions, systemic inflammatory diseases, acute heart failure, acute hemorrhage, and other causes of acute intravascular volume depletion (i.e. diarrheal diseases). The most common toxic disorders that cause ATN are nephrotoxic drugs (e.g. aminoglycosides in all species; tetracyclines in horses) and myoglobinuric nephrosis. Although morbidity and mortality are high with ATN, tubular injury frequently is reversible, depending on whether necrotic cells and intratubular casts are removed and renal tubular cells can regenerate. The prognosis is grim if there is renal cortical necrosis.

Treatment of intrinsic AKI includes: (1) treating the initiating disease; (2) correcting prerenal factors (dehydration); (3) determining if the patient is anuric, oliguric, or polyuric. If the patient is polyuric, the prognosis is better but it does not always mean GFR will improve. For the polyuric patient, fluid therapy is the gold standard except for cases of AKI. Fluid therapy re-establishes renal blood flow and glomerular pressure, flushes out tubular debris to further improve glomerular function, and corrects electrolyte abnormalities. (4) If the patient is oliguric – or, even worse, anuric – after correction of pre-renal factors, the prognosis is guarded or poor and fluid therapy alone is not the gold standard.

References

Divers TJ, Whitlock RH, Byars TD *et al.* (1987) Acute renal failure in six horses resulting from haemodynamic causes. *Equine Vet J* **19**(3):178–184.

El-Ashker MR (2011) Acute kidney injury mediated by oxidative stress in Egyptian horses with exertional rhabdomyolysis. *Vet Res Commun* **35**(5):311–320.

CASE 34

1 The primary diagnosis was not obvious, but what complication should be a concern since this is a Miniature horse that is unable to eat properly and has a 3-day-old foal by its side? There is a high risk for hyperlipemia. Continued nursing by the foal would create a further energy drain on the mare and further increase the risk of hyperlipemia and hepatic lipidosis.

2 The spun down plasma is shown (34.4). What is this an indication of? Hyperlipemia. The serum triglycerides (TGs) were 1,929 mg/dl (21.8 mmol/l).

3 What is the diagnosis based on the labwork? Based on the myoglobinuria and increase in CK and AST, a primary myopathy or musculoskeletal disease was likely preventing the mare from eating and keeping her tongue in her mouth. This caused a negative energy balance, hyperlipemia, and probable hepatic lipidosis (based on the markedly elevated liver enzymes, total and direct bilirubin, ammonia, and serum bile acids).

4 How should the hyperlipemia be treated? Treatment should involve correction of the condition that led to a negative energy balance and provision of supportive care. The most important aspect of supportive care is nutritional support. In this case, enteral nutrition was administered via an indwelling stomach tube. This mare was fed a 12% protein (whey based, which is high in branch chain amino acids), 73% carbohydrate (mainly glucose and galactose), 1% fat equine commercial liquid meal q6h in order to exceed the maintenance caloric requirements of approximately 35 Kcal/kg. At each feeding the mare was also given a small amount of a carnitine supplement, which is thought to stimulate protein synthesis, accelerate fatty acid oxidation, lower lactic acid production, and reduce hepatic fat through lipoprotein production. A small amount of high fiber, 10% protein complete feed was added to the liquid enteral feed. The mare was given a multiple vitamin mixture IV once daily for 3 days. Since the mare was hyperglycemic, she was administered 0.1 IU regular insulin IM q6h at the time of feeding. Additional treatments were 1 quart of mineral oil via nasogastric tube on day 1 and maintenance rate (60 ml/kg/day) IV fluids with acetate as a buffer, 20 mEq/l KCl, and dextrose added to make a 2.5% dextrose solution.

5 Since there was strong evidence that a myopathy might have been the primary disorder, what additional diagnostic test should be performed? Blood selenium should be measured. Testing revealed zero selenium (whole blood) at admission. The laboratory reported this as 'incompatible with life'. The mare had no routine vitamin E/selenium supplementation, hay was harvested from a selenium deficient area, and the concentrate feed was made by a local feed store. This information led to a diagnosis of masseter myopathy caused by selenium deficiency.

Follow up/discussion
The mare was given an IM injection of vitamin E and selenium. After 3 days the mare was brighter and more active. She was able to hold her tongue in her mouth (34.5) but still could not chew. The dark urine had resolved. At that time she was given a second injection of vitamin E/selenium. Laboratory findings on day 3 were: WBCs = 6,900/µl (6.9 × 10^9/l); PCV = 42%; TP = 6.9 g/dl (69 g/l); TGs = 75 mg/dl (0.85 mmol/l); ammonia = 57 µg/dl (40.7 mmol/l); CK = 21,984 U/l; AST = 10,780 U/l; SDH = 118 U/l; GLDH = 104 U/l; GGT = 146 U/l; glucose = 140 mg/dl

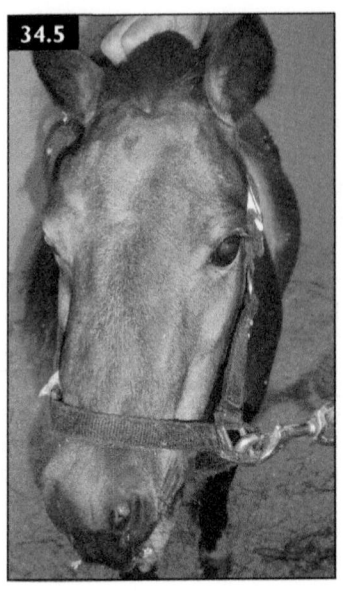

34.5

(7.8 mmol/l), total bilirubin = 2.6 mg/dl (44.5 µmol/l) (direct 0.3 mg/dl [5.13 µmol/l], indirect 2.3 mg/dl [39.3 µmol/l]).

Selenium deficiency can cause nutritional muscular dystrophy. Clinical signs are variable because any skeletal muscles may be preferentially affected in an individual animal. The clinical disease is slightly more common in young, growing animals. In foals, limb muscles and the pharynx seem to be most commonly involved. In weanlings, limb muscles and the tongue seem to be most common, although cardiac involvement can occur and result in acute death. Masseter myopathy is one form of the disease in adult horses. On necropsy, cardiac, tongue, and skeletal muscles often have chalky striations and areas of necrosis. Selenium is an essential component of the enzyme glutathione peroxidase and is present in approximately 25 other selenoproteins found within the body. Glutathione peroxidase is important in reducing lipid peroxidases and hydrogen peroxides to less harmful substances.

Methods of selenium supplementation include inorganic and organic feed supplements and injectable inorganic forms. For white muscle diseases, treatment with injectable selenium is preferred. Vitamin E has very little, if anything, to do with white muscle disease in animals.

The preferred antemortem test to determine selenium status is measurement of whole blood selenium. Glutathoine peroxidase (GP) measurements can be confusing and should only be attempted on whole blood if selenium was already given to the patient before white muscle disease was considered (it takes approximately 7–10 days for selenium to be incorporated into GP). Muscle biopsy is often non-diagnostic in the authors' experience. Treatment of white muscle disease involves IM injections of selenium and appropriate nursing care. Selenium treatment should be repeated in 2–3 days and the serum selenium rechecked. Normal horses should consume 0.2 µg selenium/gram of diet per day.

Hepatic lipidosis is a common disorder of ponies, miniature horses, and occasionally horses with Cushing's disease or equine metabolic syndrome. There is usually another disorder that causes the equine to mobilize fat, which quickly results in hyperlipemia (gross discoloration of plasma with TGs >500mg/dl [5.65 mmol/l]) and hepatic lipidosis. Treatment should be directed towards any predisposing disease and nutritional support provided to combat the hyperlipemia/hepatic lipidosis. Enteral nutrition with a high carbohydrate, moderate protein concentration of essential amino acids (especially branch chain), and some additional fiber is recommended. Parenteral nutrition with amino acids, B vitamins, glucose, and insulin can be life-saving in some cases. Crystalloids given IV would also be needed to maintain hydration and organ perfusion. Insulin treatment is beneficial in many cases. Blood glucose will need to be monitored in the first couple of hours. The IV crystalloid fluids should have glucose added if insulin is used; the amount of glucose will depend on blood glucose measurements, the amount of glucose/starch enteral feeding, and the type and amount of insulin being administered. Make sure potassium is in the normal range before starting insulin and add KCL 20 mEq/l to the fluids. Some cases have a dramatic response to treatment with TGs dropping from 2,000 mg/dl (22.6 mmol/l) to almost normal in 36 hours, although some cases will die acutely from a ruptured liver in spite of aggressive treatment. Hepatic lipidosis from hyperlipemia may occur in neonatal foals, especially miniature equine foals.

References
Conwell R (2010) Hyperlipaemia in a pregnant mare with suspected masseter myodegeneration. *Vet Rec* **166(4)**:116–117.
Step DL, Divers TJ, Cooper B *et al.* (1991) Severe masseter myonecrosis in a horse. *J Am Vet Med Assoc* **198(1)**:117–119.

CASE 35

1 **What would be the next diagnostic procedure?** Either endoscopy of the nasopharynx or radiography of this region.
2 **What is the likely cause of the black nasal discharge?** A large melanoma that has become necrotic, most likely in the guttural pouch (35.3).
3 **What are the treatment options?** Cisplatin in sesame seed oil with or without initial surgical excision is the typical treatment of choice for cutaneous

melanomas, but would not be appropriate for this pony due to the size and location of the tumor. Cimetidine therapy and autogenous vaccines (immunotherapy) have been used in horses, but the success rate is low. A canine approved xenogeneic DNA plasmid vaccine expressing human tyrosinase is currently being evaluated.
4 **What is the most common anatomic site for this condition?** Perianal. Other commonly affected areas are the tail, the head, and the throatlatch region. In gray horses, melanomas are slow to metastasize unlike the early metastasis seen in

non-gray horses. Melanomas increase in frequency as gray horses age, beginning at 5 years of age.

Discussion
Different clinical syndromes and pathologic scoring systems have been proposed for melanomas in horses. Three clinical histopathologic syndromes are recognized. (1) Melanocytic nevi are solitary superficial masses often seen in young adults (gray and non-gray) or even young foals. They may occur as solitary masses on the leg or trunk, but not under the tail. Surgical excision is often curative. (2) Dermal melanoma/melanomatosis are mostly seen in middle-aged to older gray horses. Microscopic examination of the tumor is characteristic for the disorder and dermal melanomas may metastasize to become melanomatosis. The most common site for primary tumors are under the tail, around the perineum, and the parotid salivary gland. (3) Anaplastic malignant melanomas in aged horses of any color, which often metastasize within a year after diagnosis.

References
Moore JS, Shaw C, Shaw E *et al.* (2013) Melanoma in horses: current perspectives. *Equine Vet. Educ Am Edn* **25**(3):144–150.
Phillips JC, Lembcke LM (2013) Equine melanocytic tumors. *Vet Clin North Am Equine Pract* **29**(3):673–687.

CASE 36

1 What historical information in this case makes it not surprising that tranquilization treatment for choke was unsuccessful? The fact that the horse had choked three times in the past year would suggest that there might be a structural abnormality to the esophagus, eventually making the choke more difficult to relieve.
2 What are the greatest medical concerns regarding complications of the choke in this horse? Aspiration pneumonia and septic shock.
3 What is the radiographic interpretation? The radiograph shows moderate to severe infiltrative ventral lung disease, compatible with aspiration pneumonia.
4 Describe the endoscopic findings from the cardia region of the esophagus. The esophagus appears to be edematous, with areas of erosion and proliferation (possibly a scar) around a dilated area that has accumulated fluid. The lesions look severe in this case. However, it is not unusual for very inflamed esophageal lesions to endoscopically resolve in 2–3 days when only the mucosa is damaged.
5 In addition to immediately removing all feed and water, administering a tranquilizer to relax the esophagus and lower the horse's head, and flushing, what additional parasympatholytic drug could be considered in a horse with choke? N-butylscopolammonium bromide (Bucopan®). It would be most effective for

choke in the distal one-third of the esophagus where smooth muscle is present, as opposed to the striated muscle in the proximal two-thirds. Regardless, the drug seems to be of some benefit (clinical impression) even in more proximal obstructions.

6 What is the predominant anion in equine saliva? Chloride. Therefore, horses with esophageal choke do not develop the pronounced metabolic acidosis seen in cattle that cannot swallow. Horses may develop a lactic acidosis due to dehydration and, in this case, toxemia.

Follow up/discussion

This horse was euthanized because of the severe aspiration pneumonia. Many horses with prolonged esophageal choke have ultrasound or radiographic findings suggestive of aspiration pneumonia but do not exhibit the marked increase in RR and evidence of toxemia seen in this horse.

Esophageal obstruction, most often acute, results from obstruction of the esophageal lumen with food (e.g. dried beet pulp, hay, pellets), wood chips, or bedding. This problem occurs most commonly among horses with ravenous eating habits, especially older horses being fed pelleted feed. Other risk factors include immediately feeding a nervous and excited horse on arrival at a hospital and feeding exhausted horses at rest stops. Choke occasionally occurs when a heavily sedated horse is allowed to eat. Most cases of esophageal choke occur in adult horses but the condition may occur in younger animals, even foals. Geriatric horses are predisposed to choke because of decreased saliva production and sometimes poor mastication of feed.

The most common clinical signs are excessive salivation, retching, coughing with saliva expectoration, and food dripping from the nares. In most instances, if the obstruction is in the cervical region (the most common sites for obstruction are the proximal esophagus and just cranial to the thoracic inlet) and of recent origin, enlargement of the esophagus can be palpated on the left neck. Over time, swelling and muscle spasm in this region make it difficult to delineate the mass. It is more likely that the obstruction is in the cervical portion of the esophagus if the patient retches immediately after attempting to swallow. There is often a 10–12-second delay between swallowing and the onset of retching if the obstruction is in the distal esophagus. After successfully relieving esophageal obstruction, the horse should be slowly introduced back to feed by initially feeding a gruel or mash. The horse should be monitored for 1–3 days for evidence of pneumonia. The diet, feeding environment, and teeth should be evaluated for abnormalities that may have predisposed to the choke. The workup for recurrent choke should include a neurologic evaluation, esophageal endoscopy, and possibly a contrast (e.g. barium) swallow radiographic study. Barium contrast radiographs are typically not very helpful unless the barium is given under pressure via a tube that has an inflatable balloon.

Some horses must be anesthetized to relieve more difficult or prolonged choke. In these cases, the nasoesophageal tube should be placed before anesthesia because passage of a tube into the esophagus after the horse is anesthetized can be difficult.

Reference

Chiavaccini L, Hassel DM (2010) Clinical features and prognostic variables in 109 horses with esophageal obstruction (1992–2009). *J Vet Intern Med* 24(5): 1147–1152.

CASE 37

1 What is the differential diagnosis for the neurologic signs in this case? Temporohyoid osteoarthropathy (THO) would be the primary rule out based on the acute onset of clinical signs, the age of the horse, and no known history of specific trauma. However, trauma resulting in injury to the tympanic bulla area and/or petrous temporal bone should be considered. EPM with an acute onset could be possible and can be associated with stress factors. Signs of spinal cord ataxia (which may be difficult to evaluate in a horse with such severe balance deficits) and some depression might be expected with EPM. Infectious or neoplastic involvement of the guttural pouch/tympanic bulla area should be considered and might even be related to the more chronic ipsilateral Horner's syndrome.

2 What should the diagnostic plan include (in order of importance)? (1) Endoscopic examination of the left guttural pouch (37.2) to help confirm THO or infectious disease of the pouch. In this case, a raised

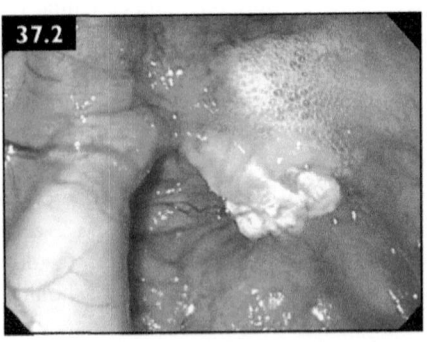

plaque, most likely fungal, was observed in the dorsal left guttural pouch. There was no evidence of THO. There was a large aneurysm and soft tissue swelling of and around the left maxillary artery. (2) Radiographs of the tympanic bulla area. These are not easy to perform in a standing horse and can be difficult to interpret. The severe head tilt does permit easier visualization of the individual tympanic bullae. GA for ventrodorsal radiographs and CT is preferred, but would not be initially indicated in this case (if available, standing CT would be appropriate). (3) Serum and/or CSF testing for EPM.

3 What are the most common causes of Horner's syndrome (mild miosis, enophthalmos, hyperemic mucous membranes, sweating of the face and cranial neck, mild ptosis) in the horse? Perivascular injection in the jugular furrow is the most common cause. Guttural pouch mycosis may cause abnormal sweating ipsilateral from the 2nd cervical vertebra cranially. Cranial cervical masses, although rare, may cause Horner's syndrome and sweating from the 7th cervical vertebra cranially if both the sympathetic trunk and caudal cervical nerve are affected, as seen in this separate case (**37.3**). Horses with cranial thoracic neoplasia can have abnormal sweating from the shoulder to the nose on the affected side.

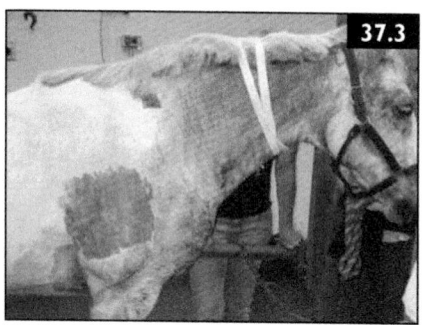

4 Can the lesion identified explain the clinical signs? The fungal infection could be the cause of the Horner's syndrome, but would not be the cause of the vestibular signs unless there was an extension of the infection/inflammation into the tympanic bulla.

5 Since there was soft tissue swelling caudal to the guttural pouch and clinical evidence (i.e. the vestibular disease) of pathology of the tympanic bulla, what additional procedure could be performed? Aspiration or biopsy of this region.

Follow up/discussion

Two aspirates were performed and both were cytologically described as pyogranulomatous inflammation. Subsequently, a 4 × 4 cm mass was observed during ultrasound examination of the thoracic inlet. Histopathologic evaluation of a biopsy of this mass was interpreted as carcinoma and the horse was euthanized. At necropsy, squamous cell carcinoma was found caudal to the guttural pouch and numerous, variously sized carcinomas were found at the thoracic inlet. There were also some very small (< 1 cm) tumors in the lungs.

In this case, the cause of the Horner's syndrome might have been either the guttural pouch fungal infection, the carcinoma and inflammatory reaction just caudal to the pouch or, more likely, the tumor at the thoracic inlet, since sweating occurred as far distal as the 3rd cervical vertebra. The origin of the tumor could not be determined and the guttural pouch infection was likely secondary to the pathology caudal to it.

Reference

Dobesova O, Schwarz B, Velde K *et al.* (2012) Guttural pouch mycosis in horses: a retrospective study of 28 cases. *Vet Rec* **171**(22):561.

CASE 38

1 What would be potential causes for the acute progressive bilateral uveitis? Differential diagnoses include *Leptospira*, herpes, parasites, immune-mediated, traumatic, drug-induced, *Salmonella*, *Streptococcus*, and *Borrelia*.

2 What is the interpretation of the cytology of the vitreous? There are large numbers of neutrophils, which indicate inflammation. The pink swirls are either protein, stain precipitant, or an infectious agent. In this case, the size of all of the swirls are the same, suggesting an infectious spirochete (detected by Dr. Heather Priest, clinical pathologist, Cornell University).

3 What diagnostic test should be performed next? PCR for both *Leptospira* and *Borrelia*. *Borrelia* was positive, *Leptospira* was negative.

4 How could the muscle wasting over the horse's top-line be explained? By a primary myopathy such as streptococcal immune myopathy, severe calorie deficiency, neoplasia, or lower motor neuron (spinal cord or nerve root) disease. In this case, there was a chronic, multifocal, lymphohistiocytic ganglioradiculitis and neuritis with presumptive neuronal degeneration in the spinal nerves supplying the top-line muscle.

Discussion

Rapidly progressive severe bilateral uveitis would be unusual for the typical equine recurrent uveitis syndrome. Acute leptospirosis was initially considered to be the most likely cause, since *Borrelia* had not previously been reported to cause such a similar and severe ocular disease. It can only be speculated whether early treatment with oral minocycline or IV tetracycline might have been effective in halting the disease progression. Orally administered minocycline would be preferred over doxycycline since the former is more likely to provide appropriate antibiotic concentration in the ocular chambers. The muscle wasting over the top-line and lymphohistiocytic ganglioradiculoneuritis have been observed in other horses with Lyme disease.

References

Curling A (2011) Equine recurrent uveitis: classification, etiology, and pathogenesis. *Compend Contin Educ Pract Vet* 33(6):E1–4.

Imai DM, Barr BC, Daft B *et al.* (2011) Lyme neuroborreliosis in 2 horses. *Vet Pathol* 48(6):1151–1157.

Priest HL, Irby NL, Schlafer DH *et al.* (2012) Diagnosis of *Borrelia*-associated uveitis in two horses. *Vet Ophthalmol* 15(6):398–405.

CASE 39

1 What are the differential diagnoses for this mare? Include hydrops allantois or amnion with or without fetal abnormalities, twin pregnancy, ruptured prepubic tendon, and ventral body wall herniation.

2 After 24 hours, the mare became recumbent due to the progressively enlarging uterus. What would be the treatment options? Because of the rapid progression in the accumulation of uterine fluid, parturition should be induced. An attempt to induce labor by manually dilating the cervix would likely be necessary in this case since the distended and stretched uterus may not respond to oxytocin. A cesarean section would be difficult because the mare's abdominal wall muscles have also been severely stretched.

3 What is the mechanism of hypotensive shock, and what would be the treatment? Hypotension may occur because rapid removal of the uterine fluid leads to vasodilation of the mesenteric vessels, which were previously compressed by the abnormally large uterus. The rapid vasodilation of these vessels expands the intravascular space, causing hypotension. The best treatment is hypertonic saline to rapidly expand the intravascular volume. While waiting for the mare's blood pressure to improve, it is important to keep her calm and on a good surface (grass in this case) in case she tries to stand while hypotensive.

4 What would be a probable uterine complication following delivery? Retained placenta.

5 In less severe cases than the mare in this case, what can be done to help prevent abdominal wall injury/hernia? Abdominal support wraps can be of benefit.

Reference
Slovis NM, Lu KG, Wolfsdorf KE *et al.* (2013) How to manage hydrops allantois/hydrops amnion in a mare. *AAEP Proceedings* **59**:34–39.

CASE 40

1 Interpret the BAL results. These are normal BAL results.

2 What is the most likely diagnosis? Excessive tracheal mucus causing poor performance.

3 What would be the preferred treatment(s) for this horse? Treatments, including antibiotics, bronchodilators, and even corticosteroid therapy, are usually of minimal benefit in 2- and 3-year-old racehorses with this syndrome. If a cough is present and the tracheal mucus contains a large number of neutrophils, it is not unusual to culture *Streptococcus zooepidemicus,* in which case there may be a clinical response to antibiotics. A Gram stain from another horse with this syndrome that had a large number of gram-positive cocci (*S. zooepidemicus*) seen on TTW is shown (**40.2**). There was a good response to antimicrobial treatment in this horse. The only proven therapy for the excessive mucus is orally administered α-interferon at 50–100 U/horse/day. The treatment is generally given for 1–2 weeks. A 1-week rest period and reduction of airway irritants is also recommended. Rapid airflow, especially cold air, may exacerbate the excessive mucus production. Some horses may recover without treatment within 2–3 weeks, but many will have a more prolonged course (months) of excessive

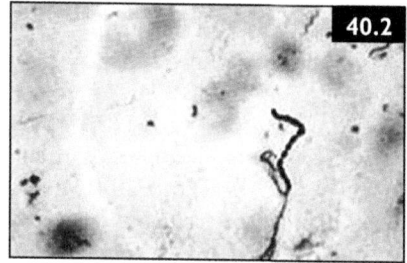

mucus following exercise with poor performance unless α-interferon treatment is given. Interferon is not effective in all cases, especially if affected horses are not also given a rest period away from airway irritants.

Discussion

Excessive tracheal mucus following exercise is a well-documented but poorly understood syndrome causing poor performance in both young racing horses and mid- to older-age event horses. In younger racehorses, there is no evidence that the excessive mucus is related to recurrent airway obstruction (RAO). However, the excessive mucus could be related to RAO in some older horses. It is not unusual to have a normal BAL several days after excessive tracheal mucus was observed following a race. There is no consistent relationship between inflammatory tracheal wash results and excessive mucus production. In fact, many successful racehorses and horses with normal treadmill performance have inflammatory tracheal wash fluid without an obvious increase in mucus. A large amount of tracheal mucus is associated with hypoxemia and poor performance. Moderate to severe tracheal mucus accumulation, without increased tracheal neutrophils, is a risk factor for poor racing performance. Therefore, functionally significant airway inflammation may best be confirmed by the presence of mucus immediately following a race rather than an increased number of neutrophils in the tracheal wash fluid.

The diagnosis of excessive mucus as a cause for poor performance can only be made by ruling out other causes of poor performance and finding excessive mucus (streams and/or pools of mucus) on endoscopic examination performed 30 to 60 minutes post racing. Additional clinical findings are scarce in affected horses, aside from the history of poor performance. A small percentage may have a history of abnormal cough or white nasal discharge, although this is not as common as in older horses with RAO. Most young racehorses with excessive mucus have normal thoracic auscultation, even when forced to take deep breaths. Although there may be some relationship between excessive tracheal mucus and inflammatory small airway disease, larger airways may be the main contributors to the excessive mucus production. It is yet to be determined if the mucus is normal in consistency and if it is triggered by a recent viral infection.

References

Holcombe SJ, Robinson NE, Derksen FJ *et al.* (2006) Effect of tracheal mucus and tracheal cytology on racing performance in Thoroughbred racehorses. *Equine Vet J* 38(4):300–304.

Moore I, Horney B, Day K *et al.* (2004) Treatment of inflammatory airway disease in young standardbreds with interferon alpha. *Can Vet J* 45(7):594–601.

Moore BR, Krakowka S, Cummings JM *et al.* (1996) Changes in the airway inflammatory cell populations in standardbred racehorses after interferon-alpha administration. *Vet Immunol Immunopathol* 49(4):347–358.

Moore BR, Krakowka S, Mcvey DS *et al.* (1997) Inflammatory markers in bronchoalveolar lavage fluid of standardbred racehorses with inflammatory airway disease: response to interferon-alpha. *Equine Vet J* **29**(2):142–47.

Ramzan PH, Parkin TD, Shepherd MC (2008) Lower respiratory tract disease in Thoroughbred racehorses: analysis of endoscopic data from a UK training yard. *Equine Vet J* **40**(1):7–13.

Richard EA, Fortier GD, Pitel PH *et al.* (2010) Sub-clinical diseases affecting performance in Standardbred trotters: diagnostic methods and predictive parameters. *Vet J* **184**:282–289.

Widmer A, Doherr MG, Tessler C *et al.* (2009) Association of increased tracheal mucus accumulation with poor willingness to perform in show-jumpers and dressage horses. *Vet J* **182**(3):430–435.

CASE 41

1 What differentials should be included at this time? The top differential is myocarditis because of the fever, tachycardia, and non-painful pitting edema. Although there were no cardiac murmurs, septic endocarditis would be an additional differential. EIA should be considered because of the pale mucous membranes, ventral edema, and absence of a Coggins test. Internal hemorrhage and thoracic neoplasia should also be considered.

2 What blood test would aid in confirming the main differential? Cardiac troponin I, which was mildly elevated (0.17 ng/ml [0.17 µg/l]).

3 Interpret the ECG. What other test should be performed? The ECG demonstrates supraventricular tachycardia with atrial bigeminy, which is a rather benign arrhythmia in which a regular (or sinus) beat is followed by a premature atrial beat. These premature beats are the result of electrical impulses sent from parts of the atrium other than the sinus. This arrhythmia is generally transient and rarely requires specific treatment. An echocardiogram showing excellent contractility, even hyperdynamic, would virtually rule out myocarditis. In this mare there was also no evidence of any valvular lesions or pericardial effusion.

4 How could EIA be ruled out? A negative Coggins test. In addition, absence of thrombocytopenia and a PCV of 28% makes acute EIA unlikely.

5 What are the laboratory abnormalities most likely caused by? The azotemia and increased lactate were most likely due to the poor cardiac output and tissue perfusion.

6 What would be a reasonable treatment plan? Administer IV crystalloids in the hope of combating the poor perfusion.

7 The mare was euthanized and several encapsulated hematomas were found at the base of the heart (41.2, 41.3). What is the most likely diagnosis? Aortic/pulmonary artery fistulation and rupture, which has been reported in Friesian horses.

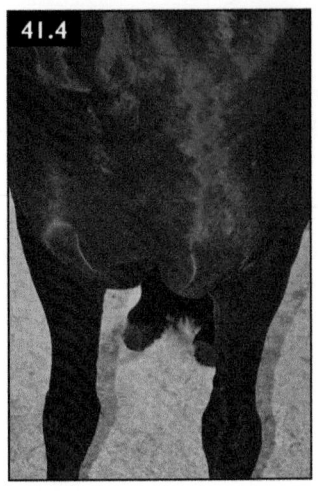
41.4

Follow up/discussion

During the 12 hours following treatment, the mare developed a remarkable volume (gallons) of ascites, pleural effusion, increased jugular distension, and a more marked ventral edema (**41.4**) consistent with interference with venous return to the heart, most likely due to a heart-based mass.

A unique form of aortic aneurysm is now recognized in Friesian horses and the authors believe the mare in this case had this recently reported syndrome. The location of the aortic aneurysm in Friesians appears to be consistent and distinctively more distal than lesions of the aortic root that are described more sporadically in breeding stallions of multiple breeds. In Friesians, the rupture is reported to occur at the level of the aortic arch near the ligamentum arteriosum and not in the sinus of Valsalva. In most of the reported Friesian cases, the rupture led to fistulation of the pulmonary artery, which is in close anatomic proximity. Clinical signs reported in the recent case series included recurrent colic, anorexia, repeated recumbency, coughing or respiratory distress, fever of unknown origin, peripheral edema, tachycardia, tachypnea, jugular pulses, and cardiac murmur. Four had a history of epistaxis. Three horses with pronounced jugular pulses also exhibited bounding carotid 'hammer' pulses. The described murmurs were highly variable, including systolic, diastolic, and right- and left-sided characterizations. Bloodwork in these cases was unremarkable. Three Friesians exhibited a cardiac arrhythmia on auscultation, but characterization by ECG was not described. Echocardiographic abnormalities included a dilated pulmonary artery (n = 5/16) and pleural effusion (n = 8/16). These were both noted on repeat echocardiogram of this patient. Pericardial effusion, not seen in this mare, was also reported (n = 3/16) in the large case series. Of the 11 horses with murmurs auscultated at presentation, only three showed valvular insufficiencies on echocardiogram.

Necropsy in the Friesian case series found a consistent pattern of transverse rupture in close proximity to the ligamentum arteriosum. A majority of the reported horses (n = 18/29) had aortopulmonary fistulation, featuring a transverse rupture of the adjacent pulmonary artery. Prominent cuffs of perivascular hemorrhage around the aorta and pulmonary trunk were also a common finding. The hemorrhage ranged from fresh to organized hematomas with fibrotic capsules. Encapsulated hematomas were a feature of the gross pathology in our patient but a discrete rupture of the great vessels was not identified.

Outcomes of the reported Friesian aortic rupture syndrome ranged from acute death with evidence of hemothorax to formation of an apparently stable fistula. The chronic cases exhibited clinical signs of right-sided heart failure for weeks to months, which was corroborated by eventual necropsy findings of hepatic congestion and/or fibrosis. Due to the unique location, and previously scarce description in the literature, aortic rupture in the Friesian can be difficult to recognize in areas where the breed is less common. The standard approach to the postmortem cardiac examination must be modified by moving the transection of the descending aorta distally to the level of the diaphragm during removal of the pluck. Ultimately, the prognosis with aortic rupture is poor with an omnipresent risk of sudden death. Early diagnosis of distal aortic rupture remains a challenge.

Reference

Ploeg M, Saey V, de Bruijn C *et al.* (2013) Aortic rupture and aorto-pulmonary fistulation in the friesian horse: characterisation of the clinical and gross post mortem findings in 24 cases. *Equine Vet J* **45(1)**:101–106.

CASE 42

1 What other test could be performed to determine if *Borrelia* might be associated with the disease? A PCR could be performed on the tissue from the skin biopsy; in this case it was positive for *Borrelia* spp.

2 What treatment is recommended for Lyme disease? Doxycycline or tetracycline is the preferred treatment in horses with clinical signs (as in this case) associated with *Borrelia* infection. This horse was treated with doxycycline (10 mg/kg PO q12h) for 30 days and the lesion completely resolved.

3 How is this different from cutaneous lymphoma? Cutaneous lymphoma in horses is typically seen in the subcutaneous tissue and may invade muscle; Cutaneous lymphoma lesions are deeper than those typically seen with pseudolymphoma.

References

Meyer J, Delay J, Bienzle D (2006) Clinical, laboratory and histopathologic features of equine lymphoma. *Vet Path* **43**:914–924.

Sears K, Divers T, Neff R *et al.* (2012) A case of *Borrelia*-associated cutaneous lymphoma in the horse. *Vet Derm* **23(2)**:153–156.

CASE 43

1 What would be the next diagnostic procedure(s)? Abdominal palpation per rectum and abdominal ultrasound.

2 What is the tentative diagnosis based on the history, laboratory findings, and clinical and gastroscopic examination? Gastric squamous cell carcinoma.

3 What other procedures might allow a diagnosis if gastroscopy is not available?
Neoplastic cells may be identified in the peritoneal fluid in approximately 50% of cases and in the thoracic fluid in <25%.

Discussion
Gastric squamous cell carcinomas occur in horses >8 years of age and arise from the non-glandular cardia in almost all cases. They most commonly metastasize to the peritoneum, allowing neoplastic cells to be seen on cytology of peritoneal fluid and, less commonly, on pleural fluid samples. Weight loss, decreased appetite, colic after eating, and clinical evidence suggesting esophageal dysfunction are common. Anemia is common, and TP can be high, low, or normal. As seen in this case, hypercalcemia caused by elevated PTH-related protein (PTHrp) due to the malignancy can occur.

Reference
Deegen E, Venner M (2000) Diagnosis of stomach carcinoma in the horse. *Dtsch Tierarztl Wachenschr* **107**(12):472–476.

CASE 44
1 What would be the primary differential diagnosis in this case? Coronary band dystrophy, nutritional deficiencies (e.g. biotin, zinc, vitamin A), or toxicity (e.g. selenium). The former seemed most likely since the lesions were restricted to the coronary bands.
2 How could the diagnosis be confirmed? Biopsy.
3 What are the possible treatments? Systemic immunosuppressive drugs including steroids and/or azathioprine or tacrolimus.

Follow up/discussion
Biopsy in this case revealed a hyperplastic stratum granulosum of the epidermis with mild surface crusting. No evidence of infection was noted. Keratinocytes were separated by fibrillar eosinophilic matrices. Coronary band dystrophy was both the clinical and the histopathologic diagnosis. Systemic immunosuppressive drugs have proven to be ineffective, or only modestly effective, in treating the disease. This case responded very well (even 2 years later) to topical tacrolimus.

Coronary band dystrophy is a sporadic disease of unknown cause in the adult horse. It appears to be a localized defect of keratinization that affects the specialized epithelium of the coronary band. All four limbs are generally affected, although each entire coronary band may not be involved. There is generally minimal lameness associated with the disease unless the diseased coronary band and hoof wall become deformed. The disease appears to be more common in Warmblood and Draft horses. Chorioptic mange, which may cause pruritus and dermatitis of the fetlock and coronary band, has been found concurrently with coronary band dystrophy in some Draft horses. Successful treatment of the chorioptic mange does not resolve the coronary band dystrophy in these cases.

Reference
Menzies-Gow NJ, Bond R, Patterson-Kane JC *et al.* (2002) Coronary band dystrophy in two horses. *Vet Rec* **150(21):**665–668.

CASE 45

1 What are possible causes for the mild-to-moderate dysphagia in this foal? Include cleft palate, fourth right branchial arch defect (which is not truly dysphagia, but regurgitation from the esophagus), neonatal encephalopathy affecting the swallowing centers in the brain, and selenium deficiency (selenium was measured and was normal in this foal). There was no clinical laboratory evidence to support any of these causes.
2 How should this foal be treated? Muzzle the foal and place a nasogastric feeding tube so that the mare's milk can be given via this route. The mare should be milked out regularly. Antibiotics were administered to the foal as treatment for mild aspiration pneumonia. The foal was allowed to nurse intermittently (every 24–48 hours) to see if the problem had resolved.
3 What further treatment and diagnostics should be recommended for the mare? Along with continuing to 'milk out' the mare, antibiotics should be initiated. In this case (with a valuable mare), IV penicillin and gentamicin were used because of their combined broad-spectrum activity. A Gram stain and culture of the milk should be performed.

Follow up/discussion
In this case, gram-negative rods were seen and a moderate growth of *Pseudomonas* sp. was cultured. The organism was resistant to all tested drugs except amikacin and imipenem. The mare's two left glands were infused with 250 mg amikacin mixed in 5 ml of isotonic bicarbonate. A 1 inch 30 gauge plastic cannula was used for each infusion. Within 24 hours of the intramammary infusion, the mare was significantly improved and the milk returned to normal appearance and had a negative CMT within 5 days. After 2 weeks and multiple attempts at getting the foal to either nurse the mare or nurse a bottle, aspiration of milk persisted and the foal was trained to drink milk from a bucket. No aspiration (confirmed by endoscopic examination) occurred with the bucket feeding.

Dysphagia in foals does not always have an anatomic cause and may be related to 'dysfunctional' swallowing. Many of these foals do not show signs of dysmaturity or neonatal encephalopathy and some are able to swallow without aspiration by days 2–4, but some are not able to nurse without aspiration after several days or even weeks. In those cases, it is best to pan or bucket feed the milk at ground level as aspiration does not occur with this method of feeding. It is assumed that there is some mild neuromuscular dysfunction that permits aspiration in the foal when its head is under the mare or nursing a bottle. The foals can usually drink from a bucket at ground level without problems. There is rarely any noticeable laryngeal dysfunction.

Mastitis in mares is not common but can be predisposed to by inadequate nursing by the foal and, in this case, continual bumping of the udder with the foal's muzzle. Mastitis can also occur in non-lactating mares. *Streptococcus zooepidemicus* is the most common causative bacteria. *Actinobacillus* and *Staphylococcus* are other common pathogens, and *Proteus* and *Pseudomonas* are less common pathogens. The prognosis is good in most cases following systemic antimicrobial therapy, frequent milking, and hot packing. Intramammary infusion may be necessary in some cases.

References

Holcombe SJ, Hurcombe SD, Barr BS *et al.* (2012) Dysphagia associated with presumed pharyngeal dysfunction in 16 neonatal foals. *Equine Vet J Suppl* (41):105–108.

McCue PM, Wilson WD (1989) Equine mastitis: a review of 28 cases. *Equine Vet J* 21(5):3.

CASE 46

1 List the clinical findings that support a diagnosis of infected ulcerative keratitis (melting corneal ulcer) (46.1). Blepharospasm, mucopurulent ocular discharge, conjunctival hyperemia, scleral injection, active neovascularization, corneal edema, cellular infiltrate, keratomalacia, stromal loss, hypopyon, miosis, and aqueous flare.

CASE 47

1 What is the explanation for the swollen and red conjunctiva in both eyes? It might be due to hepatogenous photosensitization. In bay horses, sometimes only the conjunctiva and vulva of females is involved.

2 Based on this information, does it seem reasonable that the horse has chronic bacterial cholangitis? Why? Yes. The marked elevation in GGT, fever, and neutrophilia with chronic weight loss would support this diagnosis.

3 What laboratory findings are not consistent with chronic bacterial cholangitis? The abnormally low TP and globulins do not support this diagnosis. In most chronic equine liver disease, regardless of cause, TP and globulins are increased.

4 What further diagnostic test(s) should be performed? An ultrasound of the liver (47.2), coagulation profile (PT and PTT), blood ammonia, liver biopsy and culture, and quantitative IgG and IgM.

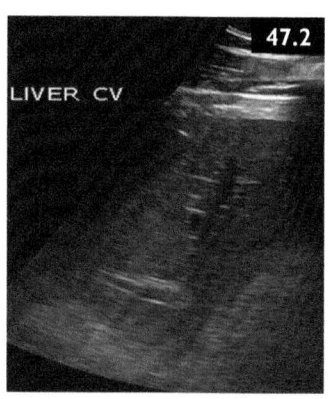

PT and PTT were prolonged, platelet count was normal, blood ammonia was normal. Immunoglobulin testing revealed an IgG of 488 mg/dl (4.9 g/l).

5 Since the PT and PTT were prolonged, should the liver biopsy be avoided? No. It should be performed to better evaluate the disease process and possible etiologic agents. When the platelet count is normal, abnormal bleeding associated with a liver biopsy is rare, even in the face of prolonged PT and PTT.

6 Undoubtedly the horse had hepatitis, which was presumably bacterial in origin, but what predisposing factors should be considered based on the laboratory testing? The possibility of an immune deficiency should be considered in this mare, either common variable immunodeficiency or secondary to a neoplasia.

Follow up/discussion
Biopsy revealed moderate-to-severe, chronic, multifocal, neutrophilic hepatitis, compatible with a chronic bacterial infection. Culture of the liver sample grew a coagulase-negative *Staphylococcus* sp. of unknown significance. Flow cytology testing of the blood revealed a marked absence of B lymphocytes. Because of this, the horse was euthanized and there was a marked absence of B lymphocytes in the lymph nodes and spleen, with a normal T lymphocyte appearance. A final diagnosis of common variable immunodeficiency syndrome was made.

Common variable immunodeficiency syndrome was first described in 2002. All affected horses have been adults, with an average age of 10 years, and multiple breeds and both genders affected. Chronic infections, most commonly *Staphylococcus*, meningitis, pneumonia with variable bacteria including *Rhodococcus equi* and fungal pathogens, occur in these horses. Hepatitis has also been reported. Common variable immunodeficiency should be strongly considered when an adult horse has chronic infection(s) with lymphopenia and hypoglobulinemia. Flow cytology testing of whole blood can help confirm the diagnosis. Although many affected horses initially respond to appropriate antimicrobial therapy, the long-term prognosis is grave. The cause of the disease is unknown.

Reference
Flaminio MJ, Tallmadge RL, Salles-Gomes CO *et al.* (2009) Common variable immunodeficiency in horses is characterized by B cell depletion in primary and secondary lymphoid tissues. *J Clin Immunol* **29(1):**107–116.

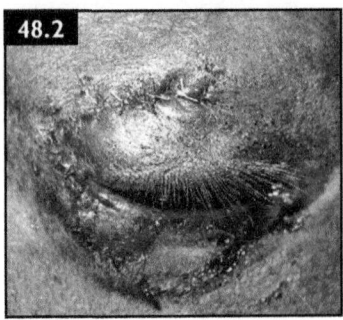

CASE 48
1 What diagnostics and therapy are indicated? The eye should be evaluated for corneal ulcers and intraocular trauma (uveitis, hyphema). Surgical reconstruction of the eyelid should be performed promptly; a two-layer closure using 4-0 to 5-0 polyglactin 910 with a figure-of-eight pattern to appose the eyelid margins is recommended (**48.2**). Broad-spectrum topical antibiotic ointment and systemic antibiotics are also indicated.

CASE 49
1 Where is the lesion(s) in the nervous system? The dysphagia, bilateral laryngeal paralysis, and absent cough reflex suggest bilateral vagal nerve dysfunction. The atrophy of both masseter muscles suggests bilateral dysfunction of the motor branch of the trigeminal nerve. The ataxia in all four limbs suggests disease in the descending long tracts in the spinal cord. The depression suggests disease of the brain or brainstem. If one lesion could explain the signs, that lesion would be in the caudal ventral brainstem.
2 What would be the next diagnostic procedure? A spinal fluid collection, as this is the most direct route to the CNS. CT or MRI of the brainstem might be helpful in determining the presence or absence of a mass in the area. The owner declined these procedures due to the poor prognosis for recovery in the horse.

Follow up/discussion
Postmortem diagnosis was lymphoma (small cell) involving the meninges from the cerebral hemisphere at the rostral commissures and continuing caudally to spinal cord segment C2. Lymphoma involving the CNS of horses is rare and primary lymphoma of the CNS is even more so. When diagnosed, the tumor most commonly involves the meninges or intramedullary capillaries and adjoining parenchyma. Lymphoma may be either T-cell lymphoma or T-cell rich B-cell lymphoma.

Other differentials for this case include EPM, granulomatous meningitis, and neuroborreliosis. If the horse had been a gray horse, melanoma would have been a differential as a similar case has been documented. Waxing and waning depression as a clinical sign is not unusual in a horse with a tumor near the brain. This might be a result of waxing and waning in blood flow to the brain. The masseter atrophy without diminished focal sensation and palpebral response indicated preferential damage to the motor nucleus of the trigeminal nerve. The absence of a cough reflex was likely a result of damage to CN X.

References
Kenney DG, Divers TJ (1993) A brainstem melanoma associated with loss of the cough reflex, resulting in aspiration pneumonia in a horse. *Prog Vet Neurol* 4:11–13.

Morrison L, Freel K, Henderson I *et al.* (2008) Lymphoproliferative disease with features of lymphoma in the central nervous system of a horse *J Comp Path* 139(4):256–261.

CASE 50
1 What is the most likely cause of the hemoabdomen and peripheral anemia? The most likely cause is continued hemorrhage from the testicular artery or testicular veins following castration. Other common causes of hemoabdomen include abdominal trauma and neoplasia; in females, hemorrhage from the reproductive tract is also common.

2 How should this case be managed? The patient should be kept as quiet as possible. IV fluids should be administered to maintain circulatory blood volume and treat hypovolemia. A plasma or whole blood transfusion should be considered if the PCV falls to <15% in subacute or chronic hemorrhage or if there is persistent acute hemorrhage or elevation of blood lactate concentration. There is no need to drain the intra-abdominal blood in this case; however, if the peritoneal effusion causes abdominal distress, it can be drained, collected (aseptically), and autotransfused back into the horse. Epsilon-aminocaproic acid or tranexamic acid can be administered as antifibronlytic agents to control haemorrhage.

3 If it is decided to administer a blood transfusion, how much blood should be collected and administered? The following equation can be used to estimate the volume of blood (in liters) needed for whole blood transfusion from a donor horse:

$$\frac{\text{Desired PCV} - \text{PCV of recipient}}{\text{PCV of donor}} \times (0.08 \times \text{body weight in kg of recipient})$$

However, this entire volume may not be necessary to re-establish effective circulating blood volume and achieve adequate tissue perfusion and oxygenation.

Discussion

Castration is a routine procedure in veterinary medicine, but complications are not uncommon. Hemorrhage from the testicular artery(ies) is usually external (at the surgical site) and is one of the most common castration-related complications. Occasionally, hemorrhage may occur internally, creating a hemoperitoneum. Ligature failure is an important cause for post-castration hemorrhage and choosing appropriate suture material and size is important. A recent study found that both a double knot loop and a single knot loop were effective for ligating the spermatic cord, and the greatest knot security was obtained with three throws of the surgeon's knot to prevent knot slippage.

Clinical signs and physical examination findings for hemoperitoneum typically include dullness, anorexia, depression, mild to moderate colic, pale to white (or, less commonly, jaundiced) mucous membranes, tachycardia, and a low PCV/TP. Abdominocentesis produces obviously hemorrhagic fluid, potentially with a higher PCV than the peripheral PCV. Ultrasound may reveal the characteristic swirling pattern of echogenic fluid in the abdomen indicative of hemorrhage. While this is not a definitive diagnostic, it offers very strong non-invasive support for a hemoperitoneum. Additionally, it may help gauge the amount of fluid within the peritoneal cavity and/or potentially help avoid a confounding splenic puncture during abdominocentesis.

Cases presenting as above do well with being kept quiet and correction of volume depletion, followed by fluid maintenance. The abdominal blood will autotransfuse, but occasionally a blood transfusion may be necessary. Antibiotics and anti-inflammatory drugs may be warranted, since blood is an excellent growth medium for bacteria. They will also help prevent adhesion formation in the abdomen. If the bleeding does not resolve or acutely worsens, surgical intervention to locate and ligate the source of the bleeding may be needed. This can be done either with exploratory laparotomy or by laparoscopy in the standing animal. Laparoscopy is preferable in castration cases where the source of bleeding is known to be the testicular artery(ies), as it avoids additional complications associated with GA.

References

Conwell RC, Hillyer MH, Mair TS *et al.* (2010) Haemoperitoneum in horses: a retrospective review of 54 cases. *Vet Rec* **167**:514–518.

Holmes JM, Nath LC, Muurlink MA (2013) Laparoscopic cauterisation of the testicular arteries to manage haemoperitoneum in a gelding. *Equine Vet Educ* **25**:297–300.

Kilcoyne I, Watson JL, Kass PH *et al.* (2013) Incidence, management, and outcome of complications of castration in equids: 324 cases (1998–2008). *J Am Vet Med Assoc* **242(6):**820–825.

Rijkenhuizen ABM, Sommerauer S, Fasching M *et al.* (2013) How securely is the testicular artery occluded in the spermatic chord by using a ligature? *Equine Vet J* **45:**649–652.

CASE 51

1 What caused this disease? Most likely by the hock being traumatized as the horse left the starting gate. Bruising of the skin could have allowed bacteria to invade the damaged soft tissue, leading to a rapidly progressive cellulitis.

2 What organism is most commonly associated with this injury, and what antimicrobial should be selected? *Staphylococcus aureus* from the skin, as in this case. Most cases are not MRSA, but rather highly necrotizing staphylococcal infections. Enrofloxacin would be a reasonable antibiotic choice, but should be based on the results of culture and sensitivity testing. This drug is effective against most *Staphylococcus* spp., can be given IV or PO, and has good tissue penetration.

3 Is it better to give the selected antibiotic as a large dose once a day or split the dose every 12 hours? Enrofloxacin, a flouroquinolone antibiotic, is a concentration-dependent antimicrobial (i.e. the higher the peak concentration, the greater the bacterial killing). Therefore, once daily dosing at 7.5mg/kg is usually recommended for equines more than 9 months of age.

4 What supportive care should the horse receive? Should include NSAIDs, tetanus toxoid administration, hydrotherapy, and a support wrap on the opposite limb.

5 What is the prognosis? There is a good prognosis for life because of the rapid surgical intervention, which was very important. The prognosis for return to racing is only fair due to the marked soft tissue damage in the area. The proximity to the superficial flexor tendon and the plantar tarsal ligament increases the likelihood of scarring in these structures.

6 What other complication might occur in the opposite limb if the horse remains non-weight bearing on the diseased limb for a prolonged period? Contralateral or support-limb laminitis, although this tends to be more common in the forelimbs than the hindlimbs.

Reference

Markel MD, Wheat JD, Jang SS (1986) Cellulitis associated with coagulase-positive staphylococci in racehorses: nine cases (1975–1984). *J Am Vet Med Assoc* **189(12):**1600–1603.

CASE 52

1 What is this structure? A rectal prolapse.

2 What are the different classifications of this disease? Four types of rectal prolapse are recognized: type 1: prolapse of rectal mucosa through the anus; type 2: full-thickness prolapse of all/part of the rectal ampulla; type 3: as for type 2 with additional prolapse of a variable length of small colon intussuscepted into the rectum (but not through the anus); type 4: peritoneal rectum and a variable length of small colon intussuscepted through the anus.

3 What are the common causes of this disease? Diarrhea (as in this case), colic, intestinal parasites, proctitis, dystocia, retained fetal membranes.

4 How should this condition be treated? Examination confirmed that this was a type 1 rectal prolapse. Sedation and butylscopalamine should be administered to decrease straining. Local anesthetic solution can be applied topically to the prolapse or caudal epidural anesthesia can be undertaken. Sugar or glycerine may be applied to the surface of the prolapse to reduce some of the edema. In most cases of type 1 prolapse, the prolapse can be reduced and kept in position with a purse-string suture around the anus for 1–2 days; this should be loosened every 2–4 hours to allow defecation/manual removal of feces. The underlying condition should also be treated as appropriate.

Discussion

Rectal prolapse is a relatively common complication of any condition in which tenesmus occurs. This includes diseases of the GI and genitourinary systems, and it is a well-known complication of parturition in mares. While the different types of prolapse increase in severity and emergency with number, all prolapses should be treated as a potential emergency. A type 1 can become a type 4 without attention, and prognosis and ease of treatment decrease as the grade increases.

In all cases of prolapse the most important aspect of therapy is to treat the underlying cause (i.e. the cause of the tenesmus). As the type increases, the risk of compromise to the involved tissues increases, possibly requiring surgical intervention (for types 3 and 4); with appropriate treatment, the prognosis is good for type 2 and 3 rectal prolapses. However, type 4 prolapses in particular have a poor prognosis due to the likelihood of trauma (including avulsion) to the relatively short mesentery of the small colon and resulting vascular compromise to that segment of the gut.

References

Gibson K, O'Hara A, Huxtable C (2001) Focal eosinophilic proctitis with associated rectal prolapse in a pony. *Aust Vet J* **79**(10):679–681.

Ragle CA, Southwood LL, Galuppo LD *et al.* (1997) Laparoscopic diagnosis of ischemic necrosis of the descending colon after rectal prolapse and rupture of the mesocolon in two postpartum mares. *J Am Vet Med Assoc* **210(11):**1646–1648.

CASE 53

1 What is the interpretation of the blood smear? There are large pleomorphic WBCs with pink cytoplasmic granules. The most likely diagnosis is a myelogenous leukemia. The platelets appear to be decreased.

2 What additional procedure(s) could be performed to help confirm the diagnosis? Bone marrow biopsy or aspirate. Although the platelet count was low, there had been no bleeding episodes, petechiae, or hematomas at the site of venipuncture, so risk of hemorrhage associated with the procedure would be low. A slide of the bone marrow is shown (**53.3**, courtesy Dr. T. Stokol). The cellularity is increased due to the large mononuclear neoplastic cells (>80% of the marrow cells). Non-neoplastic myeloid cells (mature and band neutrophils) are rare. The erythroid cells are metarubricytes and appear dysplastic. Megakaryocytes are decreased with mature and immature forms seen.

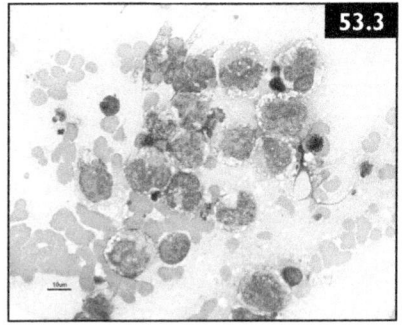

Discussion

Myelogenous leukemia in horses is an uncommon disorder that can present with a variety of clinical signs depending on the involvement of various organs. The most common clinical findings are fever, weight loss, and lethargy, which were all present in this case. Although neoplastic cells were found in the stomach, large intestine, spleen, liver, and kidney, there was no evidence of dysfunction in these organs. Severe involvement of the lungs in other cases has been seen, which results in labored breathing. Flow cytometry can be used to help determine the lineage of the abnormal cells. The abnormal RBCs in this patient also supported a mylogenous leukemia since red cells develop from the myeloid cell line and not the lymphoid line. Myeloid leukemia has a grave prognosis; more grave than lymphoid leukemia. There are several individual case reports of myelogenous leukemia in horses. Most of the horses had fever, while some had leukostasis with concurrent fungal infection.

Reference

Spier SJ, Madewell BR, Zinkl JG *et al.* (1986) Acute myelomonocytic leukemia in a horse. *J Am Vet Med Assoc* **188(8):**861–863.

CASE 54

1 What is the most likely diagnosis? Nutritional myodegeneration (white muscle disease).

2 What is the etiology and pathogenesis of this condition? The condition is associated with vitamin E and/or selenium deficiency, resulting in myodegeneration of skeletal and cardiac muscle. Oxygen free radicals (byproducts of normal cellular metabolism) can react with unsaturated fatty acids to form toxic lipid peroxidases, which can disrupt cell membranes and proteins, leading to a loss of cell integrity. Vitamin E is an antioxidant that scavenges free radicals, preventing cell disruption. Selenium is a component of glutathione peroxidase and converts hydrogen peroxide and lipoperoxidases into water and less harmful alcohols. Vitamin E and selenium deficiency allows these toxic events to occur, leading to myodegeneration.

Discussion

White muscle disease or NMD (nutritional muscular dystrophy or nutritional myodegeneration) is a non-inflammatory degenerative disease affecting the skeletal and/or cardiac muscles primarily of young foals (<30 days old) in areas around the world where the soil is selenium (Se) deficient. It may present as an acute or subacute disease, with the most common clinical signs and diagnostic findings including muscle weakness (acute or progressive) with or without skeletal muscle degeneration, severely increased serum CK, hyponatremia, hyperkalemia, hypochloremia, and myoglobinuria with low Se and, sometimes, vitamin E concentrations. Dysphagia is not uncommon, potentially resulting in secondary aspiration pneumonia. Foals with cardiac involvement may have tachycardia, arrhythmias, and myocardial degeneration. If both cardiac and skeletal muscles are significantly affected, the result may be sudden death. Histologic findings from muscle biopsy include hyaline degeneration and lysis and fragmentation of muscle fibers, but biopsy alone is not diagnostic. Diagnosis is made based on clinical signs, increased muscle enzymes, decreased concentrations of Se, and response to therapy.

Treatment is with an IM injection of a combination vitamin E/Se product at a dose not exceeding 200 µg/kg Se (which may be repeated once or twice), in combination with supportive therapy and daily vitamin E supplementation. Particular attention must be paid and support given to the renal system, as high concentrations of myoglobin released from affected muscles are nephrotoxic and may result in AKI. Hyperkalemia, as the result of potassium released from damaged myocytes, may result in sometimes fatal cardiac arrhythmias. Therapy aimed at potassium excretion or relocation to the intracellular fluid should be instituted and, if an arrhythmia is detected, correction of hyperkalemia is necessary to determine if there is primary myocardial involvement. Serial blood chemistries should be performed to track the progress of therapy with regards to normalizing electrolytes and CK.

Prognosis is guarded, with poor prognostic indicators including the acute form, inability of the foal to support its weight, severe myoglobinuria, and cardiac involvement. Many farms in Se-deficient areas choose to prophylactically treat all neonatal foals with an IM injection of vit E/Se. Again, care should be taken not to exceed 200 µg/kg as Se toxicity is possible with clinical signs including depression, ataxia, muscle weakness, blindness, diarrhea, and dyspnea. While an association between NMD and Se has been documented in foals, the same association does not appear to exist for myopathies in adult horses, as a recent study found no increased risk of myopathy associated with low Se in horses over 2 years of age.

References

Katz L, O'Dwyer S, Pollock P (2009) Nutritional muscular dystrophy in a four-day-old Connemara foal. *Ir Vet J* **62**(2):119–124.

Streeter RM, Divers TJ, Mittel L *et al.* (2012) Selenium deficiency associations with gender, breed, serum vitamin E and creatine kinase, clinical signs and diagnoses in horses of different age groups: a retrospective examination 1996–2011. *Equine Vet J Suppl* **44**(43):31–35.

CASE 55

1 Based on this information, what is the most likely etiologic diagnosis? *Rhodococcus equi* infection. The age, acute onset of severe respiratory effort, and swollen joints (which are believed to be due to immune-mediated synovitis) are characteristic of *R. equi*. Uveitis with fibrin deposition in the aqueous causes the green appearance of the eyes and is most commonly seen with either *R. equi* or *Salmonella* septicemia.

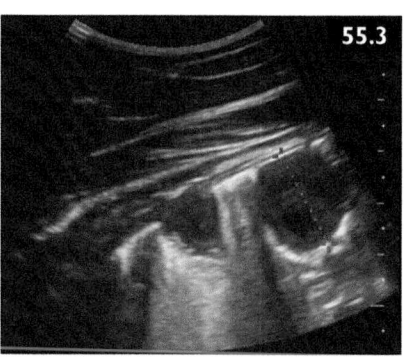

2 How could the diagnosis be supported and then confirmed? Ultrasound examination (**55.3**) of the chest often reveals focal abscessation as opposed to the more diffuse parenchymal 'consolidation' seen with other bacterial causes of foal pneumonia. CBC and fibrinogen are often more elevated with *R. equi* than with most other causes of bacterial pneumonia in foals. Transtracheal wash (TTW) with culture and cytology would be the most definitive test when used with the other clinical and laboratory data to confirm the diagnosis.

3 List some possible explanations for the poor response to seemingly appropriate treatments. (1) *Rhodococcus* might not have been the causative agent. This seems unlikely based on all the above information. (2) There was a second pathogen associated with the disease. This was possible but a second organism was not cultured and there was little improvement after 5 days of gentamicin therapy was added to the clarithromycin and rifampicin. (3) The *R. equi* may have been resistant to clarithromycin and rifampicin. This was a real possibility, as 4–5% of clinical cases have been confirmed to have this resistant pattern. Gentamicin is often recommended in those cases. In this foal, the *R. equi* grown in culture was sensitive to all the antibiotics being used. (5) A drug-induced fever. This is possible in *R. equi* cases being treated during hot weather, but the weather was cool in this case. (6) An extrapulmonary manifestation of *R. equi*. This was the reason for the poor response in this case.

Follow up/discussion
CBC and fibrinogen in this foal were relatively unremarkable for a clinical *R. equi* pneumonia (WBC = 8,100 cells/µl [8.1 × 10⁹/l] and fibrinogen was 600 mg/dl [6.0 g/l]), suggesting that there is not a good correlation between CBC and fibrinogen elevations and severity of disease. In this case, a TTW was performed through the endoscope and was interpreted as marked septic inflammation with intracellular gram-positive rods compatible with *Rhodococcus* sp.

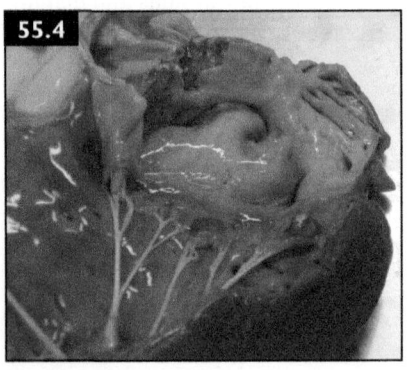

Mesenteric abscesses are the most common severe extrapulmonary disease. In this case, the foal had septic (*R. equi*) endocarditis of the mitral valve identified at necropsy (**55.4**). No murmur had been ausculted ante-mortem.

References
Carlson K, Kuskie K, Chaffin K *et al.* (2010) Antimicrobial activity of tulathromycin and 14 other antimicrobials against virulent *Rhodococcus equi* in vitro. *Vet Ther* 11(2):E1–9.
Reuss SM, Chaffin MK, Cohen ND (2009) Extrapulmonary disorders associated with *Rhodococcus equi* infection in foals: 150 cases (1987–2007). *J Am Vet Med Assoc* 235(7):855–863.

CASE 56
1 **What is the most likely diagnosis?** Equine sarcoids.
2 **What is the etiology of this condition?** A sarcoid is a tumor of fibroblastic cell origin, is the most common equine skin neoplasm, and is largely unique to horses. Bovine papillomaviruses (BPVs) appear to be involved in the etiopathogenesis,

and insect transmission may be important. Young to middle-aged horses are most commonly affected, and there may be a genetic predisposition. Lesions are often persistent and invasive, although they do not metastasize.

3 What treatment options are available? Sarcoids are difficult to treat despite a range of treatment options. Small verrucous and occult forms will sometimes remain quiescent for many years, and observation without treatment may be recommended. Wide surgical excision (traditional sharp surgery or laser surgery) is effective, but the location of many lesions often makes this impossible, and recurrence rates remain high. Laser ablation, using a CO_2 laser, is also effective in small lesions. Alternatives to surgery, or adjunctive therapies after tumor debulking, include cryosurgery, Bacillus Calmette–Guérin (BCG) or autogenous vaccine and intralesional radiation and intralesional chemotherapy (e.g. cisplatin); each has reported variable success rates. Newer topical therapies include imiquimod, an immune-enhancer in a gel form, which has recently been reported to be very effective for some sarcoids, especially superficial forms. Acylovir application has also been reported to be effective for smaller sarcoids. Topical and intralesional 5-flurouracil has been used; however, little information is available. A heavy metal and thio-uracil preparation (AW4-Ludes) from Liverpool, UK, is reported as often effective, but there are limited published studies of its efficacy. Bloodroot extract (*Sanguinaria canadensis*) has antineoplastic activity and creams containing bloodroot extract and zinc chloride have been used as escharotic agents in the treatment of equine sarcoids.

Discussion

Sarcoids can occur in various types, including: verrucous (wart-like): scaly, 'wart'-like papules and plaques; fibroblastic: ulcerated, sessile or pedunculated nodules; occult (flat): alopecic, with scaling and hypopigmentation (most common on the face, neck, and inguinal areas); nodular: nodules; malevolent: extension of nodular, fibroblastic or often mixed forms via the lymphatics; and mixed: combinations of the above.

While benign, sarcoids may significantly impact the health or use of horses due to their potentially aggressive local growth, depending on the number and location of lesions. The most common areas for sarcoid development are the head (periorbital structures, external ear, and lips), neck, lower limbs, and ventrum/groin area. Quarter Horses, Arabians, and Appaloosa breeds may be predisposed. Differentials include granulation tissue, granuloma, papilloma, fibroma/fibrosarcoma, cutaneous lymphoma, squamous cell carcinoma, habronemiasis, mast cell tumor, eosinophilic granuloma, melanoma, and folliculitis. Definitive diagnosis is by histopathology; however, careful consideration must be given to the possibility of local re-growth (often with a more aggressive mass) or inciting aggressive changes with incompletely removed sarcoids. Additionally, some masses may be covered with granulation tissue and shallow biopsies (including some punch biopsies) may only sample the overlying tissue and miss the deeper sarcoid tissue. Histopathologic characteristics include epidermal acanthosis, hyperkeratosis, and hyperplasia with long rete pegs into the dermal tissue containing immature fibroblasts, often in a whorled pattern. Schwannomas may be histologically similar to sarcoids, requiring immunohistochemistry to differentiate them from a sarcoid. Additionally, some experience on the part of the pathologist is necessary to distinguish some sarcoids from fibroma or fibrosarcoma.

BPV-1 and -2 (recently reclassified as delta papillomaviruses) have been implicated as the causative agent. Papillomaviruses are small, non-enveloped, double-stranded DNA viruses with strict tropism for cutaneous or mucosal epithelium and typically are very species-specific. Both BPV-1 and BPV-2 are an exception to this, as viral DNA can be detected in up to 100% of equine sarcoids. Injury resulting in the exposure of the basal epidermal cells at the basement membrane is necessary for infection since the virus cannot penetrate intact skin. Viral DNA has been isolated from both biting and non-biting flies removed from horses bearing sarcoids, implicating flies as possible vectors. Unlike many viruses, papillomaviruses in general do not kill the host cell, but rather shed viral particles as part of normal desquamation of the epithelium. Studies have also identified viral DNA in other species in a variety of other cells including fibroblasts and blood cells, and *in-utero* transmission has been documented in cattle. Viral DNA has also been isolated from WBCs in horses. Recently, BPV viral DNA isolated from sarcoids was found to contain several mutations when compared with BPV viral DNA isolated from lesions in cattle. These mutations were found to correlate with enhanced function within equine versus bovine fibroblasts, indicating the possibility of an equine-specific BPV-1 variant.

References

Bergvall KE (2013) Sarcoids. *Vet Clin North Am Equine Pract* 29(3):657–671.
Nasir L, Brandt S (2013) Papillomavirus associated diseases of the horse. *Vet Microbiol* 167(1-2):159–167.
Taylor S, Halderson G (2013) A review of equine sarcoid. *Equine Vet Educ* 25(4):210–216.

CASE 57

1 What possible tumor types should be considered? Common types of skin tumor occurring in mature horses include sarcoid, melanoma, squamous cell carcinoma, and mast cell tumor. This tumor has a physical appearance most consistent with either sarcoid or melanoma.

2 What should be the next course of action? Fine needle aspirates are often inconclusive in equine skin tumors. Biopsy (e.g. TruCut® biopsy needle) may be attempted, but there is concern that biopsy may aggravate some types of tumor and result in more aggressive growth/spread.

3 Anaplastic malignant melanoma was diagnosed. What further advice or treatment should be recommended? Anaplastic malignant melanoma on the tail of non-gray horses tends to be very aggressive with a high risk of metastasis, therefore amputation of the tail should be considered.

Follow up/discussion
The mass and a margin of normal skin was excised using laser surgery under standing sedation and caudal epidural anesthesia. Histopathologic examination identified the mass as an anaplastic malignant melanoma with signs of locally infiltrative growth and vascular/lymphatic vessel invasion in some areas.

Melanoma is typically associated with older gray horses but can affect horses of any age or coat color. Melanocytic tumors may be classified as gray horse dermal melanoma, gray horse dermal melanomatosis, melanocytoma, or anaplastic malignant melanoma. Anaplastic malignant melanoma is the most aggressive and often results in death within 1 year of diagnosis for non-gray horses. While many melanocytic tumors are considered benign, any melanocytic tumor has the potential to become malignant and up to 66% become malignant at some point. Regardless of coat color, the most common dermal sites of melanocytic tumor development include the tail base, perineum, sheath, parotid area, subauricular lymph nodes, and commissures of the lips. Particularly in gray horses, melanoma may develop in other organs (e.g. spine, GI tract, guttural pouch, salivary gland) without any evidence of dermal melanoma. Rapidly growing tumors are very firm and may outgrow their blood supply and develop central necrosis, which may drain externally and result in secondary bacterial infections. Left untreated, these rapidly growing masses are likely to metastasize to other areas of the body within a relatively short period of time. An excisional or Tru-cut biopsy is necessary to determine the likelihood of malignancy and differentiate melanoma from other dermal tumors. Biopsy should especially be performed in horses with a solitary mass, in non-gray horses, or with a mass in an uncommon location.

Several types of therapies have been reported, including intralesional, topical, and systemic, but these tend to be effective in only a very small proportion of cases, and typically only for horses with small (<2–4 cm) tumors. These therapies may be of use in cases where the tumors are few, small, and very slowly progressive. The more aggressive, rapidly growing tumors are unlikely to respond to any of the available medical treatments and surgery is indicated whenever possible. Complete removal of large or aggressive tumors is difficult as the masses infiltrate deeper tissues, making it difficult to obtain clean margins on all surfaces. This, coupled with the poor prognosis for survival in non-gray horses, makes amputation the recommended treatment for anaplastic masses on the tail.

References

LeRoy BE, Knight MC, Eggleston R *et al.* (2005) Tail-base mass from a "horse of a different color". *Vet Clin Pathol* **34(1)**:69–71.

Moore JS, Shaw C, Shaw E *et al.* (2013) Melanoma in horses: current perspectives. *Equine Vet Educ* **25(3)**:144–150.

Phillips JC, Lembcke LM (2013) Equine melanocytic tumors. *Vet Clin North Am Equine Pract* **29(3)**:673–687.

Valentine BA, Calderwood Mays MB, Cheramie HS (2014) Anaplastic malignant melanoma of the tail in non-grey horses. *Equine Vet Educ* **26(3)**:156–158.

CASE 58

1 List some differential diagnoses for infectious causes that could be responsible for the diarrhea in this filly. Infectious causes would include *Clostridium difficile, C. perfringens,* rotavirus, coronavirus, *Salmonella* spp., Potomac horse fever, *Rhodococcus equi, Lawsonia intracellularis,* and parasitic conditions.

2 Based on the hypoproteinemia, history, physical examination, and ultrasound findings, what is the most likely diagnosis? These signs are most consistent with *L. intracellularis* infection.

3 How should this infection be treated? Treatment with oxytetracycline IV is curative in about 90% of horses, but approximately 10% will not respond well to treatment. Oxytetracycline can be nephrotoxic and should only be used in euhydrated animals. Good supportive care is critical.

Follow up/discussion

The filly tested negative for *Cryptosporidium, C. difficile, C. perfringens,* rotavirus, coronavirus, Potomac horse fever, *R. equi,* fecal parasites, and *Salmonella.* She was PCR positive for *L. intracellularis* and had a high antibody titer.

This filly was initially treated at a different clinic with flunixin meglumine, oxytetracycline, rifampicin, gastrogaurd, pentoxyfylline, a prebiotic, plasma transfusions, and hetastarch. Clopidogrel was also added when the filly developed a fully thrombosed left jugular vein and partially thrombosed right jugular vein. Treatment for *Lawsonia* was continued at our hospital and treatment for AKI was initiated. A lateral thoracic catheter was placed and she was started on IV fluids and electrolytes. The filly was also treated with chloramphenicol, clopidogrel, pentoxifylline, and gastroprotectants. Her jugular veins were treated with hot packs and topical diclofenac. After 3 days of aggressive fluid therapy the filly's electrolytes showed improvement. She was urinating, defecating, eating, and drinking well. Unfortunately, she became febrile (105.3°F [40.7°C]), showed a decreased ability to clot when blood was drawn from her facial vein, developed hematuria, petechiation of her mucous membranes, epistaxis, and an increased respiratory effort. Prothrombin time was prolonged (32.5 seconds, control 14.5) and APTT was shortened (27.7 seconds, control 52.1). Despite treatment with oxygen and furosemide, she died shortly after. Necropsy findings are shown (**58.3**).

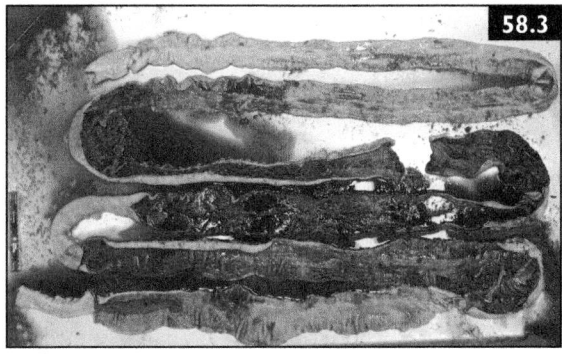

Lawsonia intracellularis is an intracellular pathogen that invades the cells that line the distal small intestines and may cause a proliferative enteropathy. This pathology prevents proper absorption of nutrients and leads to a loss of protein (initially albumin and other small proteins such as clotting factors) through the GI tract. This protein loss can be quite severe and the most common clinical finding is therefore ventral edema due to low oncotic pressure and weight loss. The presence of diarrhea, colic, and fever is more variable. The age of clinically affected horses is 4–11 months, with disease most frequently occurring around weaning time. Horses of all ages may intermittently shed the organism in the manure and have serum antibodies against the organism, but clinical disease in adults is generally sporadic. Renal failure is a common complication and in some cases may be partially due to the oxytetracycline therapy combined with the use of NSAIDs. The foal must be very well hydrated to prevent AKI.

More recently, an acute demise with renal failure and presumed disseminated coagulopathy has been described and is similar to what occurred in this filly. The loss of protein can be to the point that it depletes the clotting factors and sloughing of the intestinal mucosal lining, which leads to hemorrhage and further decreases the clotting factors. This is likely what caused the filly in this case to bleed into her GI tract, lungs, subcutaneous tissues, and brain the morning she died.

It is thought that *Lawsonia* is carried by wild rabbits. Several tests are available to check for *Lawsonia*, and foals can be screened for low TP and treated before it causes severe disease on farms where *Lawsonia* is endemic. A porcine vaccine has been used per rectum as a strategy to control disease on endemic farms.

References

Dauvillier J, Picandet V, Harel J *et al.* (2006) Diagnostic and epidemiological features of *Lawsonia intracellularis* enteropathy in 2 foals. *Can Vet J* **47**:689–691.

Page AE, Fallon LH, Bryant UK *et al.* (2012) Acute deterioration and death with necrotizing enteritis associated with *Lawsonia intracellularis* in 4 weanling horses. *J Vet Intern Med* **26**(6):1476–1480.

Pusterla N, Gebhart C (2013) *Lawsonia intracellularis* infection and proliferative enteropathy in foals. *Vet Microbiol* **167**(1–2):34–41.

Slovis NM, Elam J, Estrada M *et al.* (2014) Infectious agents associated with diarrhoea in neonatal foals in central Kentucky: a comprehensive molecular study. *Equine Vet J* 46(3):311–316.

CASE 59

1 What is the most likely diagnosis? Umbilical infection (bacterial) or omphalitis.
2 What is the recommended treatment for this foal? Surgical removal of the umbilical remnants because of enlargement of both the umbilical 'stump' (which was nearly 4 cm in this foal) and the right umbilical artery (ultrasound size was >1 cm; not shown). If the infection had involved only the umbilical stump, drainage would have been possible.

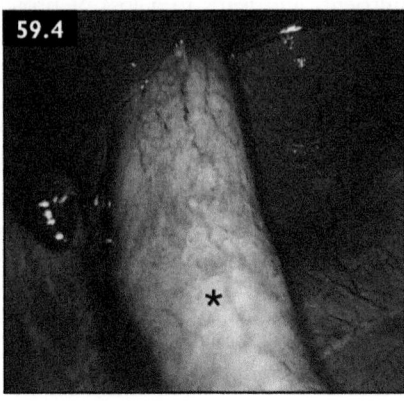

59.4

3 What is a common and serious complication with this condition in foals? Infection of one or more joints is the most common serious complication of foals >7 days of age with umbilical infections.

4 An ultrasound image of the cranio-ventral abdomen of another foal with fever is shown (59.3). What is the mixed echogenic structure coursing within the liver (identified by the broken line)? Infection of the umbilical vein (*), which was removed by laparoscopic surgery (intraoperative photo, 59.4).

Follow up/discussion
Surgery was performed in this foal and the right umbilical artery, part of the urinary bladder, and the umbilical stump were clamped, ligated, and removed. The enlarged umbilical stump was cultured and *Arcanobacterium pyogenes* and *Streptococcus zooepidemicus* were isolated. *A. pyogenes* was likely responsible for the dense exudate found in the urachus and the associated hyperechoic fluid seen on ultrasound. The foal was treated with K^+ penicillin 20,000 IU/kg IV, amikacin 20 mg/kg IV, and omeprazole 4 mg/kg PO for 24 hours prior to surgery and for 3 days following surgery.

Infection of the umbilical remnants is common in foals and may be predisposed to by poor umbilical hygiene, premature severance of the cord (inhibiting normal contractile closure), and failure of passive transfer. Although infection of the umbilical structures is not the most common route of infection for acute severe sepsis in newborn foals (oropharyngeal and GI translocation are the most common), it is a very common cause of localized sepsis with low-grade bacteremia that can result in septic joints. This type of infection is most common between 6 and

21 days of age. The infected umbilicus should always be cultured to determine the offending bacteria, either after removal or from urachal drainage. Mixed bacterial infections are common, with *E. coli* and beta-hemolytic *Streptococcus* spp. being the most common. Less commonly, *Clostridium* spp. (including *C. tetani*), *Listeria* spp., *Arcanobacterium* spp., and many other bacteria may be cultured. Ultrasound examination is important in determining the umbilical structures involved and the extent of the involvement. It is not unusual for the infection to extend all the way to, and into, the liver if the umbilical vein is involved. The umbilical vein of young foals should be <0.8 cm, each umbilical artery <1.0 cm and the combined urachus/umbilical arteries <2.1 cm. Prevention should focus on minimizing the risk factors listed above and dipping the navel with 1–4% chlorhexidine solution 2–3 times daily until the umbilicus is closed and dry.

Reference

Neil KM (2011) Disorders of the umbilicus and urachus. In: *Equine Reproduction*, 2nd edn. (eds. AO McKinnon, EL Squires, WE Vaala, DD Varner) Wiley-Blackwell, Ames, pp. 632–645.

CASE 60

1 What is this condition? Paraphimosis.
2 What are the possible causes of this condition? Potential causes include: preputial edema caused by genital trauma or systemic disease (e.g. dourine and purpura hemorrhagica); damage to penile innervation – can occur with spinal disease, trauma and infectious diseases; debilitation; phenothiazine-derivative tranquilizers (generally not a problem in geldings).
3 How should this condition be treated? Treatment depends on identifying and treating the underlying cause. Local therapies may include: replacing the prolapsed penis within the external preputial lamina and retaining it with sutures, towel clamps, or nylon netting placed across the preputial orifice; compressing the prolapsed penis against the abdomen with wraps if it cannot be replaced within the external preputial lamina; applying antimicrobial ointments to prevent epithelial excoriation and infection; administering NSAIDs to reduce swelling. If the paraphimosis is long-standing, the following measures may be taken: if preputial scarring prohibits replacement of the penis within the prepuce, a segmental posthectomy may be performed; if the penis is paralyzed, permanent placement of the penis in the prepuce (phallopexy) or amputation of the penis may be indicated.

Discussion

Paraphimosis may occur in both intact and castrated male horses and is considered an emergency. The longer the penis remains outside the prepuce, the more edematous (dependent edema) it becomes, therefore making replacement into the prepuce more difficult and increasing the likelihood of permanent paralysis. There are many different causes of paraphimosis and medical/supportive therapy is largely the same regardless of cause, with a few exceptions: prevention of retraction of the penis or occlusion of the urethra due to masses or habronemiasis probably requires surgical intervention as opposed to medical management. If the cause is spinal cord injury/disease, surgical intervention is more likely to be necessary. Drainage of hematomas or seromas may be necessary to achieve replacement, but very careful consideration should be given to potential bacterial infection. Frequent cleaning and monitoring of such sites along with topical and/or systemic antimicrobial therapy should be instituted. Ultrasound may assist in the identification of seromas and hematomas as well as assist in identifying the extent and potential communication with the corpus cavernosum. If communication exists, drainage or evacuation is not advised.

Treatment is largely supportive and aimed at providing physical support to the penis to help prevent dependent edema and the risk of permanent paralysis, and to reduce any swelling of the prepuce or penis that is preventing replacement/ retraction of the penis into the preputial cavity. The earlier therapy is instituted, the more rapid and complete the recovery is likely to be. That being said, supportive therapy is labor intensive and may be needed for many weeks before normal function returns. Care should be taken, especially early in the course of therapy, to periodically assess the urinary bladder and ensure that urethral blockage has not occurred, necessitating urinary catheterization. Urine scalding is a concern in cases where either the penis cannot immediately be replaced into the preputial cavity and must be supported against the ventral body wall, or the penis has been replaced but is being held inside the prepuce using a purse string suture. The use of barrier ointments (e.g. petroleum jelly) and frequent cleansing of the skin should help protect the skin and prevent scalding, but topical or parenteral antibiotic therapy may need to be instituted if scalding and secondary bacterial infection occur.

Reference
Beltaire KA, Tanco VM, Bedford-Guaus SJ (2011) Theriogenology question of the month. Trauma-induced paraphimosis. *J Am Vet Med Assoc* **238**(2):161–164.

CASE 61
1 What type of disease process (i.e. strangulation, complete obstruction, partial obstruction) is causing the colic? The duration of abdominal pain, normal abdominal appearance, fair appetite, normal manure, and normal abdominal palpation per rectum all suggest that this is either a non-gastrointestinal problem or

a focal, non-obstructive/non-strangulating intestinal problem. Clinical and ultrasound examination of the thorax, peritoneum, and urinary system and evaluation of liver enzymes rule out most non-intestinal causes of colic. In this case there was no evidence of a non-intestinal cause for the colic.

2 What initial diagnostic tests should be run? Measure plasma protein to help rule out colonic ulcers. The flunixin administration could be one explanation for the continued colic. In this case the

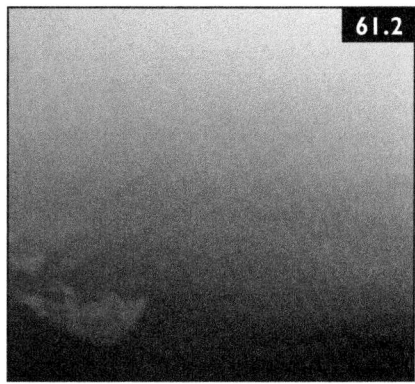

plasma protein concentration was normal (6.8 g/dl [68 g/l]). Feces should be mixed with water in a plastic bag or rectal sleeve and suspended to determine if it contains sand. This was performed and no sand was observed but the history and clinical findings, including the unusual colonic 'gushing' sound, were all compatible with sand colic. The horse was reported to have had two episodes of diarrhea without fever in the past 3 weeks. Diarrhea due to bowel irritation and motility disturbances is common in horses with sand-induced clinical disease.

3 What other diagnostic test(s) could be performed? Abdominal radiographs are useful (**61.2**).

4 What is the interpretation of the diagnostics? Although the amount of sand on the radiograph is neither impressive nor diagnostic, the location of the sand (cranioventral and close to the body wall), its depth (>5 cm) and its opacity being greater than the rib could support a diagnosis of sand colic.

5 What is the preferred treatment for the type of colic diagnosed? The preferred treatment for sand colic is psyllium (1 g/kg) via a nasogastric tube for 1–3 weeks. For horses with extremely large amounts of colonic sand accumulation, surgery may be required. For those horses, abdominocentesis should be performed with care, as even a teat cannula may penetrate the heavy, sand-laden bowel.

6 What are the most common risk factors for colic in horses in sandy areas? Although there is no definitive evidence for this, feeding or spilling grain on the sand and overgrazing sandy pastures are two presumed risk factors. There must also be individual susceptibility to the condition, perhaps based on differences in prehension and GI motility, as not every horse on sandy pastures develops sand colic.

7 Although ultrasound examination was normal in this case, what sonographic abnormality might be seen that would be supportive of the colic type diagnosed? A flattening of a colonic haustra can be seen on transabdominal ultrasound of the ventral abdomen in horses with colonic sand accumulation.

References

Granot N, Milgram J, Bdolah-Abram T *et al.* (2008) Surgical management of sand colic impactions in horses: a retrospective study of 41 cases. *Aust Vet J* 86(10):404–407.

Husted L, Andersen MS, Borggaard OK *et al.* (2005) Risk factors for faecal sand excretion in Icelandic horses. *Equine Vet J* 37(4):351–355.

Kendall A, Ley C, Egenvall A *et al.* (2008) Radiographic parameters for diagnosing sand colic in horses. *Acta Vet Scand* 50:17.

CASE 62

1 What clinical feature can be seen that would be compatible with this diagnosis?
Bilateral ptosis.

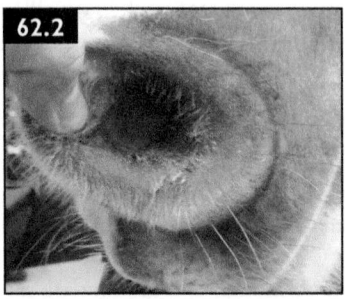

2 What other common clinical signs would be expected in subacute grass sickness? Depression/somnolence, anorexia, dysphagia, colic, absence of gut sounds, tachycardia, congestion and dryness of mucous membranes, dehydration, patchy sweating, muscle fasciculations, crusty nasal discharge/rhinitis sicca (**62.2**)

3 What abnormalities should be detected on rectal examination? Dry, tacky rectal mucosa; 'empty' rectum with few fecal pellets that are hard and have mucosal casts on the surface (**62.3**). Large intestine (especially large colon) that is shrunken down onto hard, dry, impacted contents. Distended, fluid-filled loops of small intestine may be palpable, extending back towards the pelvic inlet.

Discussion

Equine grass sickness (a.k.a. equine dysautonomia) is a frustrating disease most often associated with a poor prognosis. Clinical signs generally are related to degeneration and loss of enteric neurons, particularly in the myoenteric and submucosal plexuses of the small intestine, with a particular affinity for the ileum. Peripheral parasympathetic and postganglionic sympathetic nerves are often involved as well. It has been recognized in most Northern European countries, including the UK, as well as Chile, Argentina, the Falkland Islands, and Colombia. Rare cases have been identified in North America. Grass sickness may be classified as acute, subacute, or chronic.

With acute grass sickness, horses rapidly develop generalized GI stasis. As a result, there may be dysphagia, abdominal enlargement, and reflux of gastric/small intestinal contents. Secondary large intestinal impactions occur as gut contents dehydrate in the static bowel. Ptosis and rhinitis sicca may also be present. Fine muscle tremors, particularly of the triceps, shoulder, and flank, that do not stop when the muscle is rested or in recumbency, help differentiate grass sickness from equine motor neuron disease or botulism.

Subacute grass sickness typically presents as the case above. The clinical signs are often slightly less severe than with acute grass sickness, but subacute horses have loss of body condition and a 'tucked-up' appearance to the abdomen. They also often exhibit generalized postural muscle weakness, which may manifest as frequent shifting of body weight, resting of limbs, and a particular stance with all four feet placed well under the body. Horses also tend to hang their heads and may use the stall walls to help support their weight.

Chronic grass sickness leaves the horse with marked cachexia and generalized myasthenia, accompanied by milder signs of GI stasis. As with acute and subacute grass sickness, rhinitis is often among the clinical signs and may be severe enough to cause obstruction.

In all three categories of grass sickness, some degree of GI stasis and rhinitis sicca is typically present, which may help differentiate grass sickness from other neurologic or GI diseases. Prognosis for all three categories is poor and most horses are euthanized or succumb to the disease within 1 week of diagnosis. A few chronic grass sickness cases may recover with intensive nursing care. Antemortem diagnosis is made primarily based on clinical signs because adequate intestinal biopsies to prove dysautonomia are difficult to obtain and, given the poor prognosis, are viewed by many owners and veterinarians as an unnecessary procedure for a horse in a weakened state. Both etiology and pathogenesis have eluded science thus far. Current thought is that it may be a toxicoinfectious form of botulism whereby the ingestion and overgrowth of *Clostridium botulinum* in the gut is followed by absorption of clostridial neurotoxins. As the disease seems to be geographically consistent despite movement of horses between countries/continents, and appears not to be transmissible from horse to horse, an ingested bacterial origin does offer the best explanation. However, research findings testing this theory have been inconsistent.

References

Newton JR, Wylie CE, Proudman CJ *et al.* (2010) Equine grass sickness: are we any nearer to answers on cause and prevention after a century of research? *Equine Vet J* **42**(6):477–481.

Schwarz B (2013) Equine grass sickness: what's new? *Vet Rec* **172**(15):393–394.

CASE 63

1 What procedures should be performed to help diagnose the cause of the swollen neck? Ultrasound examination and needle aspirate for cytology and culture (aerobic and anaerobic). Gas was seen on ultrasound examination (**63.2**) and was 'bubbling' along with serum following sterile aspiration at the site (**63.3**). Gram stain of the aspirate revealed large numbers of *Clostridium perfringens*-like organisms.

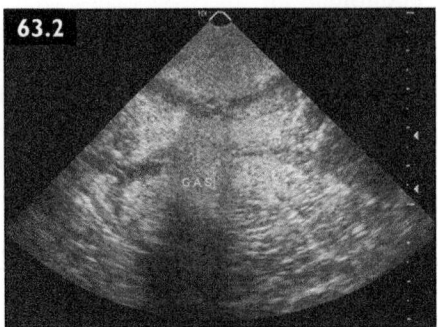
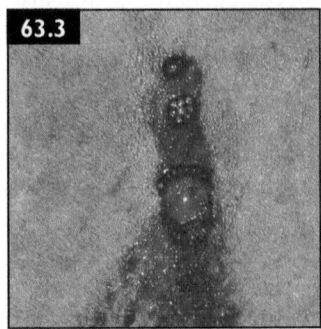

2 What further treatment should be provided following diagnostics? Treatment with antibiotics, IV penicillin, and oral metronidazole should be started as well as surgical fenestration.

3 Would you recommended the standard 22,000 IU/kg K⁺ penicillin IV q6h or would you give a larger dose in light of the life-threatening possibility of this disease? Beta-lactam drugs are time-dependent drugs and administering large doses that go well beyond the effective maximum inhibiting concentration (MIC) for bacteria is unlikely to improve efficacy and may result in toxicity. A case of clostridial myositis treated with K⁺ penicillin at 44,000 IU/kg (IV q6h) developed fatal *Clostridium difficile* colitis. Administering a 'normal dose' of K⁺ penicillin every 4 hours for the first day of treatment could be rationalized.

Follow up/discussion

The disease in this hore was too extensive to believe that antibiotics alone would be successful, so surgical drainage was performed. Although *C. perfringens* is sensitive to penicillin, metronidazole therapy was added to improve the spectrum against other clostridial species.

Spores of *Clostridium* (mostly *perfringens* type A) can lay dormant in the equine muscle, similar to black-leg (*C. chauvoei*) in cattle. When injected IM, flunixin meglumine may cause tissue injury (similar to liver flukes migrating in the liver causing *C. novyi*/black disease), which provides the right environment for the spore(s) to germinate and release toxins. The alpha toxin is most likely responsible for the rapid muscle necrosis. Gas is not always observed on ultrasound examination.

An affected horse can become toxic and die within 1–2 days. If there is noticeable swelling at the site of injection in a horse with evidence of systemic inflammation and toxemia, surgical drainage and antibiotic administration should be performed immediately. Owners should always be warned of the possibility and the importance of early diagnosis of clostridial myositis when flunixin meglumine is administered IM. Other non-antibiotic drugs, such as multi-B vitamins, prostaglandins, and ivermectin, may also cause clostridial myositis when administered IM. Hyperbaric oxygen or simply infusing oxygen into the infected site might be efficacious. Supportive therapy should include: fluids; IV or orally administered NSAIDS; anticytokine medications (pentoxifylline); antiplatelet therapy (clopidogrel); and *C. perfringens* antiserum.

References

Aggelidakis J, Lasithiotakis K, Topalidou A *et al.* (2011) Limb salvage after gas gangrene: a case report and review of the literature. *World J Emerg Surg* 6:28.

Jeanes LV, Magdesian KG, Madigan JE *et al.* (2001) Clostridial myositis in horses. *Compend Contin Educ Pract Vet* 23(6):577–587.

Peek SF, Semrad SD, Perkins GA (2003) Clostridial myonecrosis in horses (37 cases 1985–2000). *Equine Vet J* 35:86–92.

CASE 64

1 What is the tentative diagnosis and differential diagnoses? Botriomycosis (a.k.a. bacterial granuloma, staphylococcal pseudomycetoma, deep pyoderma). Differentials include exuberant granulation tissue/foreign body reaction, sarcoid, *Habronema musca* infestation, pythiosis, and other tumors.

2 What is the pathogenesis of this condition? Botriomycosis is a bacterial pyogranulomatous disease often associated with skin wounds. Multiple miliary interlinking microabscesses form from *Staphylococcus* spp. infection.

3 How should this condition be treated? Surgical excision with debridement of all affected tissue and concurrent antibiotic therapy. Repeated surgical procedures are sometimes required.

Discussion

Bacterial pseudomycetoma (botriomycosis) is uncommon and typically presents as a solitary, non-pruritic, pyogranulomatous lesion on the limbs, lips, chin, or scrotum of affected horses. While the pathogenesis is not fully known, the lesions do appear to begin with trauma to the skin and most often are associated with coagulase-positive *Staphylococcus* spp. Surgical removal is likely curative. Occasionally, a lesion will be too large for complete surgical removal, in which case surgical debulking combined with systemic antimicrobials should be successful.

On occasion, a more widespread disease with multiple lesions spread over the body occurs, particularly in horses with pituitary pars intermedia dysfunction. Surgical removal is not likely to be an option in these cases because of the number of lesions. Systemic, long-term antimicrobial therapy may be attempted, but many cases relapse once therapy is stopped.

References
Scott D, Miller W (2011) *Equine Dermatology*, 2nd edn. Elsevier, Maryland Heights, pp. 161–163.
Valentine BA, Plant JD (2009) Non-neoplastic nodular and proliferative lesions. In: *Current Therapy in Equine Medicine 6*. (eds. NE Robinson, KA Sprayberry) Saunders, St. Louis, pp. 681–686.

CASE 65
1 What is the cause of this horse's severe shock? Intestinal rupture.
2 What possible underlying causes are suspected? Common causes of intestinal rupture following parturition include cecal perforation and ischemia of the small colon due to tearing of the mesocolon. Cecal perforation usually occurs immediately after parturition. Necrosis and rupture of the small colon following tearing of the mesocolon usually takes several hours to days; this is the most likely diagnosis in this mare.

Discussion
While the vast majority of mares have uncomplicated pregnancies, foalings, and post-partum periods, there are many well-recognized peripartum complications that may arise including:

- Prepubic tendon rupture. This is apparently more common in Draft mares, but may occur in any breed. Rupture typically occurs in late pregnancy as the result of increased abdominal weight, often from excess fluid, beyond what is normal for a pregnancy. Mares are often noted to have excessive ventral edema and the prognosis is poor for the mare once rupture occurs.
- Uterine torsion. Torsion generally occurs in the last trimester, most commonly near or at term, and can be anywhere between a 180 and a 540 degree rotation of the uterus. Rapid recognition and replacement of the uterus to a normal position is imperative, as the torsion compromises the blood supply of both the uterus and the fetus. Chronic and/or severe rotation may result in uterine rupture and/or death of the foal.
- Dystocia. Abnormal delivery, most often due to malpositioning of the fetus at parturition, is a well-recognized complication. Occasionally, dystocia will be the result of factors associated with the mare (e.g. narrow pelvic canal, poorly

dilated cervix). Prompt recognition and intervention are imperative, as mares have explosive deliveries and significant, potentially fatal soft tissue damage may occur. Additionally, placental detachment begins early in parturition, creating fetal hypoxia and stress, potentially leading to death of the foal if parturition is prolonged.

- Lacerations/tears. Tears may occur in one or more of the reproductive and/or pelvic organs, including the uterus, cervix, vagina, vestibule, or vulva, with vaginal tears often creating a communication dorsally with the rectum and/or anus. Depending on the location and tissues involved, tears may be an emergency and carry a poor prognosis. The more caudal the tear, the better the prognosis.
- Retained placenta. This constitutes a medical emergency in the horse, with very little retained tissue being necessary for it to become a nidus for infection leading to severe endotoxemia, laminitis, and potentially death. Treatment should be initiated as soon as there is suspicion that any placental tissue may remain.
- Uterine artery rupture. Uterine arteries may rupture pre-partum, but typically rupture during parturition either into the broad ligament or into the abdomen. Ruptures into the abdomen typically are rapidly fatal, but ruptures into the broad ligament may carry a better prognosis if the mare is kept quiet and calm while the resulting hematoma resolves.
- Gastrointestinal. Contusions or rupture of bowel (small or large) that becomes trapped in the pelvis during parturition, development of large colon torsion, large colon impaction, and constipation (from avoidance of defecation due to trauma to the caudal reproductive tract) may develop post-partum.

Reference
Schweizer CM, Cable CS, Squires EL (2006) *Breeder's Guide to Mare, Foal & Stallion Care.* Blood Horse Publications, Lexington, pp. 203–220.

CASE 66
1 Interpret the abdominal ultrasound. There is marked thickening of the small intestinal wall, most likely the muscular layer.
2 What diagnostics should be performed next? Rectal biopsy and duodenal biopsy (via gastroscopy).
3 How should this horse be treated? With IV fluids and frequent feeding of small amounts of low-bulk pelleted feed (Purina Equine Senior®) once the colic episode has resolved.

Follow up/discussion
Biopsy results were histologically normal. The stomach had a normal appearance on gastroscopy with no ulcerations noted.

Attempts to feed any hay caused the colic signs to recur. Attempts to graze the horse would also result in colic, although not as consistently as with hay. Prior to receiving the biopsy results, the stallion received three daily doses of dexamethasone (50 mg/day), but the treatment dose was tapered and discontinued after 7 days. Although fecal examination for parasites was negative, the horse was given moxidectin and praziquantel.

Six months later the signs still persisted, so a laparotomy was performed and a full-thickness small intestinal biopsy carried out. This revealed severe and diffuse muscular hypertrophy of the small intestine.

Muscular hypertrophy of the small intestine is a thickening of the smooth muscle layer that has been reported in a number of species including the horse, pig, cat, and human. There are two recognized presentations: secondary hypertrophy, which occurs as a compensation for a chronic, aboral stenosis, and idiopathic, in which there is no obvious inciting constriction. The underlying cause of idiopathic muscular hypertrophy is unknown, but food allergies, parasites, and underlying inflammatory conditions have been hypothesized. In either case, hypertrophy of the muscularis layer occurs, causing a narrowing of the intestinal lumen, which may eventually lead to partial or complete obstruction. Though rare, muscular hypertrophy in horses most commonly appears to cause a thickening of the ileum; however, more diffuse presentations have been reported. Clinical signs generally include mild-to-moderate intermittent colic, which may recur over the course of months or years, is responsive to analgesics, may be associated with meals, and may progressively become more serious. Weight loss and anorexia are also reported. Blood work and peritoneal fluid analysis are usually unremarkable unless constriction of the intestinal lumen has led to secondary rupture of the GI tract. The horse in this case has done well for 3 years with a special low-bulk diet similar to the original one that was initially instituted, and he has been successfully breeding as well.

Reference
Chaffin MK, Fuenteabla IC, Schumacher J *et al.* (1992) Idiopathic muscular hypertrophy of the equine small intestine: 11cases (1980–1991). *Equine Vet J* **24**(5):372–378.

CASE 67

1 What would be the most likely neuroanatomic location for a lesion? Focal cervical lesion.

2 What is the interpretation of the radiograph? There is some radiographic evidence of osteochondrosis at C_4–C_5 and C_5–C_6 with C_5–C_6 being more severely affected. The intravertebral measurement of the vertebral canal to vertebral body ratio at this site is 0.44 (normal >0.5).

3 What is the most likely diagnosis based on the radiograph and the clinical picture? Compressive cervical myelopathy caused by abnormal development of cervical vertebrae and soft tissues (ligaments).

Follow up/discussion

At necropsy there was evidence of dynamic compression at both C_3–C_4 and C_5–C_6. The lesion at C_5–C_6 was most severe and likely explained why the ataxia seen in the forelimbs was as severe as in the hindlimbs, and why the horse would 'buckle' when he lowered his head.

Cervical stenotic myelopathy (CSM), also referred to as cervical compressive myelopathy or 'wobbler syndrome', results from impingement by the cervical vertebrae and/or surrounding soft tissues on the spinal cord resulting in compression. Ante-mortem diagnosis of CSM is made by signalment (most common in 1–3 year old Throughbreds), neurologic evaluation, exclusion of other diseases, and imaging examination. The information provided by survey radiographs can be moderately to highly diagnostic if there is obvious malalignment of the vertebrae or the intravetebral canal ratio is abnormal (indicating stenosis). These can be misleading since soft tissue compression would not be detected and, conversely, some horses with the aforementioned radiographic abnormalities do not have a compressive lesion. Myelography is a more accurate test but both false positives and false negatives still can occur. Vertebral canal stenosis is generally considered to be the most important feature of cervical stenotic myelopathy, which can be missed by a myelogram if the compression is not dorsal or ventral because only the height of the dorsal dye column is evaluated with this technique. Additionally, complications such as transient blindness or seizures, may occur following the myelogram. Clinical signs are ataxia of all four limbs, generally worse in the hind- than the forelimbs. If the compression is at C5–C7, the forelimbs may be the same – or even worse – than the hindlimbs. Focal hyperesthesia, sweating, and muscle spasms or atrophy may occur at the site of the compression. It is common for the horse to have pain or difficulty bending its neck. Surgical fusion of the vertebrae on either side of the compression will allow atrophy of the articular facets and eventually alleviate the compression. Likewise, if the compression is dynamic (associated with flexion of the neck), surgery will alleviate the compression. If the surgery is beneficial, a period of up to 1 year is needed to determine how much the gait has improved.

Reference

Levine JM, Scrivani PV, Divers TJ *et al.* (2010) Multicenter case-control study of signalment, diagnostic features, and outcome associated with cervical vertebral malformation-malarticulation in horses. *J Am Vet Med Assoc* **237**(7): 812–822.

CASE 68

1 What is the diagnosis? Ascarid (*Parascaris equorum*) impaction of the small intestine.

2 How is the occurrence of this disease explained in foals that are routinely administered anthelmintic drugs? There is widespread resistance of *P. equorum* to ivermectin and moxidection in some geographical areas. Testing for possible anthelmintic resistance can be undertaken using a fecal egg count reduction test.

3 Describe the life cycle of this parasite. *P. equorum* has a direct, migratory life cycle. The adults reside unattached in the small intestine. Eggs are passed in the feces. The eggs are resilient and are the infective stage. The host ingests the embryonated egg and, under optimal conditions, the larval stage 2 (L2) can develop in 14 days. The L2 larvae then pass through the intestinal wall and transform into L3. The larvae then migrate to the liver via the hepatic portal vein, where they remain for approximately a week. They then enter the caudal vena cava and travel to the alveoli of the lungs, where they transform to L4. The larvae travel up the bronchi to the trachea and are then coughed up and swallowed. Finally, they return to the stomach and small intestine, where they mature into adults. The prepatent period for *P. equorum* is 12–16 weeks.

4 What other clinical signs can infestation with this parasite cause? Poor growth/failure to thrive and coughing.

Discussion

P. equorum has long been a concern, but with the recent development of anthelmintic resistance it has come back to the forefront as a top concern. Ascarids are well known for their potential to cause disease in foals. This includes not only the risk of intestinal impaction or perforation, but also disease resulting from liver damage or pneumonia from the parasite's larval migration. Disease is well described and fecal egg shedding is high in foals (up to about 1 year of age).

Intestinal disease results from very heavy adult worm burdens in one of two ways: either alive (the sheer mass and number of these very large parasites blocks the gut) or dead, following routine deworming (the large die-off of undigested parasites creates a blockage). If the ascarid mass is large enough, it may cause perforation of the gut and severe peritonitis, which is nearly always fatal. Ascarid impactions are difficult to treat medically and often require surgical intervention. Deworming of a foal suspected of having a heavy ascarid burden should be done with care to avoid impaction from sudden mass die-off of parasites. Interestingly, it is very rare to see ascarid disease or evidence of infection (i.e. egg shedding) in an adult non-Cushing's disease horse, suggesting that horses develop a natural immunity to this parasite as they age. Nonetheless, it is still of great concern in foals.

P. equorum has come to the forefront of discussion along with cyathostomins in recent years because of the emergence of anthelmintic resistance, and the associated increase in related foal disease. Care must be taken when designing a parasite control program for farms with young animals to preserve anthelmintic susceptibility whenever possible. This involves utilizing fecal egg counts, reduction tests, and egg reappearance times as well as careful consideration of general management practices (i.e. manure and pasture management).

References

Lester HE, Matthews JB (2014) Fecal worm egg count analysis for targeting anthelmintic treatment in horses: points to consider. *Equine Vet J* **46**:139–145.

Peregrine AS, Molento MB, Kaplan RM *et al.* (2014) Anthelmintic resistance in important parasites of horses: does it really matter? *Vet Parasit* **201**:1–8.

von Samson-Himmelstjerna G (2012) Anthelmintic resistance in equine parasites: detection, potential clinical relevance and implications for control. *Vet Parasit* **185**:2–8.

CASE 69

1 Based on the history and appearance of the lesions, what is the most likely diagnosis? Erythema multiforme (EM).

2 What procedure(s) should be performed to help confirm the diagnosis? A skin biopsy.

3 What treatments should recommended for this horse? The owners were concerned about steroid-induced laminitis and the disorder was present for many months. Therefore, treating with azathioprine (3 mg/kg PO q24h) for at least 3 weeks, and then trying tapering the dose, would be recommended.

Discussion

EM is an acute and often self-limiting urticarial dermatitis. The urticarial lesions are usually symmetrical, frequently raised at the lesion periphery, and do not pit like classic equine urticaria. Scaling, crusting, and alopecia would not be expected with EM, which helps separate this disease from pemphigus. The lesions can be painful, are not usually pruritic, and are most common on the neck, thorax, topline, and rump. Mucocutaneous involvement may be seen in some cases. EM causes apoptosis and necrosis of keratinocytes. It is believed to be caused by an immunologic (T-cell) reaction to a variety of antigens (e.g. bacteria, viruses, drugs). There is a recent case report on a horse with EM that was believed to be caused by EHV-5. In rare cases, the hives develop necrosis in the center of the lesions on the back or rump area (**69.2**). Horses with minor forms of EM may recover in a few months with no treatment, while

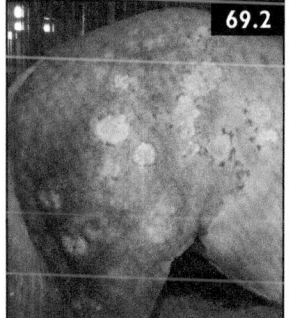

69.2

others persist and require long-term immunosuppressive therapy. Corticosteroids and azathioprine have been the preferred treatments. The author has had cases of EM that responded poorly to corticosteroids but had a good response to azathioprine at 3 mg/kg.

References

Divers TJ (2010) Azathioprine – a useful treatment for immune-mediated disorders in the horse? *Equine Vet Educ* **22**(10):501–502.

Herder V, Barsnick R, Walliser U *et al.* (2011) Equid herpesvirus 5-associated dermatitis in a horse resembling herpes-associated erythema multiforme. *Vet Microbiol* **155**:420–424.

Oryan A, Ghane M, Ahmadi N (2010) Erythema multiforme and its clinicopathological disorders in a horse. *Comp Clin Pathol* **19**(2):179–184

Scott DW, Miller WH (1998) Erythema multiforme in the horse: literature review and report of 9 cases (1988–1996). *Equine Pract* **20**(6):6–9.

White SD, Affolter VK, Dewey J *et al.* (2009) Cutaneous vasculitis in equines: a retrospective study of 72 cases. *Vet Dermatol* **20**(5–6):600–606.

CASE 70

1 Based on the history, what are the additional differentials? Include *Clostridium difficile* colitis, other antimicrobial-induced colitis, salmonellosis, and right dorsal colitis.

2 What diagnostics should be performed to determine the cause? Test the feces for *C. difficile* toxin (positive), *C. perfringens* enterotoxin (negative), and *Salmonella* (negative). Ultrasonography of the right dorsal colon (0.4 cm thick) to help rule out right dorsal colitis.

3 What treatment should be recommended for the colitis? Discontinue the penicillin therapy (4 days of treatment and adequate drainage were likely adequate), begin metronidazole (20 mg/kg PO q6h), administer Bio-Sponge® for enteric toxin neutralization, and provide supportive care (IV crystalloids, polymyxin B, and continue flunixin meglumine at 0.3 mg/kg q8h).

4 What are the most common life-threatening complications associated with any inflammatory colitis in the adult horse? The biggest complication would be systemic inflammatory response syndrome-associated laminitis. Additional complications would include AKI, septic shock, thrombophlebitis, and DIC.

5 What is the single most important treatment that should be incorporated to help prevent this complication? Cryotherapy of the distal limbs (**70.1**).

6 The horse stabilized but continued to have very watery diarrhea and a poor appetite, and the plasma protein decreased (but remained low normal). Is there any other treatment that could be highly effective in treating this diarrhea? Yes, fecal

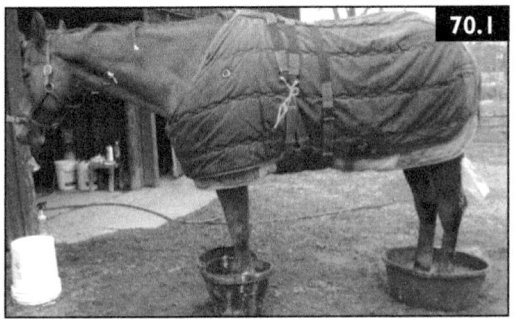

transfaunation should be performed. Cecal or fecal contents can be obtained from a healthy horse and administered via a nasogastric tube. In this case, cecal contents were obtained from a systemically healthy horse that was euthanized for orthopedic reasons. Three quarts of cecal contents were given once. The next day the horse had an excellent appetite and the manure is shown (**70.2**).

Discussion
This case demonstrates some important points. First, antibiotics are a risk factor for *C. difficile* colitis in horses. *C. difficile* colitis occurs most commonly during or following administration of antibiotics, often occurring 3–4 days after the initiation of antibiotic therapy. An additional risk factor is witholding feed (i.e. for anesthesia). Colic, abdominal bloat, and ileus may be more common with *C. difficile* colitis than with salmonellosis or Potomac horse fever (author's personal opinion), and *C. difficile* colitis may have a higher incidence at referral centers and broodmare farms than other equine facilities. Specific treatments include oral metronidazole, Bio-Sponge®, and gastric administration of fecal cocktails. Metronidazole-resistant strains do occur, and have mostly been found on the West coast of the US. The use of other antibiotics (IV or IM) to protect against bacterial translocation is controversial; if used, they should be antimicrobials and at dosages that have the least possible effect on normal intestinal flora. A variable (depending on the location) percentage of healthy horses carry toxigenic strains of *C. difficile*. The second important point from this case is that bacteriotherapy might be an important treatment for *C. difficile* infections. In this case, cecal transfaunation may have saved the horse's life.

References
Guo B, Harstall C, Louie T *et al.* (2012) Systematic review: faecal transplantation for the treatment of *Clostridium difficile*-associated disease. *Aliment Pharmacol Ther* **35**(8):865–875.

Medina-Torres CE, Weese JS, Staempfli HR (2011) Prevalence of *Clostridium difficile* in horses. *Vet Microbiol* **152**(1–2):212–215.

Reesink HL, Divers TJ, Bookbinder LC *et al.* (2012) Measurement of digital laminar and venous temperatures as a means of comparing three methods of topically applied cold treatment for digits of horses. *Am J Vet Res* **73**(6): 860–866.

Thean S, Elliott B, Riley TV (2011) *Clostridium difficile* in horses in Australia: a preliminary study. *J Med Microbiol* **60**(Pt 8):1188–1192.

Van Eps AW, Pollitt CC (2009) Equine laminitis model: cryotherapy reduces the severity of lesions evaluated seven days after induction with oligofructose. *Equine Vet J* **41**(8):741–746.

CASE 71

1 What are the differential diagnoses in this case? Include mandibular fracture, tooth root abscess, lumpy jaw/osteomyelitis, and neoplasia.

2 What diagnostic procedures should be considered? Complete oral examination, radiographs +/– ultrasound of the draining tract, biopsy, and culture should be performed.

3 What is the radiographic diagnosis (71.3)? The radiograph shows a possible mandibular fracture of the lower left diastema with a sequestrum or osteomyelitis.

Follow up/discusion

A large biopsy revealed remodeling of bone. A sample was obtained for culture and the horse was treated with trimethoprim sulfa for 3 weeks with no response. Culture results reported 2 weeks later demonstrated an *Actinomyces* spp.

Based on the culture results, the horse was treated with sodium iodide (40 g IV q24h) for 4 days and then potassium iodide (60 g PO q24h) for 2 weeks with minimal improvement. The mare was then anesthetized and her second premolar (307) was removed via buccotomy and the open socket was packed with polyvinyl siloxane impression material. An elliptical skin incision was made over the draining tract and the mandible at the site was curetted. The surgical site was packed with betadine-soaked gauze and allowed to heal by second intention. Trimethoprim/sulfa and potassium iodide treatments were continued. She was rechecked 6 weeks later and healing had occurred.

Mandibular cheek teeth abscesses are not as common as maxillary tooth abscesses. Swelling of the mandible or mandibular area is the most common presenting complaint. Indications for extraction include dental fractures, end-stage periodontal disease, severe tooth decay, and periapical abscess. In many cases, the diseased tooth becomes loose and can be more easily extracted. Complications associated with extraction are not unusual. Radiographs are usually sufficient to identify the diseased tooth, but CT may be required in some cases. Tooth pulp involvement, lamina dura destruction, bone fragments, and cortical bone destruction are more conspicuous with CT. *Actinomyces* is a common bacterial pathogen in chronic infections of the mouth and, occasionally, the mandibular lymph nodes. Bone biopsy to rule out neoplasia is not unreasonable, as the mandible is a common site of neoplasia in the horse. Sodium iodide is commonly used for treating *Actinomyces* infection in horses and cattle. The treatment is safe as long as there is no perivascular injection. Fever associated with treatment is common, but horses do not have the other typical signs of iodism seem in cattle (lacrimation, nasal discharge, and dandruff-like skin reaction).

References

Bailey GD, Love DN (1991) Oral associated bacterial infection in horses: studies on the normal anaerobic flora from the pharyngeal tonsillar surface and its association with lower respiratory tract and paraoral infections. *Vet Microbiol* **26**(4):367–379.

Vos NJ (2007) Actinomycosis of the mandible, mimicking a malignancy in a horse. *Can Vet J* **48**(12):1261–1263.

CASE 72

1 What are the three most likely differential diagnoses? Hypocalcemia, myopathy, and tetanus.

2 Given the high index of suspicion, what would be the first treatment? Given the high index of suspicion, calcium supplementation would be recommended. Calcium should be given slowly IV, and careful monitoring of the HR during treatment is recommended.

3 What else should be taken into consideration with this case, given the signalment and history? Given that the mare was overweight and possibly late-term pregnant, it would be important to check for concurrent hyperlipemia.

Follow up/discusion

In this case, palpation of the muscles was not painful, as it would have been with a myopathy, and tetanus was unlikely because there was no evidence of third eyelid prolapse or elevated tail head and the mare was well-vaccinated against tetanus. Hypocalcemia was considered most likely given the improvement when recumbent and the classic goose-stepping hindlimb gait. Hypocalcemia can occur in late pregnancy or early lactation, usually due to decreased calcium intake/delayed calcium mobilization and increased body requirements for calcium.

Hyperlipemia is common in overweight ponies, especially during late pregnancy or early lactation when energy demands are higher. In this case, the serum was lipemic and triglycerides were increased at 665 mg/dl (7.51 mmol/l). Aggressive treatment with IV fluids, enteral feeding via a nasogastric tube, and correction of the hypocalcemia improved the mare's appetite and her triglyceride levels returned to normal after 4 days.

CASE 73

1 What is the diagnosis? The history, laboratory findings, results of peritoneal fluid analysis, and rectal palpation are all consistent with a diagnosis of an abdominal abscess. The ultrasonographic appearance is consistent with an abscess, showing an ill-defined mass with heterogeneous echogenicity and hyperechoic foci.

2 What is the most likely cause? Although there are several possible causes of abdominal abscesses (including foreign bodies, external trauma, neoplasia, intestinal perforation), in the absence of any other relevant history, metastatic spread of *Streptococcus equi* subsp. *equi* ('bastard strangles') should be considered.

3 What other diagnostics should be recommended? Bacterial culture of peritoneal fluid should be performed. Serologic evaluation for antibodies to *Streptococcus equi* subsp. *equi* may be helpful. Endoscopic examination of the guttural pouches and culture of guttural pouch lavage samples should be considered to identify a carrier state. Laparoscopic examination or exploratory celiotomy may be considered.

4 What treatment options should be considered? Prolonged penicillin therapy may be successful in eliminating infection in small abscesses. For larger abscesses, surgical drainage and lavage may be required.

Discussion

Determining the cause of chronic colic is often difficult and frustrating. The interval between bouts of colic may be quite prolonged (weeks or months) and the intensity of pain exhibited may be variable between episodes. While the source of pain generally stems from the GI tract, the cause for that pain may be intra- or extraluminal and disease of other abdominal or pelvic organs, or of the peritoneal cavity itself, should be considered. Attempts at diagnosis for chronic colic may be unrewarding owing to the transient nature of the pain and inaccessibility of much of the equine abdomen to rectal palpation and ultrasound. However, in foals/weanlings, young horses, and some small ponies, ultrasound can offer a much more thorough evaluation of the abdomen and takes the place of rectal palpation, which is not possible in these smaller animals. Other initial diagnostic procedures are similar to those used for acute colic, expanding to a more systemic evaluation. If the bouts of colic worsen either progressively or acutely, or other diagnostics do not reveal a likely cause, surgical or laparoscopic exploration +/- biopsies may be indicated. As with all cases, but particularly for chronic colic, a thorough history is very important as a precipitating cause may be identified.

Abdominal abcesses are not an uncommon cause of chronic colic and can occur without any obvious recent history or precipitating event (e.g. trauma). They generally are accompanied by varying degrees of peritonitis, as is this case. Abscesses of any cause can be difficult to treat, most likely due to poor penetration of the abscess by antibiotics (whether given PO or IV). Binding and potential inactivation of the antimicrobial by the WBCs within the abscess may also be a factor. If antibiotic therapy is attempted, it should be based on culture and sensitivity results from abdominal fluid (or abscess if it is adhered to the body wall or otherwise easily accessible, as in surgery). Furthermore, antibiotics are likely to be needed for over 1 month (up to 6 months) to be effective. Depending on the size, number, and location of abscesses, surgical drainage and flushing may be possible or necessary.

Abdominal or thoracic abscess formation is a well known but uncommon complication of *S. equi* infection. In most cases there is swelling and abscess formation of the submandibular, submaxillary, and/or retropharyngeal lymph nodes following an acute fever and onset of nasal discharge. Older horses, such as the one in this case, are more likely to have at least partial immunity from previous exposure or possibly vaccination (although there was no mention of vaccination here), which may lessen the severity of clinical signs, possibly going unnoticed by the owner. While the mechanism is not understood, it is also possible for a horse to become infected with *S. equi* and be asymptomatic. These horses shed the bacteria and become a source of infection for other horses. This horse may fit into one of these categories, hence the lack of history regarding possible causes.

References
Arnold CE, Chaffin MK (2012) Abdominal abscesses in adult horses: 61 cases (1993–008). *J Am Vet Med Assoc* **241(12)**:1659–1665.
Berlin D, Kelmer G, Steinman A *et al.* (2013) Successful medical management of intra-abdominal abscesses in 4 adult horses. *Can Vet J* **54**:157–161.
Mair TS, Sherlock CE (2011) Surgical drainage and post-operative lavage of large abdominal abscesses in six mature horses. *Equine Vet J* Suppl. **39**:123–127.
Rumbaugh GE, Smith BP, Carlson GP (1978) Internal abdominal abscesses in the horse: a study of 25 cases. *J Am Vet Med Assoc* **198**:1045–1048.

CASE 74

1 What are the most likely differential diagnoses? Given the large size of the foal and presence of a swelling on the left thorax, thoracic trauma associated with foaling was highly likely. Clinical examination findings of pale mucous membranes and decreased lung sounds should make you suspicious of hemorrhage into a body cavity, most likely the thorax. Rib fractures occur commonly during foaling and most are undetected because they do not cause clinical signs. However, clinical signs can be seen if the fracture becomes displaced and hemorrhage occurs. Interestingly, the left side is more commonly affected.

2 What diagnostic tests should be chosen first? Thoracic ultrasound (74.2) is very useful in diagnosing rib fractures, and can also reveal any hematomas,

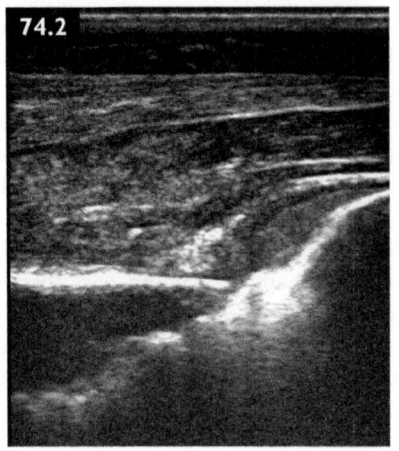

74.2

hemothorax, or lung pathology, as were seen in this case. Fractures typically occur around the costochondral junction, and the 3rd–8th ribs are most commonly affected. Radiographs and careful palpation of the rib cage can also be used. Hematology confirmed anemia with a PCV of 22% and TS of 4.6 g/dl (46 g/l), consistent with blood loss anemia of >12 hours' duration.

3 What treatment options should be discussed with the owner? Treatment options depend on the number of ribs affected and the degree of displacement. Generally, non-displaced fractures can be treated conservatively with stall rest. Surgical stabilization is warranted if the fractures are severely displaced, are causing damage to internal structures (heart/lungs), or if a flail chest is present (free floating segment).

References
Jean D, Laverty S, Halley J *et al*. (1999) Thoracic trauma in newborn foals. *Equine Vet J* **31**(2):149–152.
Jean D, Picandet V, Macieira S *et al*. (2007) Detection of rib trauma in newborn foals in an equine critical care unit: a comparison of ultrasonography, radiography and physical examination. *Equine Vet J* **39**(2):158–163.

CASE 75
1 What diagnostic procedure(s) should be performed next? Radiography and ultrasonography of the stifle joint to determine if there is evidence of septic osteoarthritis or other bony abnormalities such as cartilage defects or fracture. Radiographs showed soft tissue swelling of the joint with no bony abnormalities.
2 What is the interpretation of the synovial fluid analysis and cytology? It is consistent with marked suppurative inflammation. Although no infectious agents were identified on Wright's- or Gram-stained smears, infectious causes could not be ruled out. This is especially important to consider based on the history. Bacteria are typically seen in synovial fluid from a septic joint unless the sepsis is due to hematogenous infection, in which case the organisms may reside mostly in the synovial membranes.
3 What else should be done with the synovial fluid? Aerobic culture and sensitivity.
4 If there could be a reasonable chance of sepsis in this case, and the owner wants the horse to return to performance soundness, should arthroscopic surgery and flushing the joint, in addition to local and systemic antibiotics, be recommended? Yes.

Arthroscopic surgery was performed under GA to flush fibrin from the medial and lateral components of the joint. The mare was started on IV penicillin and gentamicin therapy.

5 What are the options for therapy in light of the culture results? Treatment for the yeast infection could include local flushes with miconazole, enilconazole, vorconazole, or nystatin solution. Amphotericin may be irritating to the synovium so should be avoided. If these antifungal drugs are not available, dilute betadine (1%) could be used as a flush. Options for systemic therapy include: fluconazole (reasonably well absorbed in horses and economical); voriconazole (better absorbed and a high range of sensitivity against various fungal organisms, but expensive); or itraconazole (may not have as good bioavailability but has good distribution once absorbed). Ketoconazole has poor bioavailability and is not recommended. Antibiotics should be discontinued if there is no bacterial growth or at least limited to a narrow-spectrum antibiotic that may be effective against *Staphlococcus*.

Follow up/discusion

The mare was treated with fluconazole systemically (14 mg/kg as a loading dose PO and then 5 mg/kg q24h thereafter) and 200 mg voriconazole and 250 mg amikacin diluted in saline injected intra-articularly every day for 5 days. Ten days later the joint was less swollen, the WBCs in the joint fluid had decreased to 1,700 PMNs/µl (non-degenerative) and the culture was negative. The horse was discharged and treatment with fluconazole was continued for 2 more weeks. Two follow-up examinations, at 1 and 3 months, found the horse to be improving but still mildly lame at the jog and joint fluid indicated ongoing synovitis without sepsis. The joint was injected with hyaluronic acid at monthly intervals for 3 months.

Fungal arthritis is a rare condition in horses, although infections may be seen in areas endemic for coccidioimycosis and blastomycosis, often in the intervertebral joints. *Candida* arthritis has been reported in foals, likely caused by systemic invasion from intestinal colonization. An accurate prognosis is difficult to provide due to the limited number of cases reported. The combination of systemic and local antifungal therapy was successful in treating the arthritis in this case.

References

Gamaletsou MN, Kontoyiannis DP, Sipsas NV *et al.* (2012) *Candida* osteomyelitis: analysis of 207 pediatric and adult cases (1970–2011). *Clin Infect Dis* **55**(10):1338–1351.

Higgins JC, Pusterla N, Pappagianis D (2007) Comparison of *Coccidioides immitis* serological antibody titres between forms of clinical coccidioidomycosis in horses. *Vet J* **173**(1):118–123.

Swerczek TW, Donahue JM, Hunt RJ (2001) *Scedosporium prolificans* infection associated with arthritis and osteomyelitis in a horse. *J Am Vet Med Assoc* **218**(11):1800–1802.

CASE 76

1 What is the most likely explanation for the high fibrinogen, low serum iron, and elevated serum globulins? These are markers of acute (fibrinogen and iron) and chronic (all three) inflammation.

2 What is the most likely diagnosis in this case? Chronic progressive lymphedema.

3 How can this condition be treated? There is no treatment cure but controlling secondary bacterial infection and the chorioptic mange can be helpful. Clipping the leg feathers carefully and soaking the distal limbs in diluted betadine or chlorhexidine for a couple of days in hopes of removing a large amount of scales and crusts is often a helpful start. After that, the leg should be kept clean and dry, and all four limbs should be sprayed from the carpus or tarsus distally with fipronil spray (125 ml/leg) twice 3–4 weeks apart. (**Note:** This product cannot be legally used in horses in some countries.) Trimethoprim sulfa could be used to help control bacterial infections.

Discussion

Chronic pastern dermatitis, also known as chronic progressive lymphedema, is a skin disease that affects Draft horses. It has been found in numerous horses within the Shire, Clydesdale, Belgian Draft, and gypsy cob breeds. The disorder is caused by abnormal lymphatics in all four limbs. It starts at an early age as edema of the limb and is progressive. Early lesions include skin thickening, slight crusting, and skin folds at the pastern, which are often missed because of heavy feathering. Secondary infections ('greasy heel' and 'scratches') are a frequent occurrence. The disease causes painful lower leg swelling, nodule formation, excessive skin folds that can progress to verruciform lesions, and skin ulceration, interfering with movement. Treatments are intended to improve the clinical signs and slow the progression of the disease. Every effort should be made to control the common secondary bacterial infections and chorioptic mange. Aggressive efforts should be made in the early part of the disease process to decrease edema by compression bandages, massages, and forced exercise in order to reduce edema and soften fibrosis. Following maximum edema reduction, the horse should be kept in equine compression stockings. Chronic progressive lymphedema appears to be hereditary but no gene has yet been discovered that would allow genetic testing.

References

DeCock HE, Affolter VK, Wisner ER *et al.* (2003) Progressive swelling, hyperkeratosis, and fibrosis of distal limbs in Clydesdales, Shires, and Belgian draft horses, suggestive of primary lymphedema. *Lymphat Res Biol* 1(3):191–199.

Powell H (2009) Therapy for horses with chronic progressive lymphoedema. *Vet Rec* 165(25):758.

Powell H, Affolter VK (2012) Combined decongestive therapy including equine manual lymph drainage to assist management of chronic progressive lymphoedema in draught horses. *Equine Vet Educ* 24:81–89.

CASE 77

1 What five diagnostic tests should be performed next? Hematology and biochemisty; ultrasonography; duodenoscopy; biopsy; glucose absorption test.

2 List the most likely differential diagnoses. The most likely diagnoses given the history and clinical signs are inflammatory bowel disease with systemic involvement (e.g. multisystemic eosinophilic epitheliotropic disease [MEED]), generalized lymphoma, or sarcoidosis.

3 What advice should the owner be given regarding the likely response to treatment? The response to treatment for inflammatory bowel disease in horses has typically been variable but often poor. Treatment involves immunosuppression with dexamethasone given parenterally (IV or IM), as oral absorption of medication is poor in these cases. Treatment should persist for several months as the dose of dexamethasone is tapered slowly. Concurrent treatment with gastroprotectants (sucralfate and omeprazole) is recommended given the severity of ulceration typically present in the stomach and duodenum. Owners should be warned of the likelihood of a poor response and the need for long-term or indefinite treatment depending on the response.

Follow up/discussion
In this case, there was mild anemia, hypoalbuminemia, and mild increases in liver enzymes (GGT, GLDH, SDH, AST, ALP). Abdominal ultrasound revealed thickening of the small intestinal wall, especially the duodenum, with some areas measuring 12 mm thick (normal <4 mm). Feed was withheld for 12 hours and duodenoscopy (77.2) revealed diffuse ulceration and thickening of the duodenal muscosa, with large areas of fibrinonecrotic membrane. Biopsies of the abnormal areas (duodenum, skin, and lymph node) confirmed the diagnosis of MEED. A glucose absorption test was performed and revealed poor absorption <50% (normal >80% or more increase in blood glucose).

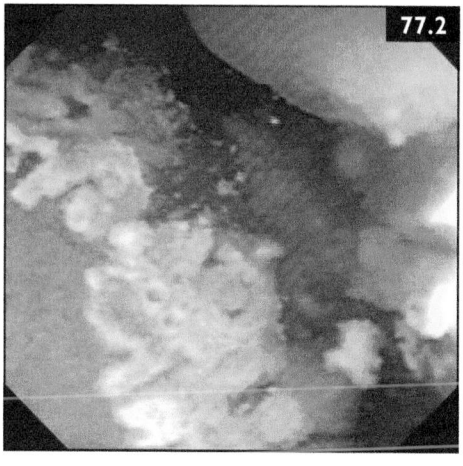

77.2

Reference
Bosseler L, Verryken K, Bauwens C *et al.* (2013) Equine multisystemic eosinophilic epitheliotrophic disease: a case report and review of literature. *N Z Vet J* **61**(3):177–182.

CASE 78

1 List some differentials for the chronic colic and weight loss in the face of an otherwise normal physical examination and normal laboratory testing? Include abdominal gastrointestinal neoplasia and gastric ulceration. The possibility of abdominal neoplasia was more likely due to the duration of signs, age, and use of the mare. An intra-abdominal abscess was unlikely due to normal CBC, fibrinogen, and TS. Parasitism was also unlikely because of the excellent deworming history.

2 What other procedures should be performed to arrive at a diagnosis? Abdominocentesis with peritoneal fluid evaluation and percutaneous biopsy of the mass.

Follow up/discussion

The peritoneal fluid was a mixed inflammatory fluid (6,500 cells/µl (6.5 × 10⁹/l) with a TP concentration of 2.4 g/dl (24 g/l) and a predominance of neutrophils, reactive macrophages, and mesothelial cells. There were no neoplastic cells observed but erythrophagia was noted. Biopsy (**78.3**) characterized it as streaming spindle to stellate cells with no malignant features. The histopathologic diagnosis was a mesenchymal tumor, most likely a GI stromal tumor (GIST) arising from the colon wall.

The horse did well on grass pasture and equine senior feed for several months with only occasional treatment needed for sporadic mild colic. The horse was eventually euthanized because of a more severe episode of colic. There are a few types of abdominal mesenchymal tumors in horses, including GIST, leiomyoma, and myofibroblastic sarcoma. The most common mesenchymal tumor is GIST, which may be myogenic, neurogenic, or undifferentiated. These tumors are usually

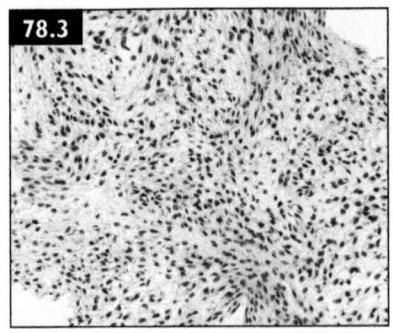

benign and do not metastasize but cause clinical problems due to their large size. The tumors are found in horses >12 years of age and the most common clinical signs are weight loss with chronic colic. The tumor can be felt by abdominal palpation per rectum in some cases and seen by abdominal ultrasound in most cases. Some horses can be managed, as was the horse in this case, until the tumor compromises quality of life by either repeated colic episodes or a severe colic episode.

References

Del Piero F, Summers BA, Cummings JF *et al.* (2001) Gastrointestinal stromal tumors in equids. *Vet Pathol* 38:689–697.

Hafner S, Harmon BG, King T (2001) Gastrointestinal stromal tumors of the equine cecum. *Vet Pathol* 38:242–46.

Muravnick KB, Parente EJ, Del Piero F (2009) An atypical equine gastrointestinal stromal tumor. *J Vet Diagn Invest* 21:387–390.

Tomlinson J (2009) Abdominal spindle cell tumor in an 18 year old mare. *Senior Seminar Paper*, Cornell University 9/02/2009 (T. Divers, advisor).

CASE 79

1 Describe the fundic examination abnormalities in this horse (79.1). Pale optic disk and vascular attenuation of retinal vessels consistent with optic nerve atrophy/degeneration.

CASE 80

1 What would be the likely differentials for these clinical signs? Include peritonitis and physical damage to the urethral sphincter or bladder associated with the dystocia, a vaginal or uterine tear, or a uterine hematoma. Another possibility would be physical damage to the small colon and peritonitis in addition to injury to the bladder and urethra from the dystocia. Herpes myelitis causes acute bladder dysfunction and fever but ataxia would be expected, which was not present in this case.

2 What other diagnostic procedures could performed? Vaginal speculum examination (bruising with marked discoloration was found); gentle palpation of the uterine body via vaginal examination (no abnormalities detected); abdominal palpation per rectum (small colon and uterus normal but the bladder was abnormally enlarged and expressed easily without the mare posturing to urinate); transrectal ultrasound examination (no obvious abnormalities seen); abdominocentesis of the increased peritoneal fluid that is easily seen next to the involuting uterus on the ultrasonogram.

3 What would the peritoneal fluid analysis in a normal post-foaling mare be expected to show? WBCs <5,000/μl (5.0×10^9/l), TP <2.0 g/dl (20 g/l), amber color. This is no different from that found in normal horses.

4 How should this mare be treated? With broad-spectrum antibiotics (in this case, IV penicillin and gentamicin plus metronidazole PO) because of the possibility of septic peritonitis.

Follow up/discusion

The peritoneal fluid revealed: WBCs = 390,000/μl (390×10^9/l) with neutrophils 87%; no bacteria; TP = 5.7 g/dl (57 g/l); RBCs = 680,000/μl (0.68×10^{12}/l), dark red color. CBC of the mare was performed: PCV, TP and fibrinogen were all normal but the neutrophil cell count was high (9,200/μl [9.2×10^9/l]) and there were 400 band neutrophils/μl (0.4×10^9/l).

A Foley catheter was placed in the bladder and bethanechol (12 mg SC q9h) was administered to combat the bladder incontinence and atony. Flunixin meglumine was continued and the mare was given 1 gallon of mineral oil via nasogastric tube as a laxative. The mare was administered oxytocin (5 U IM q4h) for 1 day to improve uterine contraction in case there was a very small perforation. The mare was bright and alert with normal vital signs and appetite within 24 hours of initiating treatment. Bladder dysfunction continued for 72 hours but then resolved. A small uterine tear plus peritonitis and injury to the nervous innervation arising from sacral spinal cord segments (S2–S4) to the bladder and urethral sphincter were presumed to be responsible for the clinical findings in the mare.

Peritonitis following foaling occurs occasionally. Causes include perforation of the uterus, vaginal tear, small colon injury (presumably from being 'caught' against the fetus and pelvis during delivery), ruptured bladder, and cecal perforation. Onset of fever and clinical signs 24 hours after foaling would be most compatible with reproductive tract or small colon injury causing peritonitis. Urinary bladder rupture may not cause fever and clinical signs may not be noticed as early as 24 hours post foaling. Cecal rupture usually occurs within the first 3 days and causes signs of acute shock immediately following the rupture. These problems must be ruled out by clinical, laboratory, and ultrasound examination, and must be surgically corrected if necessary. Medical therapy alone may be appropriate for small colon impaction and, occasionally, small dorsal tears of the uterus and/or bladder.

Reference

Javsicas LH, Giguère S, Freeman DE *et al.* (2010) Comparison of surgical and medical treatment of 49 postpartum mares with presumptive or confirmed uterine tears. *Vet Surg* **39**:254–260.

CASE 81

1 What other questions should the owner be asked? What is the reproductive history of other animals on the farm. This farm had a history of EEL affecting four other mares, and two mid- to late-term abortions. The owner reported a severe infestation of tent caterpillars at that time. One other case of pericarditis had been diagnosed 2 years previously but a cause was not identified.

2 What diagnostic tests should be performed initially? Echocardiography (81.2), which revealed severe pericardial effusion causing cardiac tamponade, with 28 liters of fluid subsequently drained off the pericardial sac. The fluid was described as septic suppurative exudate, with intracellular gram-negative rods seen. The horse was euthanized because of the poor prognosis. At necropsy, a 2 cm thick layer of fibrin was present on the epicardium, causing restriction of cardiac function (81.3).

3 What are the most likely pathogens associated with this problem? Given the combination of historical reproductive disease and bacterial pericarditis, a diagnosis of mare reproductive loss syndrome was made. This disease is secondary

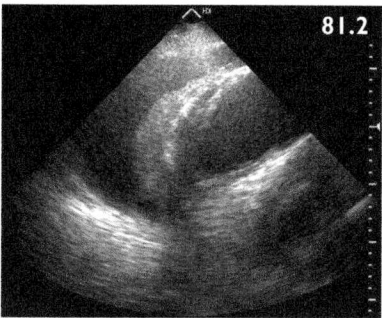
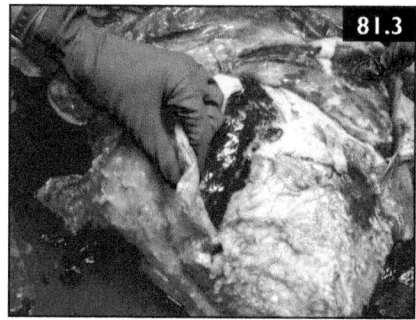

to ingestion of tent caterpillars, with subsequent migration of the caterpillar spines through internal organs including the uterus and pericardium. Clinical signs include EEL, abortion, and stillbirths. Pericarditis and endophthalmitis are less common sequelae. Commonly associated bacterial species include *Actinobacillus* spp., *Streptococcus zooepidemicus, E. coli,* and *Enterococcus.*

References
Bolin DC, Donahue JM, Vickers ML *et al.* (2005) Microbiologic and pathologic findings in an epidemic of equine pericarditis. *J Vet Diagn Invest* **17**(1):38–44.
Sebastian MM, Bernard WV, Riddle TW *et al.* (2008) Review paper: mare reproductive loss syndrome. *Vet Pathol* **45**(5):710–722.

CASE 82
1 What is equine Cushing's disease, and what clinical signs are commonly associated with it? Equine Cushing's disease is also known as pituitary pars intermedia dysfunction (PPID). It is a prominent cause of morbidity in mature to elderly horses and ponies and may affect >20% of horses aged ≥15 years. This disease is a neurodegenerative condition that progresses over time and occurs in its advanced form in old horses. Early signs of PPID include laminitis, abnormal hair-shedding patterns, muscle atrophy leading to a 'pot-belly' appearance, abnormal fat distribution (especially periorbital), and lethargy. Other signs that may be seen in association with PPID include hair color fading or changing, polydipsia/polyuria, excessive or decreased sweating, susceptibility to secondary infections, infertility, and, rarely, seizure-like activity.

2 How can the diagnosis be confirmed? Current commonly used diagnostic tests for PPID include basal plasma adrenocorticotropic hormone (ACTH) concentration, overnight dexamethasone suppression test (ODST), and the thyrotropin-releasing hormone (TRH) stimulation test (measuring ACTH).

3 Should this pony be treated, and if so, how? Detection of early PPID is important because the disorder is easily managed with pergolide treatment. Treatment is advisable to prevent laminitis, rather than wait for laminitis to develop before treating, because laminitis is so difficult to manage.

Discussion

The exact pathogenesis of PPID is unknown, but the result is loss of negative dopamine control of endocrine function of the pars intermedia, leading to overproduction of proopiomelanocortin (POMC)-derived peptides (including ACTH), which in turn produces the clinical signs noted by owners. Adenomas are commonly found within the pituitary of affected horses at necropsy and at one time were thought to be the cause of this progressive disease, but more recent studies have shown that this is unlikely and that PPID is more likely an age-related degeneration. Hirsutism is considered pathognomonic for PPID, as it does not occur with any other disease in horses and thus may be as accurate as laboratory diagnostics. Laminitis in horses with PPID is thought to be the result of insulin resistance, which has been well documented in some cases. There is no apparent breed or sex predilection. While PPID is well recognized to affect older horses, many cases are not properly diagnosed until later in the progression of disease, as many changes may appear gradually over a period of years and be erroneously considered by owners to be normal for aging horses. In particular, the loss of topline musculature or a pendulous belly, especially in a multiparous broodmare, or weight loss, muscle wasting, and lethargy in any aging horse may not be recognized as abnormal by an owner, leading to a delay in diagnosis. Owner education and encouragement to report any changes is important, as PPID is manageable and quality of life and use will be maintained best with early diagnosis before the development of laminitis. Pergolide is the therapeutic drug of choice. No cardiac abnormalities associated with pergolide have been reported in horses, although some horses will have a transient decrease in appetite at the onset of therapy. Cyproheptadine and trilostane have also been used for treatment, and can sometimes be helpful if pergolide therapy alone is ineffective. To date, no predisposing factors other than age have been identified. Similarly, no association between clinical or laboratory findings and long-term survival have been identified.

References

McFarlane D (2011) Equine pituitary pars intermedia dysfunction. *Vet Clin North Am Equine Pract* **27**(1):93–113.

McGowan TW, Pinchbeck GP, McGowan CM (2013) Prevalence, risk factors and clinical signs predictive for equine pituitary pars intermedia dysfunction in aged horses. *Equine Vet J* **45**(1):74–79.

Rohrbach BW, Stafford JR, Clermont RS *et al.* (2012) Diagnostic frequency, response to therapy, and long-term prognosis among horses and ponies with pituitary par intermedia dysfunction, 1993–2004. *J Vet Intern Med* **26**(4):1027–1034.

CASE 83

1 Interpret the radiograph. A large soft tissue mass can be seen that has spherically eroded through the dorsolateral aspect of the coffin bone (distal phalanx).
2 What is the most likely diagnosis? A keratoma.
3 What would be the preferred treatment? Surgical removal via a partial dorsolateral wall resection performed either standing or under GA.
4 Because of the severity of the lameness in the left hindlimb, what would be a major concern for the right hindlimb? Support limb laminitis.
5 What could be done to help prevent this condition? As in this case, the most important preventive for support limb laminitis is to properly treat the lame leg to decrease the lameness and improve weight-bearing. If the severe lameness had persisted, treatments such as NSAID therapy, morphine epidural, and intermittent nerve block to the lame leg could have been performed to 'rest' the opposite leg from bearing nearly 100% of the hind end weight. Another option would have been to intermittently place the horse in a sling such that some of the weight-bearing was relieved from the hindlimb. Lastly, the right rear foot could have padding or commercial boots applied to the bottom of the foot to help more equally distribute the weight between the hoof wall and sole.

Follow up/discussion

Surgery was performed and the keratoma literally 'popped out' following removal of a portion of the hoof wall. The hoof wall defect created by resection was packed with iodine-soaked sponges and bandaged. The horse was treated with trimethoprim-sulfa and phenylbutazone. The day following surgery, an egg bar shoe with clips and a treatment plate were applied to the foot. The horse's lameness improved dramatically within 36 hours of surgery and he was discharged with instructions for soaking, bandaging, and shoeing of the hoof to allow proper healing of the hoof wall.

Keratomas are uncommon benign growths of the keratin-producing epidermal cells of the inner hoof wall. They can be cylindrical to spherical in shape and are generally found in the toe or quarters of the hoof. Keratomas are slow growing and lameness occurs when the mass enlarges to the point of causing excessive pressure on the laminar tissue and coffin bone (distal phalanx). The cause of keratoma formation is not known. Chronic abscesses and draining tracts are common sequelae. Surgical removal may result in complete recovery in nearly 90% of horses, although recovery time is prolonged (months). The prognosis is not as favorable if the coffin bone (distal phalanx) becomes infected or develops a sequestrum.

Reference

Boys Smith SJ, Clegg PD, Hughes I *et al.* (2006) Complete and partial hoof wall resection for keratoma removal: postoperative complications and final outcome in 26 horses (1994–2004). *Equine Vet J* 38(2):127–133.

Answers

CASE 84

1 What is the most likely diagnosis? Thoracic neoplasia. The heavily blood-stained modified transudate shows no evidence of being septic (i.e. not compatible with septic pleuritis). The reactive mesothelial cells seen on cytology and the hyperfibrinogenemia are compatible with neoplasia.

2 How should this horse be further evaluated? Repeated thoracic radiography and ultrasononography after drainage of the pleural fluid may sometimes be helpful in identifying abnormal masses. Pleuroscopic examination of the thoracic cavity and direct biopsy of any observed lesions may be helpful in establishing a firm diagnosis. In this case, mesothelioma was diagnosed.

Discusssion

Mesothelium is the monolayer of epithelium lining the visceral and parietal surfaces of the thorax, pericardium, and abdomen. Inflammation or mechanical irritation of the mesothelium may result in cellular hypertrophy and/or hyperplasia. Malignancies arising from mesothelium are known as mesotheliomas and are very rare in horses. While there is an established association between asbestos and human mesothelioma, no etiologic agents have been identified for equine mesothelioma. There are three main histologic types of mesothelioma: biphasic, sarcomatoid, and epithelioid. Epthelioid mesotheliomas have several known variants, including tubopapillary, acinar, polygonal, microcystic, clear cell, deciduoid, desmoplastic, and lymphohistiocytic. Clinical signs depend on the location of the tumor's origin and diagnosis may be difficult, but most cases will present with significant effusion of one or more coelomic cavities. Thickening of the pleural, pericardial, or abdominal lining may or may not be visible on ultrasound examination, and in most cases hematology and biochemistry results will be unremarkable, aside from possibly hyperfibrinogenemia. Evaluation of the effusion typically reveals a modified transudate, with or without evidence of mesothelial hyperplasia or reactive mesothelial cells, which may represent either a neoplastic or marked inflammatory processes. The distinction between these two processes is very difficult based on fluid cytology alone. The main differentials for mesothelioma are metastatic adenocarcinoma and severe mesothelial reactivity. Definitive diagnosis is by biopsy of mass(es) submitted for histopathology and immunohistochemistry. Other neoplasms that may affect the thoracic cavity and viscera, either primarily or as metastatic disease, include lymphosarcoma, hemangiosarcoma, melanoma, granular cell tumor, adenocarcinoma, squamous cell carcinoma, chondrosarcoma, myxoma, thymoma/malignant thymic tumors, lipoma, and fibrosarcoma. Treatment is rarely feasible.

References

Dobromylskyj M, Copas V, Durham A *et al.* (2011) Disseminated lipid-rich peritoneal mesothelioma in a horse. *J Vet Diagn Invest* 23(3):615–618.

Fry MM, Magdesian KG, Judy CE *et al.* (2003) Antemortem diagnosis of equine mesothelioma by pleural biopsy. *Equine Vet J* 35(7):723–727.

Stoica G, Cohen N, Mendes O *et al.* (2004) Use of immunohistochemical marker calretinin in the diagnosis of a diffuse malignant metastatic mesothelioma in an equine. *J Vet Diagn Invest* 16(3):240–243.

CASE 85

1 What is the diagnosis? This is a deep stromal abscess with extension into the anterior chamber. A posterior lamellar keratoplasty was performed to excise the fungal abscess. Prolonged medical therapy may be a suitable alternative to surgery for superficial or mid-stromal abscesses.

CASE 86

1 List some differentials for the recurrent bacterial respiratory disease and fever that responded to antibiotics but recurred. There could be an anatomic or functional problem in the pharynx causing chronic aspiration or an immunodeficiency.

2 What diagnostic test or procedure should be performed next? Airway endoscopy should be performed, which was normal in this case. Serum immunoglobulins should be measured (IgG = 1,957 mg/dl [19.57 g/l]; IgM = 19 mg/dl [0.19 g/l]). This made IgM deficiency a likely diagnosis for the chronic infections.

3 List some differentials for the bilateral hyphema and mildly thickened eyelids. Thrombocytopenia associated with drug administration should be considered as a possible cause for the hyphema, but the platelet count was normal and there were no petechiations of the membranes or bleeding elsewhere. Trauma would be possible, but there was no evidence or history of this and the eyelid thickening was bilateral and non-painful. It is very rare for lymphoma to cause hyphema, but it can commonly cause eyelid thickening and might help explain the IgM deficiency.

Follow up/discussion
Because of the lack of improvement in both eyes and the suspicion of neoplasia, the yearling was euthanized and lymphosarcoma was found bilaterally in the eyelids, posterior choroid, ciliary body, iris, and gastric and tracheobronchial lymph nodes. A large amount of blood was present in both the aqueous and vitreous bilaterally.

Ocular lesions may precede or be more obvious than peripheral lymph node enlargement in horses with ocular lymphoma. Infiltration of the palpebral conjunctiva and eyelids is most common. Retrobulbar masses, which are common in cattle with lymphoma, are not common in horses. Uveitis was reported in 4/21 cases in one study, but hyphema appears to be rare. Horses with the nodular form of extraocular lymphoma have a fair to good prognosis if the neoplastic tissue is completely excised.

References
Rebhun WC, Del Piero F (1998) Ocular lesions in horses with lymphosarcoma: 21 cases (1977–1997). *J Am Vet Med Assoc* **212(6)**:852–954.
Schnoke AT, Brooks DE, Wilkie DA *et al.* (2013) Extraocular lymphoma in the horse. *Vet Ophthalmol* **16(1)**:35–42.

CASE 87

1 What is the most likely cause of the increased creatinine in this colt? The initial assessment was that the elevated creatinine may have been a result of placentitis, but the remarkably high potassium with low sodium and chloride are not expected in foals born to mares with placentitis.

2 The dark colored urine was occult blood positive on urine dipstick examination. What are the possible causes for this finding? Hemoglobinuria, hematuria with lysis of RBCs, or myoglobinuria. A complete chemistry was performed and the CK was 599,280 U/l and AST was 12,880 U/l, which suggests that the pigmenturia was due to myoglobinuria.

3 Do these urinalysis and laboratory findings help explain the electrolyte abnormalities and elevated creatinine? Yes. Severe muscle disease causes hyponatremia, hypochloremia, and hyperkalemia. Renal failure with azotemia may ensue, which may cause further changes in these serum electrolytes.

4 The mare and foal are from the northeastern US, so what would be a suspected nutritional cause for the severe myopathy in this case? Nutritional (selenium deficient) myopathy.

Follow up/discussion
The foal's blood selenium result prior to treatment was <2.5 µg/dl. Treatment for white muscle disease, renal failure, and sepsis (*Enterobacter* was cultured from a blood sample) was initiated. This consisted of antimicrobials, IV plasma and crystalloids to maintain perfusion and urine output, selenium/vitamin E administration, and nutritional support. The foal's mentation improved within 2 days but he still could not stand when lifted. A second injection of selenium/vitamin E was given on day 3 and the selenium value was 5.15 µg/ml 2 days later, therefore a third injection was given. The third injection raised the serum selenium to 12.84 µg/dl, which is normal for a foal. The foal was able to stand on his own 19 days after treatment was begun, could nurse the mare, and made a complete recovery.

White muscle disease is a distinct entity in young horses, with numerous case reports documenting the lesions and their association with low selenium. Vitamin E levels are not closely associated with white muscle disease, even though affected foals are often concurrently administered vitamin E. White muscle disease may also occur in weanlings, yearlings, and adults in association with extremely low selenium levels. Skeletal and/or cardiac muscle can be involved. The tongue and muscles of the limbs are commonly involved in foals. An unusual masseter myopathy can be seen in adult horses.

Selenium deficiency occurs most often in areas with selenium-deficient soil. The development of this disease in foals <1 week of age would suggest that the deficiency is associated with poor maternal selenium status, but this has been difficult to document. The prognosis is variable in foals, although foals with noticeably swollen limb muscles rarely survive. (**Note:** As in this case, it can be difficult to determine the cause of discolored urine by appearance alone.)

References

Perkins G, Valberg SJ, Madigan JM *et al.* (1998) Electrolyte disturbances in foals with severe rhabdomyolysis. *J Vet Intern Med* **12**(3):173–177.

Streeter RM, Divers TJ, Mittel L *et al.* (2012) Selenium deficiency associations with gender, breed, serum vitamin E and creatine kinase, clinical signs and diagnoses in horses of different age groups: a retrospective examination 1996–2011. Equine Vet J **Suppl. 43**:31–35.

CASE 88

1 What are the most common causes of entropion in foals? Loss of orbital fat (as a result of cachexia) or due to dehydration from systemic illness; entropion can also be hereditary. Premature foals are more prone to entropion.

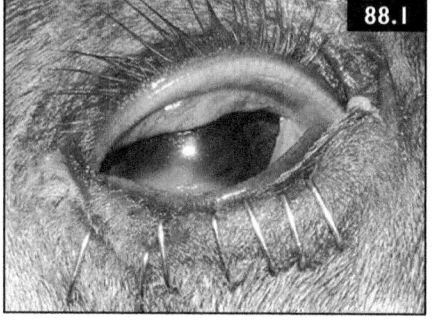

2 What treatment options are available? Include temporary tacking (suture or staples) (**88.1**); sutures are left in place for 2–3 weeks. Permanent surgical correction should be performed only if temporary tacking fails.

CASE 89

1 What conditions commonly result in acute swelling of the head? Acute cellulitis, emphysema (e.g. oropharyngeal or esophageal perforation, tracheal perforation), snake bite, vasculitis/purpura hemorrhagica, bilateral jugular thrombophlebitis.

2 What abnormalities can be seen on the first set of radiographs? They show soft tissue swelling with linear radiolucencies in the proximal neck (suggestive of emphysema, possibly secondary to the tracheotomy).

3 What abnormalities can be seen on the second set of radiographs? There are radiolucent lines running through the horizontal mandibular ramus, indicative of fracture of the mandible.

4 What condition is now suspected? A pathologic fracture secondary to neoplasia should be considered in the absence of trauma and with the evidence of tongue ulceration. Squamous cell carcinoma (SCC) was identified in biopsies of the tongue and mandible.

Discussion

SCC is one of the most common neoplasms to affect horses. It is an epithelial neoplasm and the majority of SCCs are associated with the ocular/periocular and external genitalia (of male horses) regions, particularly on light-colored skin and in older animals. While there are several neoplasms that may be found in the oral cavity, including odontogenic tumors, primary bone and soft tissue neoplasms, SCC is the most common of the soft tissue neoplasms and can involve any mucosal surface. SCC is commonly locally invasive but may also become metastatic with spread via local lymphatics to involve other nearby structures. Invasive SCC has a poor prognosis because of the high post-surgical recurrence rate and the likelihood of metastasis at the time of diagnosis. Skeletal neoplasia is very rare in horses, but can be associated with pathologic fracture as a result of destruction and necrosis of the bone by the neoplasm.

References

Jann HW, Breshears MA, Allison RW et al. (2009) Occult metastatic intestinal adenocarcinoma resulting in pathological fracture of the proximal humerus. *Equine Vet J* 41(9):915–917.
Perrier M, Schwarz T, Gonzalez O et al. (2010) Squamous cell carcinoma invading the right temporomandibular joint in a Belgian mare. *Can Vet J* 51:885–887.

CASE 90

1 What is the most likely diagnosis? Neonatal isoerythrolysis (NI) (a.k.a. isoimmune or alloimmune hemolytic anemia).

2 What is the cause of this disease? This is a hemolytic condition of neonatal foals in which the mare develops antibodies against the foal's erythrocytes (RBCs). It occurs when the foal inherits paternal RBC antigens that the mare does not have, and the mare has previously been exposed to these antigens (exposure usually occurs in a previous gestation from the same stallion). After the foal nurses and absorbs colostral antibodies, severe hemolytic anemia occurs. Factors Aa and Qa are the most immunogenic in horses, but it can happen with other blood groups. Mares that are negative for Aa or Qa are at higher risk of producing foals that will develop isoerythrolysis.

3 How should this foal be treated? A blood transfusion should be considered if the anemia is severe (PCV <15%) or if the foal is weak and showing additional signs of poor perfusion (e.g. tachycardia, cool limbs, increased blood lactate). Sources of blood for the transfusion may include the mare's blood (the RBCs must be washed at least three times and suspended in saline prior to the transfusion). If possible, other donors should be cross-matched with the foal's blood prior to transfusion. If unable to cross-match, using an Aa/Qa-negative mare is usually safe; alternatively, a gelding donor should be used. Transfused RBCs may have a short half-life depending on the compatibility of the donor cells, therefore the PCV needs to be closely monitored as further transfusions may be needed. The foal will likely remain icteric until the transfused RBCs have broken down and been eliminated. Plasma transfusions may be necessary if serum IgG levels are low. Broad-spectrum antibiotics should be given to protect against opportunist infections. Anti-ulcer medications and nutritional support may also be required.

Discussion
NI is a rare alloimmune disease in foals and is the most common cause of icterus in the neonate. Reported prevalence of clinically affected foals is 1–2%. Alloimmune disease is an acquired immune disease from exogenous sources of antibodies in contrast to autoimmune disease where the antibodies are produced by the affected animal. Acquired exogenous antibodies may result from transfusion of blood products (more common in adults) or, as is the case with NI, from absorption of colostral antibodies. While NI is the most common alloimmune disease in foals, an alloimmune thrombocytopenia has also been documented. Foals with NI typically show clinical signs of icterus, quiet, weak or obtunded mentation, and variable heart and respiratory rates within the first week of life, but milder cases may not be evident until as much as 12 days of age. In general, foals affected by NI do not have failure of passive transfer, but rare cases of partial failure of passive transfer resulting in NI have been reported. Adequate passive transfer does not necessarily prevent concurrent neonatal disease (i.e. sepsis), which may contribute to the foal's condition and clinical signs. Pathogenesis and treatment are described above. NI is well recognized and as a result, intervention typically occurs early, improving the prognosis, although some foals are more severely affected than others. Future athletic prospects are unaffected for recovered foals.

Screening tests for mares and stallions are available to help estimate the probability of NI. The mare will continue to produce NI affected foals if bred back to the same stallion or another stallion with RBC factors against which she has produced antibodies. Given this knowledge, many owners choose to attend the foalings of mares known to have produced NI affected foals and muzzle the foal during the period of gut closure to prevent nursing and absorption of alloantibodies. Commercial or frozen colostrum from non-NI mares should be administered via nasogastric tube initially, followed by bottle or bucket feeding of milk replacer or milk from another mare until gut closure (at least 24 hours). The fact that foals nurse roughly every 2 hours must be kept in mind to ensure the foal remains hydrated, receives the proper nutrition, and to help prevent secondary complications such as gastric ulcer formation. Milk replacer must be carefully mixed at the correct dilution to prevent iatrogenic hypernatremia. The mare should be milked several times (typically at each foal feeding session) during this period for udder health, continued milk production, and elimination of the colostrum potentially containing alloantibodies. After a sufficient period to allow gut closure and milking the mare to eliminate colostrum, the muzzle can be removed and the foal allowed to nurse.

References

Boyle AG, Magdesian KG, Ruby RE (2005) Neonatal isoerythrolysis in horse foals and a mule foal: 18 cases (1988–2003). *J Am Vet Med Assoc* **227**(8):1276–1283.

de Graaf-Roelfsema E, van der Kolk JH, Boerma S *et al.* (2007) Non-specific haemolytic alloantibody causing equine neonatal isoerythrolysis. *Vet Rec* **161**(6):202–204.

Polkes AC, Giguère S, Lester GD *et al.* (2008) Factors associated with outcome in foals with neonatal isoerythrolysis (72 cases, 1988–2003). *J Vet Intern Med* **22**(5):1216–1222.

CASE 91

1 Should an abdominal disorder be further pursued (i.e. repeat abdominocentesis, exploratory laparotomy) as the most likely diagnosis, or should another organ system disorder be investigated? The focus of the investigation should be directed towards the heart due to the marked tachycardia, jugular pulses, and normal findings regarding the GI system.

2 Is the increase in cTnI clinically important? The half-life of cTnI in the horse is less than 1 hour, so lack of a decrease or, in this case, an increase in cTnI is strongly suggestive of ongoing myocardial damage. The continued elevation in cTnI, pronounced tachycardia with jugular pulsations, and recurrence of the tachycardia are all suggestive of an acute myocardial injury.

3 What treatments should be considered? Myocarditis was considered the likely diagnosis and treatment with IV lidocaine, magnesium sulfate, and dexamethasone (30 mg) was initiated. Twelve hours after beginning the treatment, the HR began to decrease. Thirty-six hours later the HR was 40 bpm so the lidocaine was discontinued. Dexamethasone (30 mg PO q24h) was continued for 4 days (until the cTnI returned to normal) and then discontinued in a taper fashion over the next week. The horse continued to recover successfully and was back into full work 6 weeks later.

Discussion

Colic signs are not unusual with marked ventricular or supraventricular tachycardia, and jugular pulsations are an important diagnostic clue. The pulsations are not a true regurgitation but instead occur because the marked cardiac rate (often >180 bpm) does not allow adequate time for cardiac volume-loading, which causes increased jugular vein distension and a 'pulse appearance' with each beat of the heart. When cTnI is elevated, myocardial injury is likely and corticosteroids are frequently used to decrease suspected focal or diffuse inflammation. If there is diffuse myocarditis, echocardiography can usually demonstrate decreased function (decreased contractility/fractional shortening). Ventricular or supraventricular tachycardia with minimal or no increase in cTnI can also be relatively common in horses, and in these cases treatment with corticosteroids may not be indicated. The cause of the myocardial injury in this case was not determined. Some cases of myocarditis caused by equine influenza, EHV, other viruses, or some immune-mediated diseases (e.g. purpura hemorrhagica) are associated with a fever or respiratory signs. The lack of these signs in this horse did not support an infectious cause. The presence of urticaria at the time of the cardiac disease did raise the possibility of an allergic/immune reaction. Toxic causes, such as snake bites, ionophores, and pasture myopathy (AST = 316 U/l), were ruled out.

Reference

Schwarzwald CC, Hardy J, Buccellato M (2003) High cardiac troponin I serum concentration in a horse with multiform ventricular tachycardia and myocardial necrosis. *J Vet Intern Med* **17**(3):364–368.

CASE 92

1 **What is the most likely diagnosis based on the clinical signs?** The history and clinical findings are highly suggestive of lymphangitis, most likely of bacterial origin.

2 **What diagnostic test should be performed to help establish the diagnosis?** Ultrasound to detect distended lymphatics and rule out any focal abscess or tendon involvement. The distended lymphatics seen on ultrasound examination are shown (**92.2**).

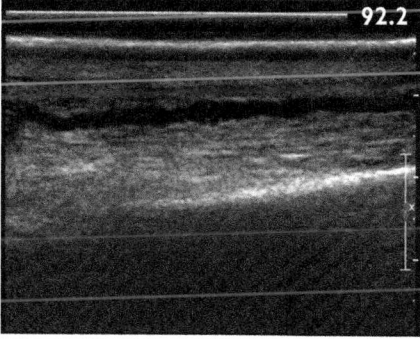

3 **What are the etiologic causes of the condition in this horse, and which one is most likely here?** *Staphylococcus aureus* is the most likely infectious organism to

cause the signs seen in this horse, although there are no published case series to answer this question. *Steptococcus* spp. would also be possible. *Corynebacterium pseudotuberculosis* is a proven cause of lymphangitis in the horse, but it generally causes a nodular pattern and is seen in limited geographic areas. The same is true for fungal causes (e.g. sporotrichosis, blastomycosis). *Clostridium perfringens* is more likely to cause cellulitis than lymphangitis.

4 How should this horse be treated? Systemic antimicrobials that would likely be effective against *S. aureus* should be initiated. In this case, IV enrofloxacin and

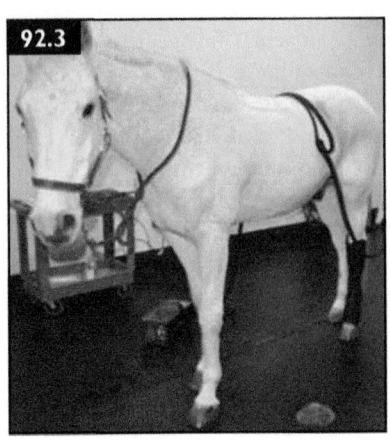

potassium penicillin were administered (enrofloxacin is a preferred treatment for *S. aureus* and penicillin was chosen in case there was beta-*Streptococcus* present, which is typically resistant to enrofloxacin). Anti-inflammatory treatment should be aggressive; high-normal doses of an NSAID, pentoxifylline, and cold water hydrotherapy with massage and pressure bandages. The Game Ready® cooling device (**92.3**) is excellent at both cooling and compressing the leg and does not require great manipulation of the limb for treatments, as might be required for jacuzzi tubs. Unfortunately, they are more expensive.

5 Due to the non-weight bearing of this limb, what concern is there for the opposite limb, and how might this be prevented? Support limb laminitis (SLL) would be a huge concern. Prevention centers around maintaining adequate blood flow, oxygen, and nutrient supply to the weight-bearing limb. The provision of adequate systemic analgesics and, sometimes, a morphine epidural in order to allow the horse to walk on the affected leg is important in preventing SLL. If the horse will not bear weight on the affected leg, a sling can be used to rest the contralateral hindlimb. However, actually loading and unloading weight by walking is preferred to placing the horse in a sling.

Follow up/discussion

Lymphangitis is an emergency: the longer the affected leg remains swollen, the more severe will be the permanent anatomic disruption of the lymphatic vessels. There is often an acute, progressive swelling of one hindlimb. Acute bacterial lymphangitis may cause limb swelling with serum oozing through the skin. Acutely affected patients have a fever and frequently are non-weight bearing, which predisposes to SLL. Diagnosis is based on clinical signs and results of ultrasound examination using a 7.5-MHz probe, which reveals numerous dilated vessels (lymphatic vessels).

If the inflammation and damage to the lymphatics cannot be quickly controlled, the limb may not return to normal size. If it does, the horse may have recurrence of non-septic swelling with any future trauma to the limb.

Reference

Nogradi N, Spier SJ, Toth B *et al.* (2012) Musculoskeletal *Corynebacterium pseudotuberculosis* infection in horses: 35 cases (1999–2009). *J Am Vet Med Assoc* **241(6):**771–777.

CASE 93

1 What is the most likely diagnosis? Purpura hemorrhagica. This is a syndrome of cutaneous vasculitis that occurs 2–4 weeks after infection with *Streptococcus equi* subsp. *equi* or *Streptococcus equi* subsp. *zooepidemicus* or after other infections. Affected animals present with painful subcutaneous edema, especially of the distal limbs and ventral abdomen. Petechiae and ecchymoses may be present on the oral, nasal, and conjunctival mucous membranes.

2 How could the diagnosis be confirmed? Skin biopsy of affected areas may confirm the presence of vasculitis of the small dermal and subcutaneous vessels. Serology may be helpful in identifying high titers to *S. equi* subsp. *equi* M protein.

3 What treatments are recommended? In horses recently affected by strangles, penicillin therapy should be instituted. Hydrotherapy can help resolve the edema. In most cases, corticosteroid (dexamethasone) therapy will be required as an immunosuppressive drug to treat the underlying vasculitis. Furosemide may be needed if there is severe edema. A tracheostomy may be needed if there is respiratory distress secondary to upper airway edema.

Discussion

Vasculitis is a common sequela to several viral and bacterial infections, but the recent history of *S. equi* in this case was strongly suggestive of purpura hemorrhagica. Unlike some bacterial or viral infections that directly damage the vasculature as part of the infective process, purpura is an immune-mediated response to infection with *S. equi*. While unproven, purpura is thought to be a type III hypersensitivity with deposition of immune complexes (likely IgA) in the walls of small vessels, which then become an attractant for WBCs, particularly neutrophils and macrophages. The release of lysosomal enzymes from these WBCs creates vascular damage, leading to increased vessel permeability and the clinical signs noted in this case.

Left untreated, many horses affected with purpura may die within 4–7 days due to complications associated with the vasculitis including pneumonia, renal failure, colic (from visceral edema), or cardiac arrhythmia. It is important to note that while petechiae and ecchymoses on mucous membranes are commonly seen with purpura, these are not typically the result of thrombocytopenia or other coagulopathies, but rather due to a severe increase in vessel permeability.

Approximately 75% of horses develop good immunity post *S. equi* infection that may last up to 5 years; however, vaccines have been developed and are available for horses. The vaccines will provide some protection, but that protection is short-lived (approximately 3 months) and therefore they are generally only recommended for horses at moderate to high risk of exposure to *S. equi*. The intranasal vaccine also carries some risk of triggering purpura or causing IM abscesses (see **case 18**) when given in temporal proximity to an IM injection. For these reasons, it is not recommended for practitioners to give the intranasal vaccine on the same day as any IM vaccine or injection. It is also not recommended to give the intranasal vaccine to any horse with an SeM titre of >1:1,600 or to give another intranasal vaccine at the same visit.

References

Kaese HJ, Valberg SJ, Hayden DW *et al.* (2005) Infarctive purpura hemorrhagica in five horses. *J Am Vet Med Assoc* **226**(11):1893–1898, 1845.

Pusterla N, Watson JL, Affolter VK *et al.* (2003) Purpura haemorrhagica in 53 horses. *Vet Rec* **153**(4):118–121.

CASE 94

1 What is the likely diagnosis? Melanoma.

2 What are the different forms of this disease? At least four manifestations of equine melanotic disease have been proposed: melanocytic nevus, discrete dermal melanoma, dermal melanomatosis, and anaplastic malignant melanoma. Discrete dermal melanoma and dermal melanomatosis, collectively referred to as dermal melanoma, represent the large majority of melanoma diagnoses in gray horses. Discrete dermal melanomas are seen in gray horses and generally exist as single masses in typical or atypical locations. They are further differentiated into benign and malignant forms, with surgical excision appearing to be curative in most circumstances. Dermal melanomatosis is a condition seen in gray horses involving multiple cutaneous masses, with at least one of the masses presenting in a 'typical' location. These typical sites include the undersurface of the tail, anal, perianal and genital regions, perineum, and lip commissures. These masses may not be amenable to surgical resection and can be associated with visceral metastasis.

3 How can these lesions be treated? Treatment of multiple dermal melanomas is difficult. Surgical debulking (including laser surgery) can be palliative, along with cryotherapy, intralesional cisplatin, or PO cimetidine to slow down the rate of growth in rapidly growing melanomas. Staged treatments may be required to treat multiple lesions. A variety of treatments for melanoma, including surgical, radiologic, and local and systemic chemotherapies or other pharmaceuticals, have been described, most with only anecdotal evidence of efficacy. Surgical removal is often the treatment of choice for horses with accessible, growing melanomas, with the best results coming from the removal of small, benign masses. Other therapies include:

- Radiation therapy. Brachytherapy, where the source of radiation is placed within or near the tumor, has been described and has shown some promise. However, availability is limited and treatment of more than one tumor may be cost-prohibitive.
- Intratumoral chemotherapy. The injection or placement (via impregnated beads) of a cytotoxic drug into the tumor or the tissue surrounding the tumor. Carboplatin and cisplatin have been, used with higher response rates reported for smaller tumors than larger tumors. Various techniques have been developed to enhance tumoral uptake of the cytotoxic agent.
- Intratumoral immunotherapy. The injection of DNA plasmids encoding for IL-12 and IL-18 remain under investigation, but show some promise.

The results for systemic therapy with cimetidine are variable, leaving this as a questionable therapy for melanoma. The H2 receptor antagonist is thought to enhance natural killer cell recognition of tumors and suppress immunosuppressive actions by T-regulatory cells.

The use of whole tumor cell autogenous and DNA-based vaccines has been investigated in horses. Recently, the canine tyrosine kinase vaccine has also been used in horses. Studies on autogenous vaccines showed efficacy but results were confounded by simultaneous additional therapies. While DNA-based vaccines are not commercially available in many parts of the world at this time, they have shown promise in early studies.

References

Moore JS, Shaw C, Shaw E *et al.* (2013) Melanoma in horses: current perspectives. *Equine Vet Educ* 25(3):144–150.
Phillips JC, Lembcke LM (2013) Equine melanocytic tumors. *Vet Clin North Am Equine Pract* 29(3):673–687.

CASE 95

1 **What is the cause of the opaque plasma?** Hyperlipemia.
2 **What is the pathogenesis of this condition?** Hyperlipemia represents an excessively rapid mobilization of the body's fat reserves in response to stress or

failure to maintain energy homeostasis. In response to negative energy balance and after depletion of glycogen reserves, non-esterified fatty acids are mobilized from fat stores and released into the circulation. The majority of non-esterified fatty acids are taken up by the liver, where they may overwhelm the oxidative, gluconeogenic, and ketogenic pathways and be esterified to form triglycerides. Triglycerides then accumulate in the liver and are exported in the circulation in the form of very low density lipoproteins (VLDLs). This process occurs at such a fast rate that the VLDLs cannot be utilized by peripheral tissues, and plasma levels become excessive. VLDLs are also taken up by cells of the reticuloendothelial system, resulting in fatty infiltration of many organs.

3 How should this pony be treated? The treatment of hyperlipemia has five different objectives: treatment of the underlying or concurrent disease; correction of dehydration, electrolyte, and acid/base imbalances; symptomatic therapies; nutritional support; and normalization of lipid metabolism. The induction of abortion or premature foaling in pregnant mares has been recommended, since this significantly reduces the demands for energy. However, prematurely delivered foals have a high mortality rate because of their immature body systems and susceptibility to infectious disease. There is also a risk of retained placenta and laminitis in the mare. There are two possible approaches to modifying lipid metabolism in hyperlipemic patients: reducing the net release of non-esterified fatty acids from adipose tissues (insulin therapy) and accelerating the removal of triglycerides from plasma to adipose tissues and skeletal muscle (heparin therapy).

Discussion
Dyslipidemia is a disorder of lipid metabolism resulting in abnormal lipid (specifically triglycerides in horses) concentrations in peripheral circulation. Hyperlipidemias are most common in miniature horses, ponies, and donkeys, but do occur in horses. Pregnant or lactating mares, obese horses, or horses with equine metabolic syndrome or pituitary pars intermedia dysfunction (PPID or equine Cushing's syndrome) are particularly at risk. Lipemia is not a clinical syndrome, but rather a descriptive term for the visible turbidity of serum due to high concentrations of triglyceride. There are four dyslipidemias in horses:

- Hypertriglyceridemia: triglyceride concentrations above the normal reference range; no clinical disease evident.
- Hyperlipidemia: serum triglyceride concentrations are elevated but <500 mg/dl (5.65 mmol/l), without grossly visible serum lipemia; usually no specific clinical signs.

- Severe hypertriglyceridemia: serum triglyceride concentrations >500 mg/dl (5.65 mmol/l) without grossly visible serum lipemia; clinical signs usually evident.
- Hyperlipemia: serum triglyceride concentrations >500 mg/dl (5.65 mmol/l) with grossly visible serum lipemia and fatty infiltration of the liver or other organs; clinical signs usually evident.

Dyslipidemias result from the catabolic processes of gluconeogenesis, glycogenolysis, and peripheral lipolysis related to periods of stress or negative energy balance. The two most common clinical signs associated with dyslipidemias, depression and anorexia, are non-specific and must be interpreted along with relevant history and other clinical findings including measurement of triglyceride concentrations in any at-risk patient. This is important as treatment of dyslipidemias relies first and foremost on correcting/managing the underlying cause or disease. Insulin resistance is a significant component of dyslipidemias and may either be induced by the underlying disease and physiologic process or have been present prior to physiologic stress/negative energy balance as the result of obesity, PPID, or metabolic syndrome. Prognosis for dyslipidemias in horses has been improving in recent years as recognition and treatment improve, but reports still indicate a 33% mortality rate.

Reference
McKenzie HC3rd (2011) Equine hyperlipidemias. *Vet Clin North Am Equine Pract* **27(1)**:59–72.

CASE 96
1 What is the most likely diagnosis, and what are the differential diagnoses? Polyneuritis equi. The history of early signs of hyperesthesia of the perineal region followed by progressive signs of tail, anal, and perineal paralysis is typical of this disease. Differential diagnoses include sacrococcygeal trauma and fractures, EPM, equine herpesvirus-1 myeloencephalopathy, verminous myeloencephalitis, and equine motor neuron disease.

2 What other clinical signs can be associated with this disease? CN involvement occurs in a proportion of cases of polyneuritis equi. Although any CN may be involved, CNs V, VII, and VIII are most commonly involved. Head tilt, ear droop, lip droop, ptosis, and dysphagia are common signs. Colic due to fecal retention may also occur.

3 How should this horse be treated? Treatment is primarily palliative. Manually evacuating the rectum and catheterizing the urinary bladder may be necessary. Systemic antibiotics may be necessary if secondary cystitis occurs. Corticosteroids usually have little or no beneficial effects.

Answers

Discussion

Polyneuritis equi (a.k.a. cauda equina neuritis) is a rare condition affecting adult horses, in most cases presenting very much like the one above. Clinical signs may be present unilaterally or bilaterally (both symmetrically and asymmetrically). In the absence of any known etiologic agent and an unknown pathogenesis, it is generally considered to be an immune-mediated disease with a presumptive diagnosis often made ante-mortem, but the definitive diagnosis is often made at necropsy. Characteristics of the lesions are well established and include granulomatous inflammation (and resultant hemorrhage and swelling) of affected nerve roots progressing to demyelination, axonal degeneration, fibrosis, and adhesion formation.

The cellular infiltrate of examined lesions largely consists of lymphocytes (both T and B lymphocytes) and macrophages, with smaller numbers of various other leukocytes. Macrophages are responsible for the damage to the myelin, as they invade not only the myelin sheath but also the basal lamina of the Schwann cells. It is likely that the axonal damage is secondary to demyelination. The macrophages are activated by T lymphocytes and antibodies have been found against myelin's P2 protein, indicating a role for the B lymphocytes present in/around the lesions.

Historically, definitive diagnosis has been made at necropsy when samples of the affected cauda equina can be obtained, but recently it has become possible to obtain an ante-mortem diagnosis by examining the nerve endings in biopsies of affected tail muscles. The tail muscles are among the muscles most frequently affected by this disease and they contain a large number of nerve endings from the cauda equina, making them an ideal choice for biopsy. Frozen and formalin-fixed samples should be submitted. While CNs can be affected, other etiologies for CN deficits should be ruled out in the absence of cauda equina clinical signs.

References

Aleman M, Katzman SA, Vaughan B *et al.* (2009) Antemortem diagnosis of polyneuritis equi. *J Vet Intern Med* 23(3):665–668.

Hahn CN (2008) Polyneuritis equi: the role of T-lymphocytes and importance of differential clinical signs. *Equine Vet J* 40(2):100.

van Galen G, Cassart D, Sandersen C *et al.* (2008) The composition of the inflammatory infiltrate in three cases of polyneuritis equi. *Equine Vet J* 40(2):185–188.

CASE 97

1 What is the likely diagnosis? Meconium impaction. Although meconium can often be palpated by digital palpation *per rectum*, in some cases the meconium may be lodged further orad in the small colon and cannot be felt.

2 What other procedures can be used to help confirm the diagnosis? In some cases, the obstruction can be observed by ultrasound. Abdominocentesis is not usually necessary, and the peritoneal fluid is normal in most cases.

3 How should this condition be treated? An enema should be administered, using either warm soapy water or a commercial phosphate enema. The clinician should be careful not to tear the rectum, which is very delicate in the neonate. If the meconium is not passed, the enema can be repeated in 1–2 hours. A retention enema using acetylcysteine can also be used with the foal lightly sedated (i.e. with diazepam); 4% acetylcysteine solution is infused into the rectum via a Foley catheter and left in the rectum for 30 minutes. Additional treatments used for meconium impactions include oral laxatives such as mineral oil given via nasogastric tube. Dioctyl sodium succinate has also been advocated, but should not be repeated because multiple doses can lead to toxicity. Although smooth plastic loops have been used to remove the meconium, manual removal of meconium using forceps is not recommended because of the risk of rectal trauma or penetration.

Discusion
Meconium is a concretion of glandular secretions from the developing fetus's GI tract, amniotic fluid, mucus, and cellular debris. It is very firm and dark in color and is the 'first feces' a neonate passes, usually beginning shortly after the first suckle. While most owners are aware that the meconium must pass and can cause impaction if it does not, many mistakenly think that a single successful defecation of meconium feces observed shortly after birth means all of the meconium has passed. This, however, is not the case because meconium passage takes several bowel movements and is a process that may take up to 12–24 hours to complete. Meconium passage is complete only when observation of the passage of 'milk feces' (soft, pasty, lighter brown feces) occurs. Meconium impaction is common and is the leading cause of signs of colic in the neonate.

In rare cases, hyperammonemia may develop, resulting in altered mentation. Impaction has been noted to affect males slightly more often than females, possibly due to a smaller diameter pelvic inlet in male neonates. Treatment for meconium impaction is as above, with preference given to warm soapy enemas because they are inexpensive, safe, and very effective. Care must be taken with commercial enemas as the phosphate may be absorbed through the rectal mucosa, causing hyperphosphatemia, which then results in hypocalcemia; phosphate enemas should not be repeated more than 3–4 times in 24 hours in an average sized foal. All enemas should be given by gravity flow through soft tubing (such as a Foley or stallion catheter) to avoid damage to the rectum. Some farms prefer to give a prophylactic soapy water enema approximately 12 hours after birth to ensure meconium passage. If signs of pain are severe, analgesics may be required. IV fluid therapy may be warranted if the foal has stopped nursing or is clinically dehydrated. Foals should be adequately hydrated prior to any NSAID administration.

Reference
Pusterla N, Magdesian KG, Maleski K *et al.* (2004) Retrospective evaluation of the use of acetylcysteine enemas in the treatment of meconium retention in foals: 44 cases (1987–2002). *Equine Vet Educ* 16(3):133–136.

CASE 98

1 What do the endoscopic images show? Evidence of severe pharyngitis.

2 Image 98.4 is from endoscopy of the left guttural pouch. What abnormality is observed on the mucosa overlying the stylohyoid bone? Lymphoid follicles can be seen on the mucosa overlying the stylohyoid bone. These are common with viral respiratory infection in horses.

3 Image 98.5 shows endoscopy of the trachea. Interpret this image. There is thick mucus and most likely purulent exudate on the ventral floor of the trachea. Cytology and culture may be indicated, although this finding is also common in acute severe upper respiratory viral infections in horses.

4 What sample should be collected to further identify the cause of this colt's clinical signs? A nasal swab for PCR for *Streptococcus equi* subsp. *equi*, influenza, equine herpesvirus-1 and -4, and equine rhinitis virus. This horse was PCR negative for *S. equi*, equine herpesvirus-1 and -4, and equine influenza, but PCR positive for equine rhinitis A virus (ERAV).

5 How should this weanling be treated? Administer NSAIDs for 3 days and feed a soft pelleted feed and soaked alfalfa cubes for 1 week. Also administer 1 mg/kg omeprazole for 5 days.

Discussion

In an experimental study, ERAV-inoculated ponies developed respiratory tract disease characterized by pyrexia, nasal discharge, adventitious lung sounds, and enlarged mandibular lymph nodes. Additionally, these animals had purulent mucus in their lower airways up to the last evaluation time 21 days after inoculation (detected endoscopically). The virus was isolated from various samples obtained from the lower and upper airways of ERAV-inoculated ponies up to 7 days after exposure; this time corresponded with an increase in serum titers of neutralizing antibodies against ERAV. None of the ponies developed clinical signs of disease after reinoculation 1 year later.

Reference

Diaz-Méndez A, Hewson J, Shewen P *et al.* (2014) Characteristics of respiratory tract disease in horses inoculated with equine rhinitis A virus. *Am J Vet Res* 75(2):169–178.

CASE 99

1 What is the recommended treatment? Iridal cysts can be deflated with an ND:YAG laser (**99.2**); cysts only require treatment if they cause pain or visual impairment.

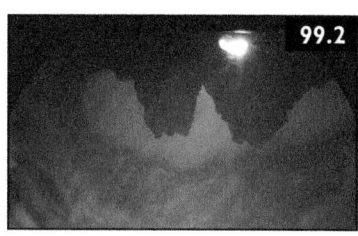

CASE 100

1 What is the most likely type of worm identified in the fecal preparation? Cyathostomin larvae (small strongyles).

2 How does the presence of these worms in the feces relate to the fecal egg count of 75 epg? The parasites in the feces are larval stages of cyathostomins. Horses can carry large burdens of larval stages of these parasites without necessarily having heavy burdens of egg-laying adult cyathostomins.

3 Describe the life cycle of this parasite and how the condition 'acute larval cyathostominosis' occurs. The cyathostomins have a direct, non-migratory life cycle. Adult stages reside in the large intestine. Eggs pass in the feces, and development to third stage infective larvae (L3) occurs on the pasture. The prepatent period is 6–20 weeks. Ingested L3 larvae migrate into the mucosa of the cecum and large colon where they continue their development. During larval development within the colonic mucosa, cyathostomins have the ability to go into a stage of arrested development within mucosal cysts. Such arrested encysted larvae stimulate little inflammatory response. Huge numbers of parasite larvae can inhabit the mucosa of susceptible horses. Mass emergence of these larvae when development is reactivated leads to severe inflammation and colitis, a condition known as 'acute larval cyathostominosis'.

Discussion

Over 50 different species of cyathostomins have been documented in horses, and in most cases these intestinal parasites do not cause clinical disease. When clinical disease occurs it is generally in younger horses (<5 years) and often sporadic with only one or a few horses in a herd being affected. Cyathostomins are unique in their ability to arrest development of encysted larval stages. Little is known about what triggers this arrest or the resumption of development, which culminates in mass exodus from the cecal and large intestinal wall in some horses, resulting in acute disease.

Theories include seasonality and host immunity factors, but it is known that larvae may remain arrested in the gut wall for as long as 2–3 years or as little as 1 month. Seasonality theories are based on the idea that the developing larvae wait to complete their development until the most opportune time for the eggs to be on pasture, thereby maximizing the chances of their offspring surviving and finding a host. This seems logical with the patterns of emergence being in late winter/early spring in northern temperate climates and in late summer/early fall in the warmer southern, subtropical regions where pastures dry up and horses are fed hay during the summer months. To date, this is only a hypothesis and the mechanism(s) by which this is accomplished are unknown. Individual differences in immune system recognition and function are also logical, although largely unproven. It has been shown that some individuals will harbor larger parasite burdens than their counterparts under identical conditions.

There is some evidence that an equilibrium may exist between the adults and larvae, with the number of adults somewhat controlling the number of emerging larvae. When the adults are killed off in large numbers by anthelmintics, it triggers massive numbers of encysted larvae to resume development to replace the adults, resulting in acute disease. It is worth remembering that the adult cyathostome creates no clinical disease in the horse. Disease results only from the emergence of the encysted larvae from the gut wall.

Treatment is supportive and symptomatic once clinical disease is evident. At this point adulticidal anthelmintics are not likely to be of use. Since adult cyathostomins do not cause intestinal disease, and eliminating the adults may trigger additional encysted larvae to emerge, administration of adulticidal anthelmintics furthers the intestinal damage, inflammation, and clinical signs. Fenbendazole and moxidectin are the only anthelmintics known to be larvicidal for encysted cyathostomins. It would be prudent to administer a 5-day course of fenbendazole or a dose of moxidectin to decrease or eliminate remaining encysted larvae during or after treatment of clinical disease, but prior to use of an adulticide. Following recovery, it is recommended that the individual is monitored using fecal egg counts and treated yearly (in the fall in the northern hemisphere and spring in the southern hemisphere) with moxidectin or fenbendazole to eliminate as many new encysted larvae as possible.

References
Andersen UV, Reinemeyer CR, Toft N *et al.* (2014) Physiologic and systemic acute phase inflammatory responses in young horses repeatedly infected with cyathostomins and *Strongylus vulgaris*. *Vet Parasitol* **201**(1-2):67–74.

Lyons ET, Drudge JH, Tolliver SC (2000) Larval cyathostomiasis. *Vet Clin North Am Equine Pract* **16**(3):501–513.

Matthews JB (2011) Facing the threat of equine parasitic disease. *Equine Vet J* **43**(2):126–132.

CASE 101
1 What is the most likely cause of the nodular lesions in the skin and mucous membranes? Besnoitiosis (*Bestnoitia bennetti*).

2 What procedures or test could be performed to confirm the diagnosis? A skin biopsy to visualize the parasite and the cystic lesions characteristic of the disease (**101.6**) or serum immunofluorescent antibody testing for *B. bennetti*.

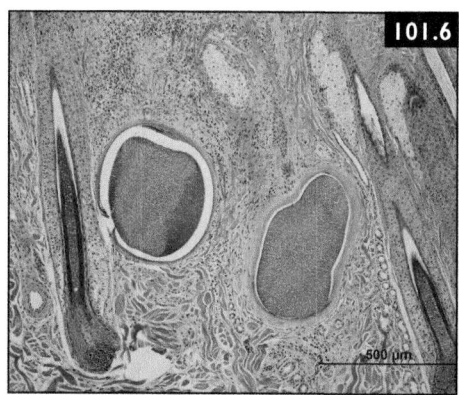

3 How is the disease spread? The mechanism of transmission is unknown, but possibly due to insects.

4 How should this disease be treated? There is no known successful treatment for the disease, although improvement with nitazoxanide has been reported.

5 Why did the donkey sneeze immediately after the sedation? Sneezing is a common response to sedation with alpha-2 agonist drugs in donkeys.

Discussion

Besnoitiosis is caused by infection with protozoal parasites (*Besnoitia* spp.), which are cyst-forming coccidians that affect multiple host species worldwide. Reported equid cases of besnoitiosis are mostly limited to donkeys. Clinical disease is characterized by the development of multifocal pinpoint parasitic cysts, approximately 1.0 mm in diameter, in the skin over the face and body, within the nares, on the internal and external pinnae, and on the limbs and perineum. Mucous membranes are also frequently affected, and one of the most unusual clinical features of besnoitiosis is the development of cysts along the limbal margin of the sclera (i.e. 'scleral pearls'). Some infected animals remain otherwise healthy, whereas others, such as the one in this case, become cachexic and debilitated as a result of the disease.

Reference

Ness SL, Peters-Kennedy J, Schares G *et al.* (2012) Investigation of an outbreak of besnoitiosis in donkeys in northeastern Pennsylvania. *J Am Vet Med Assoc* **240**(11):1329–1337.

CASE 102

1 What diagnostic procedures should be recommended in the hope of determining the cause of the clinical abnormalities? Radiographs of the neck should be taken because of the history of stumbling and the abnormalities found during neurologic examination. These were normal. Spinal fluid should be collected for cytology and to test for EPM and Lyme neuroborreliosis. The CSF cytology was normal.

2 What is the preferred way to use serology for the diagnosis of EPM or neuroborreliosis? An antibody ratio between serum and CSF should be performed. If the cytology is abnormal, albumin should also be measured in CSF and serum to estimate the change in the blood–brain barrier. Albumin measurements are not required if the cytology and CSF protein are normal. Normal horses have a serum to CSF IgG ratio of 130:1–250:1. Specific antibody ratios less than this value are suggestive of intrathecal antibody production (i.e. intrathecal infection). There were no antibodies against *Sarcocystis neurona* surface antigens (SAG 2, 3, or 4) detected in the CSF. The *Borrelia burgdorferi* OspA serum to CSF ratio was 100:1, which was a borderline positive test for neuroborreliosis. The horse had not been vaccinated for Lyme disease. The importance of these findings was unclear but neuroborreliosis was considered as a possible diagnosis.

3 What would be an appropriate treatment for this disease? Includes oxytetracycline IV or minocycline PO.

4 Why would doxycycline PO not be an appropriate treatment in this case? Doxycycline given orally does not reach detectable CSF concentrations in most horses.

5 Holding the tail to one side is a common finding with pain in what area? The back.

Follow up/discussion
The horse was treated with minocycline (4 mg/kg PO q12h) and 5,000 IU of vitamin E daily for 2 months. The horse appeared much improved after 4 weeks (102.3, 102.4) and returned to normal appearance and use at 3 months.

Muscle wasting and pain over the thoracolumbar area have been present in a few horses with high *Borrelia* spp. serum titers and some of these horses have

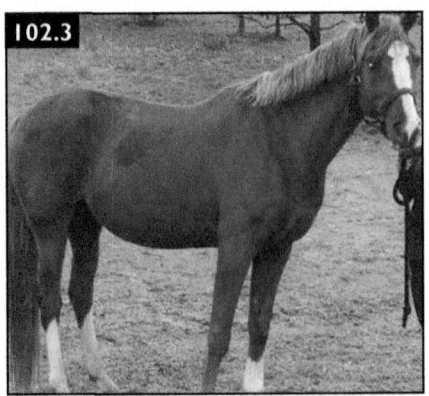

had neurologic signs (ataxia). In one report, two horses were diagnosed with Lyme neuroborreliosis and both had chronic, necrosuppurative-to-nonsuppurative, perivascular-to-diffuse meningoradiculoneuritis on necropsy examination. Hyperesthesia, lumbar pain, and muscle wasting were the initial clinical findings followed by ataxia of all four limbs, facial nerve paralysis, and finally head tremors with depression in one horse. On necropsy spirochetes were identified by

Steiner silver impregnation in both cases, predominantly in the affected dura mater of the brain and spinal cord. *Borrelia burgdorferi sensu stricto* was identified by PCR with the highest spirochetal burden in tissues with inflammation, including the spinal cord, muscle, and joint capsules. In another report, a horse with severe neck stiffness that progressed to ataxia had lymphohistiocytic meningitis and *B. burgdorferi* DNA in the CSF. That horse originally responded to doxycycline treatment but relapsed after discontinuing the treatment. Based on these few cases, ataxia and lumbar muscle wasting caused by lymphohistiocytic meningitis and radiculoneuritis with occasional fasciculations and neck stiffness may be common characteristics of neuroborreliosis in the horse. Analysis of CSF would likely show a lymphocytic pleocytosis and, although uncommon in humans with neuroborreliosis, some horses have

CSF that is PCR positive for *B. burgdorferi*. Additionally, a horse can be considered positive for neuroborreliosis if it has a serum to CSF antibody ratio less than 100:1 (along with appropriate clinical signs). It is important to keep in mind that the Lyme Multiplex test dilutes serum but not CSF, so serum antibody levels should be corrected for this dilution prior to calculating the serum to CSF ratio. Unfortunately, some horses that have tested antibody negative on both CSF and serum were found to be positive on necropsy for Lyme neuroborreliosis.

Reference

Imai DM, Barr BC, Daft B *et al.* (2011) Lyme neuroborreliosis in 2 horses. *Vet Pathol* **48(6)**:1151–1157.

CASE 103

1 What further tests are needed to confirm the suspicion of thrombocytopenia?
Thrombocytopenia is confirmed if the peripheral blood platelet count is <100,000/μl (100 × 10^9/l). Clinical bleeding is usually associated with platelet counts of <30,000/μl (30 × 10^9/l). Proper sample collection and platelet counting

are important to prevent false-positive results. Platelets of some horses will clump in EDTA anticoagulant, thereby producing artificially low platelet counts (pseudothrombocytopenia). A sample of blood placed in sodium citrate instead of EDTA should be assessed. Platelets should also be evaluated on a blood smear. A Coombs test should be considered if there is evidence of concurrent hemolytic anemia (Evans syndrome involves autoimmune hemolytic anemia with concurrent immune-mediated thrombocytopenia). Blood samples can be tested at specialized laboratories for detection of antibody-coated platelets and regenerative platelets.

2 What are the important causes of thrombocytopenia in adult horses? Thrombocytopenia can occur as a result of a severe systemic inflammatory response or as a result of immune-mediated platelet damage. Primary infections, including infection with *Anaplasma phagocytophilum* and equine infectious anemia virus, can also result in thrombocytopenia. Immune-mediated thrombocytopenia can occur secondary to viral infections, neoplasia, abscessation, or in association with drug therapy (especially trimethoprim sulfa combinations). Other cases are idiopathic.

3 How should this horse be treated? Immune-mediated thrombocytopenia should be treated with dexamethasone; risks of laminitis should be discussed prior to starting on this therapy. Azathioprine can also be used as an immunosuppressive agent. Fresh plasma or whole blood transfusions can provide a source of platelets, keeping in mind that transfused platelets have a life span of roughly 2–5 days.

Discussion

True thrombocytopenia (as opposed to pseudothrombocytopenia) results from either increased destruction or consumption (regenerative thrombocytopenia) or decreased production (non-regenerative thrombocytopenia) of platelets. Non-regenerative thrombocytopenias are less common and typically are the result of aplastic anemia, myeloproliferative disease, or neoplasia. Myelosuppression induced by drugs or toxins is possible and identification and removal of the causative agent is necessary in these cases. Microscopic examination of a bone marrow biopsy to evaluate a decrease in the numbers of megakaryocytes may be necessary to confirm a non-regenerative thrombocytopenia.

Regenerative thrombocytopenias result from either an increased peripheral destruction of platelets (as in immune-mediated thrombocytopenia) or an increase in consumption of platelets (e.g. as with severe hemorrhage, DIC). Immune-mediated thrombocytopenia can be further broken down into either primary (typically autoimmune) or secondary (e.g. following drug administration, bacterial or viral infection) categories and is considered a type II hypersensitivity. In these cases, antibodies attach to the platelets, leading to their removal from circulation for destruction, likely in the spleen. Identification followed by the treatment or removal of the underlying cause is necessary in these cases. Trimethoprim sulfa administration is the most common drug-related cause and should be considered in horses currently or recently (within the previous 14 days) receiving the drug, although any drug or vaccine has the potential to trigger this type of reaction.

Reference
McGurrin MK, Arroyo LG, Bienzle D (2004) Flow cytometric detection of platelet-bound antibody in three horses with immune-mediated thrombocytopenia. *J Am Vet Med Assoc* **224**(1):83–7, 53.

CASE 104

1 What are the differential diagnoses for diffuse subcutaneous emphysema of the neck and shoulder region? Include an axillary wound, esophageal perforation, ruptured trachea, lung rupture into the mediastinum, and caudal pharyngeal injury.

2 What diagnostic procedure(s) should be performed next? It would be important to check carefully for puncture wounds (especially in the axillary region) and evidence of any external trauma to the neck. Thoracic and cervical radiographs, endoscopy of the trachea and esophagus, and a CBC (to determine if there is an inflammatory component) should be performed. The CBC was normal.

3 What clinical findings would help rule out a ruptured esophagus? Absence of both fever and laboratory evidence of sepsis make esophageal perforation unlikely.

4 How should this case be treated? A conservative approach would involve keeping the horse quiet, administering antibiotics, and monitoring for worsening of clinical signs, especially respiratory distress. If the emphysema worsened, a tracheostomy could be performed to decrease the high pressure influx of air through a possible tracheal perforation into the subcutis. Suturing of a tracheal tear could be performed but would be difficult.

Follow up/discussion
Cervical radiographs revealed diffuse air present within the subcutaneous space and dissecting along fascial planes, including around the trachea (**104.3**). A small area of increased opacity and irregular tracheal margin was appreciated in the dorsal aspect of the trachea between the middle and distal third of the neck. Thoracic radiographs showed mild amounts of air present within the mediastinum surrounding the aorta and the esophagus (**104.4**).

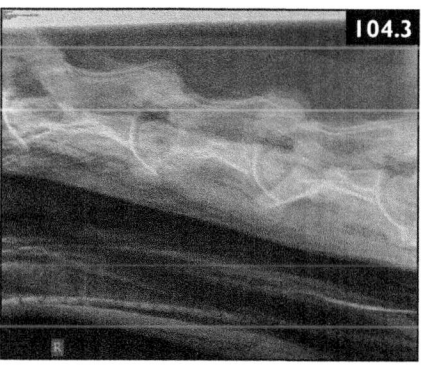

104.3

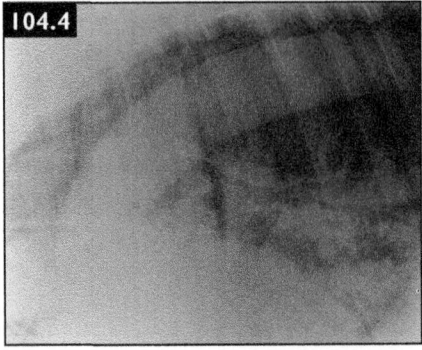

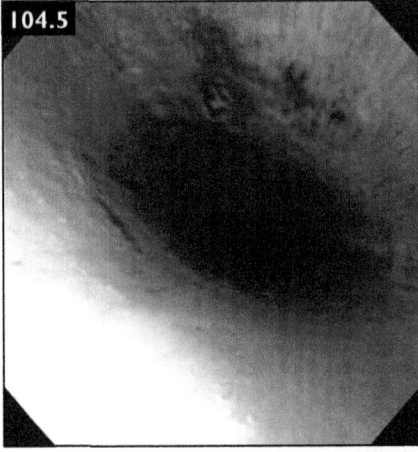

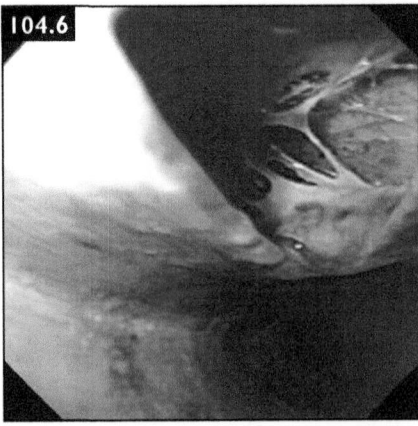

The horse was sedated with detomidine and butorphanol and endoscopic examination of the trachea was performed after lidocaine infusion into the trachea (to prevent coughing). It revealed a tear of indeterminate size on the dorsal surface of the trachea approximately 80 cm caudal to the nares. The tracheal mucosa ventral and adjacent to the defect was hyperemic and abraded (104.5, 104.6). The clipped area over the thorax (104.2) was in preparation for placement of a thoracic tube in case the mediastinum ruptured into the pleural space.

Conservative medical management was elected based on the location of the tear. The horse was closely observed for development of worsening clinical signs associated with a worsening pneumomediastinum progressing to pneumothorax. The horse was started on broad-spectrum antibiotics and flunixin meglumine. A tetanus toxoid booster was administered because the last tetanus vaccine was given over 6 months prior. Within 3 days of presentation the subcutaneous emphysema had decreased and the horse appeared more comfortable. The cause of the tracheal perforation could not be determined. The horse was stall rested for 3 weeks and kept close to other familiar horses to decrease vocalization. Approximately 2 weeks were required for resolution of the subcutaneous emphysema. Repeat tracheal endoscopy was performed 1 month later and the tracheal perforation had healed.

Tracheal rupture is an uncommon cause of subcutaneous emphysema in the horse but may occur from trauma, tracheal foreign body, or esophageal perforation that also penetrates the trachea. In this case, a traumatic incident was suspected (i.e. a kick), although no injury could be seen on the skin in the area of the perforation. Some reported complications of tracheal rupture include dyspnea from either upper airway obstruction or pneumomediastinum and pneumothorax. Medical treatment as outlined above is generally successful. Surgical repair should be considered if the tear is large and a cartilaginous ring is involved. A tracheostomy may be required if emphysema is so severe that the upper airway is being noticeably compromised. Severe subcutaneous emphysema in horses is most often caused by 'suction' wounds in the axillary region, but ruptured trachea should also be considered as a differential. Clostridial infections would be unlikely to cause such widespread emphysema and would be expected to show obvious clinical and laboratory signs of sepsis.

References

Caron JP, Townsend HG (1984) Tracheal perforation and widespread subcutaneous emphysema in a horse. *Can Vet J* **25**(9):339–341.

Fubini SL, Todhunter RJ, Vivrette SL *et al.* (1985) Tracheal rupture in two horses. *J Am Vet Med Assoc* **187**(1):69–70.

Gronvold AMR, Ihler CF, Hanche-Olsen S (2005) Conservative treatment of tracheal perforation in a 13-year-old hunter stallion. *Equine Vet Educ* **17**(3):142–145.

CASE 105

1 What is the interpretation of the radiographs? The radiographs reveal bulbous enlargement and radiopaque masses in the intra-alveolar spaces, as well as marked osteomyelitis and loss of the periodontal space.

2 Based on the history, physical examination, and radiographic findings, what is the most likely diagnosis for this case? Equine odontoclastic tooth resorption and hypercementosis (EOTRH).

3 How should this horse be treated? Currently, the preferred treatment method is removal of the affected incisors, which can be done standing with appropriate sedation and regional anesthesia. Typically, the horse becomes much more comfortable following removal of the affected teeth and will resume grazing eagerly once the extraction sites are healed.

Discussion

This relatively new dental disease was first documented in 2007 and typically affects the incisors and canines of middle-aged to older horses. There is no obvious breed predilection and the condition may be more common in males. Clinical signs may include reluctance to accept treats, head shaking, problems accepting the bit, changes in prehension, decreased appetite, and weight loss. Hallmark findings on examination include dental resorption, radiopaque masses in the intra-alveolar spaces causing bulbous enlargement, and loss of the periodontal space of the incisor. In later stages, and with more aggressive resorption, there may be disruption of alveolar bone, osteomyelitis, and even tooth fractures. Although the pathogenesis of the disease is not yet known, it has been proposed that periodontal inflammation begins the process of tooth resorption and the hypercementosis is actually an attempt by the body, albeit inappropriate, to stabilize the tooth.

References

Caldwell LA (2007) Clinical features of chronic disease of the anterior dentition in horses. *Proceedings of the 21st Annual Veterinary Dental Forum*, Minneapolis, pp. 18–21.

Earley E, Rawlinson JT (2013) A new understanding of oral and dental disorders of the equine incisor and canine teeth. *Vet Clin North Am Equine Pract* **29**:273–300.

Staszyk C, Bienert A, Simhofer H *et al.* (2008) Equine odontoclastic tooth resorption and hypercementosis. *Vet J* **178**:372–379.

CASE 106

1 List some differentials for the clinical signs found in this horse. Icterus can occur secondary to anorexia in the horse, therefore the differentials might include: viral respiratory disease (although some serous nasal discharge or cough might be expected), peritonitis (although HR may be expected to be >44), bacterial pneumonia, multisystemic viral or rickettsial infection, purpura hemorrhagica, liver disease, or leukemia.

2 What is the tentative diagnosis? Anaplasmosis (*Anaplasma phagocytophila* infection).

3 What is the preferred treatment? Oxytetracycline IV is preferred, although doxycycline or minocycline could be given PO instead.

4 How quickly can the clinical and laboratory findings be expected to improve? Twenty-four to 48 hours.

5 What additional clinical signs might occur with this disease? Myopathy or ataxia.

6 What is the best (most sensitive) method to confirm the diagnosis during this febrile stage? Whole blood PCR.

7 Would seroconversion be expected after 3 days of fever (likely 7 days after infection)? Approximately 50% of clinical cases are seropositive when tested with the point of care ELISA (SNAP test) at this stage.

Discussion

A. phagocytophila is a common and geographically expanding tick-borne disease in horses. Clinical signs include fever, jaundice, petechiations, leg edema, and depression, often with mild ataxia. In rare cases, myopathy and recumbency may occur. Diagnosis is based on presence of typical clinical signs, positive PCR during the febrile stage, finding inclusion bodies (morulae) in neutrophils in the early febrile period, and response to therapy. Most clinically affected horses have thrombocytopenia and leukopenia with an occasional left shift. Serology is only mildly helpful in confirming the disease; approximately 50% of infected horses are positive on the SNAP test towards the end of the febrile stage. The incubation period after the tick bite is approximately 5 days, with clinical signs occurring on days 5–12, and untreated horses may remain infected for ≥4 weeks. Tetracycline IV is the treatment of choice; minocycline and doxycycline are second-choice antibiotics. Routine supportive care should be administered. Steroids improve clinical signs but may prolong infection. The recovery rate is nearly 100%, even without treatment. Immunity is thought to last approximately 2 years after infection.

References

Davies RS, Madigan JE, Hodzic E *et al.* (2011) Dexamethasone-induced cytokine changes associated with diminished disease severity in horses infected with *Anaplasma phagocytophilum*. *Clin Vaccine Immunol* **18(11)**:1962–1968.

Franzén P, Aspan A, Egenvall A *et al.* (2009) Molecular evidence for persistence of *Anaplasma phagocytophilum* in the absence of clinical abnormalities in horses after recovery from acute experimental infection. *J Vet Intern Med* **23(3)**:636–642.

CASE 107

1 What medical therapy is indicated? Place a subpalpebral lavage system; topical antibiotics (gram-positive: neopolygramicidin or cefazolin; gram-negative: ciprofloxacin or ofloxacin) every 2–4 hours; topical mydriatic/cycloplegic (1% atropine drops) every 6–8 hours as needed for mydriasis then taper; taper anticollagenase (serum, EDTA, 10% acetylcysteine) every 4–6 hours; systemic NSAID (flunixin meglumine); eye saver mask.

CASE 108

1 What is the diagnosis? Atypical myopathy (AM) (a.k.a. seasonal pasture myopathy [SPM]; atypical myoglobinuria).

2 What is the cause of this condition? AM is a severe and often fatal condition that can occur in grazing horses. It results from the development of a multiple acyl-CoA dehydrogenase deficiency that blocks several steps in mitochondrial lipid metabolism. This significantly decreases the respiratory capacity of mitochondrial complexes of the electron transport system. These dysfunctions lead to the exhaustion of energy sources in muscle and subsequent rhabdomyolysis, which is principally seen in muscles that contain a high proportion of oxidative type I fibers. The disease is believed to be caused by hypoglycin A toxicity. In Europe, this toxicity is believed to be primarily associated with ingestion of seeds of the sycamore tree (*Acer pseudoplatanus*); in North America it is associated with the ingestion of seeds of the box elder tree (*Acer negundo*).

Discussion

AM (in Europe) or SPM (in the USA) is a seasonal, acute, severe, very often fatal non-exertional rhabdomyolysis of the respiratory, postural, and occasionally cardiac muscles of horses. The etiology and pathogenesis are described above and have only recently been elucidated. Previously, associations had been noted with the autumn or spring season, stormy or windy weather, trees around or in a pasture, and proximity to a stream or river. All of these coincide with seed production and preferred growing conditions of the sycamore or box elder tree. Overgrazing without supplementation of other feed is also likely a contributing factor. Onset of clinical signs is sudden, appearing within hours, and signs include extreme weakness and stiffness of the postural and respiratory muscles, particularly the neck, shoulder, diaphragm, and intercostal muscles. Rarely, the myocardial muscle is affected. Progression is commonly so rapid that horses may be found in lateral recumbency in the field, unable to rise. Death may ensue in as little as 72 hours in severely affected animals. Horses are afebrile and may be hypothermic if found on pasture, but this corrects quickly once the horse is brought indoors. The conjunctivae progressively redden, and rectal examination is typically normal except potentially for a distended bladder. Urine is dark brown. Laboratory findings, similar to other forms of rhabdomyolysis, include an extremely high CK, often rising to over 100,000 IU/l. Hypocalcemia and hyperglycemia are also found in the majority of cases and it is worth noting that blood and urine creatinine are typically normal.

Histopathology of muscle reveals hyaline myonecrosis/myodegeneration (Zenker's degeneration) of respiratory, neck, shoulder, biceps, masseter, and back/hindquarter muscles; the myocardium can also be affected in some cases. Myodegeneration of the respiratory muscles is likely the commonest cause of the reported high mortality rates. Minimal inflammatory response is present. Treatment is supportive, mainly involving analgesia and aggressive fluid therapy in a warm, well-padded environment. Multivitamin or vitamin E/Se injections have been advocated, but have not been shown to prevent death. Calcium supplementation and manual bladder expression at regular intervals may be required for some cases.

Prognosis is guarded to poor with reported mortality rates as high as 85%. Poor prognostic indicators include severe respiratory dyspnea or distress, increasing dyspnea during treatment, PaO_2 <85 mmHg, and increasing pain. Prognosis is likely dependent on individual factors of toxin tolerance as well as the concentration of toxin ingested. To date, there is no test to help determine these factors as a toxic dose has not been determined for horses and toxin production varies between trees and even between seeds produced by the same tree.

References

Valberg SJ, Sponseller BT, Hegeman AD *et al.* (2013) Seasonal pasture myopathy/ atypical myopathy in North America associated with ingestion of hypoglycin A within seeds of the box elder tree. *Equine Vet J* 45(4):419–426.

Votion DM, Serteyn D (2008) Equine atypical myopathy: a review. *Vet J* 178(2):185–90.

Votion DM, van Galen G, Sweetman L *et al.* (2014) Identification of methylenecyclopropyl acetic acid in serum of European horses with atypical myopathy. *Equine Vet J* 46(2):146–149.

CASE 109

1 What is the tentative diagnosis based on the above information? Ascarid impaction of the small intestine.

2 What are the hyperechoic areas marked (+) in the ultrasonograms? They are longitudinal and cross-sectional images of *Parascaris equorum* within the small intestinal lumen.

3 What is the greatest concern regarding complications that may occur within the next several days to months following this surgery? Peritoneal adhesions causing abdominal pain.

4 What anthelmintic would have been a better and safer treatment in this foal? Fenbendazole.

5 What is the prepatent period for *P. equorum*? 10–12 weeks.

Follow up/discusssion

At surgery the roundworms were successfully massaged into the cecum, but the small intestine became very hyperemic during the procedure. The foal recovered from anesthesia well and was nursing the mare within a few hours. The feces passed the following day are shown (**109.3**).

In one study, 72% of foals with roundworm impaction and colic had been administered anthelmintics, including pyrantel (n = 8), ivermectin (n = 7), and trichlorphon (n = 1), within 24 hours prior to the onset of colic. Of the 25 cases reviewed, 16 had simple obstructive ascarid impaction (SOAI) and nine had complicated

109.3

obstructive ascarid impaction, with the latter including concurrent volvulus (n = 6) or intussusception (n = 3). Short-term survival (defined as discharge from the hospital) occurred in 79% of foals treated for SOAI and in 64% of all foals with ascarid impaction. Long-term survival (>1 year) occurred in 33% of foals with SOAI, and the overall long-term survival was 27% for all foals with ascarid impaction. Formation of adhesions was the most frequent finding associated with death for foals that did not survive long term. In another study, manual expression of the impaction into the cecum without enterotomy had a higher survival rate than enterotomy and removal of the parasites.

References

Cribb NC, Cote NM, Bouré LP *et al.* (2006) Acute small intestinal obstruction associated with *Parascaris equorum* infection in young horses: 25 cases (1985–2004). *N Z Vet J* **54(6)**:338–343.

Tatz AJ, Segev G, Steinman A *et al.* (2012) Surgical treatment for acute small intestinal obstruction caused by *Parascaris equorum* infection in 15 horses (2002–2011). *Equine Vet J* **Suppl. 43**:111–114.

CASE 110

1 What is the tentative diagnosis? Severe rhinitis of unknown etiology.

2 What further diagnostic tests should be considered? Bacterial culture of nasal swabs. Viral infections (e.g. equine influenza, equine herpesvirus, equine viral arteritis, equine picornavirus, equine adenovirus) should be considered and appropriate samples obtained. Biopsy of the nasal mucosa should be considered in order to characterize the pathologic nature of the rhinitis (e.g. to identify evidence of bacterial invasion, vasculitis). Biopsy of an enlarged submandibular lymph node could also be considered in light of the possibility of neoplasia.

3 What is the significance of the large reactive lymphocytes identified on the blood smear? They could be indicative of abnormal/neoplastic cells (i.e. leukemia); however, they can also be present in severe inflammatory conditions. The morphologic changes of the cells in this case were not considered to be sufficient to support a diagnosis of leukemia. The thrombocytopenia could also be indicative of bone marrow suppression associated with hematopoietic neoplasia.

Follow up/discussion

Biopsy of both the nasal mucosa (obtained via the endoscope) and a submandibular lymph node showed tissue infiltration with a population of atypical round cells, the appearance of which was most consistent with a diagnosis of lymphoma (stage V). Malignant neoplasia is considered rare in the horse, but lymphoma is the most commonly diagnosed of these rarities, comprising approximately 3% of malignant diagnoses. There are multiple distinct disease patterns associated with lymphoma, including multicentric/generalized, alimentary, mediastinal, cutaneous, and solitary tumors of extranodal sites. Tumors may be of either B or T cell origin and there is no sex or breed predilection, although roughly 50% of cases are young adults. Clinical signs are variable depending on extent and location, but the most common include ventral edema, enlarged lymph nodes, respiratory distress, diarrhea, and colic. The presence or lack of pain on palpation will help distinguish between infection and possible neoplasm of solid masses. To date, there is no known etiology, but a recent case study has suggested a possible link to equine herpesvirus-5, a gamma herpesvirus.

The case presented here is of generalized or multicentric lymphoma and is the most common presentation of equine lymphoma. It generally affects adult animals and clinical signs vary depending on the tissue or organs affected. Masses may be quite large and are generally very firm.

Alimentary lymphoma typically affects younger horses (<5 years) and clinical signs are often consistent with protein-losing enteropathy, with or without diarrhea, and recurrent colic. Disease usually affects the small intestine and may manifest as either a generalized thickening of the bowel wall or distinct masses.

Mediastinal lymphoma largely affects adult horses and may present as a primary neoplasm of the thymus or thoracic lymphadenopathy. Clinical signs are consistent with space-occupying mass(es) that compress vessels and small airways. Thoracic radiographs or ultrasound may reveal pleural fluid, which if drained may allow visualization of thoracic masses.

Cutaneous lymphoma most often affects horses between 4 and 9 years of age and is characterized by single or multiple subcutaneous masses and/or enlarged peripheral lymph nodes. If growth is rapid, central necrosis and external drainage may occur. Most cutaneous lymphomas are of B cell origin, often being characterized as T cell-rich B-cell lymphomas.

Solitary extranodal tumors have been reported in several organs/locations. The clinical signs are determined by the tumor's size and location.

The prognosis is poor in all cases of equine lymphoma, with most animals progressing rapidly and either dying or being euthanized within 6 months of the onset of clinical signs. Surgery may be attempted for solitary tumors, depending on size and accessibility. Medical treatment, including intratumoral cisplatin, may be attempted depending on disease pattern and severity. Systemic chemotherapies are rarely utilized due to the large amount of drug required to treat adequately and the associated high cost in the face of a poor prognosis. Otherwise treatment is palliative, including corticosteroids, which may provide temporary relief from symptoms.

References

Doyle AJ, MacDonald VS, Bourque A (2013) Use of lomustine (CCNU) in a case of cutaneous equine lymphoma. *Can Vet J* **54(12):**1137–1141.

Mair TS, Pearson GR, Scase TJ (2011) Multiple small intestinal pseudodiverticula associated with lymphoma in three horses. *Equine Vet J* **Suppl. 39:**128–12.

Montgomery JB, Duckett WM, Bourque AC (2009) Pelvic lymphoma as a cause of urethral compression in a mare. *Can Vet J* **50(7):**751–754.

Sanz MG, Sellon DC, Potter KA (2010) Primary epitheliotropic intestinal T-cell lymphoma as a cause of diarrhea in a horse. *Can Vet J* **51(5):**522–524.

Van der Werf K, Davis E (2013) Disease remission in a horse with EHV-5-associated lymphoma. *J Vet Intern Med* **27(2):**387–389.

CASE 111

1 What is the provisional diagnosis? Infective valvular endocarditis with infarction of the spleen and liver.

2 What is the underlying cause of this disease? Infective endocarditis involves microbial infection of the endothelial surfaces of the heart. The valves are most commonly affected (i.e. valvular endocarditis), especially the aortic and left atrioventricular (mitral) valves. The source of infection is often not determined. A wide range of bacterial pathogens can be involved, but *Streptococcus* spp. and *Actinobacillus* spp. are most common. Embolization of fragments of the vegetative heart valve lesions to distant sites results in infection or infarction.

3 How could this condition have been diagnosed during life? Echocardiography allows identification and location of the heart valve lesions. Vegetative lesions appear as irregular hypoechoic to hyperechoic masses on the valves. Blood culture is used to identify the infective agent; several cultures obtained while the horse is off antibiotic treatment may be required.

Discussion

Infective endocarditis is rare but unfortunately typically carries a very poor prognosis. As noted above, the vast majority of cases are bacterial (*Actinobacillus* spp. or *Streptococcus* spp., although other bacteria have been isolated) but occasionally the infective organism is fungal (*Aspergillus* spp. or *Candida* spp.). Typically the lesions are not found in horses until they have become infective lesions, but they likely begin as a platelet-fibrin 'clot' adhered to a damaged valve, giving the infective agent a matrix to which it can attach and grow. The cause of the damage to the valve is generally unknown, but any recent illness may have contributed. The most common clinical signs are fever and a heart murmur that was not previously present, but a myriad of other clinical signs may be present as well, including all those present in this case. The mitral and aortic valves are more commonly affected and carry a worse prognosis than lesions on the pulmonic and/or tricuspid valves. As the lesions grow, small pieces may break off and lodge as emboli in other organs, in this case the spleen and kidney. The lungs may be affected if there are lesions on the pulmonic or tricuspid valves.

The prognosis is very poor due to several factors inherent to the condition. Very often there is no obvious precipitating event/cause and attempts at treating the initial clinical signs may have been made, delaying diagnosis and the start of appropriate therapy. Thus, by the time the condition is recognized it is often quite advanced. Additionally, the fibrin-platelet matrix allows the infective agent to shield itself well both from the host's immune system as well as from therapeutic agents. These factors make it notoriously difficult to treat and it is recommended to use antibiotics IV for at least 1–2 weeks before changing to antibiotics PO (for a total of up to 3 months). Antibiotic choice should be bactericidal and ideally based on blood culture and sensitivity. If this is not possible, or if the culture is negative, broad-spectrum antibiotic therapy may be necessary. It is also important to recognize that other supportive therapy may be necessary for the clinical signs exhibited on an individual basis. This may include aspirin PO for its antithrombotic properties, but because of inherent risks to the GI tract and renal system other NSAID are not recommended.

It is important to remember that prognosis remains poor in these cases, even with relatively prompt diagnosis and institution of therapy. Horses often relapse within a couple months of cessation of apparently successful therapy, or the affected valve may remodel and fibrose, predisposing the horse to heart failure and likely preventing return to work at the same level as prior to diagnosis.

References

Jesty SA, Reef VB. (2006) Septicemia and cardiovascular infections in horses. *Vet Clin North Am Equine Pract* **22(2)**:481–495, ix.

Maxson AD, Reef VB (1997) Bacterial endocarditis in horses: ten cases (1984–1995) *Equine Vet J* **29**:394–399.

Porter SR, Saegerman C, van Galen G *et al.* (2008) Vegetative endocarditis in equids (1994–2006). *J Vet Intern Med* **22(6)**:1411–1416.

CASE 112

1 What condition should be suspected? Equine grass sickness (EGS).

2 How does the cecocecal intussusception fit in with this disease? The cecocecal intussusception was probably a manifestation of abnormal intestinal motility due to EGS.

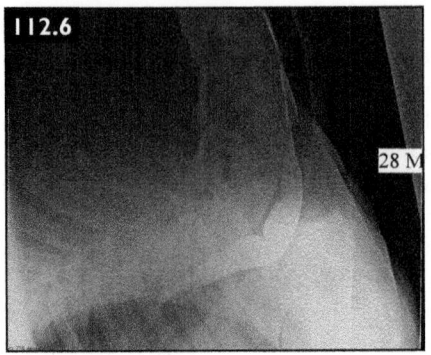

112.6

28 M

3 What clinical tests could be performed to help confirm the provisional diagnosis?
The phenylephrine eye test would show a large increase in size of the palpebral fissure; contrast esophography (barium swallow) may show delayed clearance of contrast agent in the esophagus (**112.6**); histologic examination of ileal biopsies would show damage to the enteric ganglia.

Discussion

There are three presentations of EGS: acute, subacute, and chronic. The case presented fits best into the chronic category, where the clinical signs related to GI stasis are often milder. It also demonstrates the wide variety of clinical signs and some of the difficulties associated with diagnosis of equine EGS. Grass sickness typically causes dysautonomia of the small intestine, particularly the ileum, but the large intestine may also be involved, particularly in subacute and chronic cases. Intussusception is uncommon in the horse and cecocecal intussusception is uncommon among intussusceptions (most occur in the small intestine or the ileocecal or cecocolic junctions). They are often associated with motility dysfunctions as with intussusception of other portions of bowel; ileocecal intussusception is sometimes associated with tape worm infection. Intussusceptions may be identified by ultrasound, but are sometimes only identified during exploratory laparotomy.

In most cases, a presumptive diagnosis of EGS is obtained ante-mortem based on the clinical signs and history, but definitive diagnosis of EGS currently is accomplished only through histopathologic examination of full-thickness intestinal biopsies obtained during exploratory laparotomy or postmortem examination to confirm neuronal loss. The ileum is the preferred site for biopsy as it is typically the most affected. In an effort to develop improved ante-mortem, less invasive diagnostics, rectal biopsy (using standard technique) was compared with ileal biopsy and was found to be inconsistent in comparison. Nasal biopsy has also been evaluated, but with disappointing results.

Other ancillary tests combined with the clinical picture may support a presumptive diagnosis, but are not definitive. These include fluoroscopic observation of a barium swallow or esophageal endoscopy. These methods may help evaluate esophageal smooth muscle function and, in the case of endoscopy, may also reveal the presence or absence of esophageal ulceration. These tests may be more helpful in acute or subacute cases where dysautonomia is more severe and widespread compared with chronic cases. Since bilateral ptosis is common in cases of EGS, corneal application of 0.5% phenylephrine resulting in temporary resolution of ptosis is supportive, but again, not pathognomonic of EGS. Hematology, biochemistry, urinalysis, and peritoneal fluid analysis are generally unhelpful in distinguishing EGS from other disorders, although there is some recent evidence that elevated serum amyloid A and fibrinogen concentrations may help differentiate between EGS and strangulating/non-strangulating small intestinal obstruction.

References

Copas VE, Durham AE, Stratford CH *et al.* (2013) In equine grass sickness, serum amyloid A and fibrinogen are elevated, and can aid differential diagnosis from non-inflammatory causes of colic. *Vet Rec* **172(15)**:395.

Mair TS, Kelley AM, Pearson GR (2011) Comparison of ileal and rectal biopsies in the diagnosis of equine grass sickness. *Vet Rec* **168(10)**:266.

Pirie RS, Jago RC, Hudson NP (2014) Equine grass sickness. *Equine Vet J* **46(5)**:545–553.

Schwarz B (2013) Equine grass sickness: what's new? *Vet Rec* **172(15)**:393–394.

CASE 113

1 How do you account for the large volume of peritoneal fluid? The excessive volume of peritoneal fluid probably represents fluid loss into the 'third space'. This is likely to be the result of increased vascular permeability associated with severe endotoxemia. Increased third space fluid losses decrease the effective circulating volume and lead to worsening shock and multi-organ dysfunction; this is the probable cause of the worsening azotemia.

2 What is the cause of the focal severe discoloration of the large colon? The discolored segment of intestine has been caused by intestinal infarction. This condition is termed non-strangulating intestinal infarction, and is likely to have been caused by intravascular thrombosis associated with DIC.

Discussion

Horses harbor large numbers of gram-negative bacteria in their large intestine. With colitis, the bowel wall becomes inflamed, which creates increased permeability resulting in systemic absorption of large amounts of endotoxin (a.k.a. lipopolysaccharide [LPS], a component of the outer membrane of gram-negative bacteria). Small amounts of endotoxin are absorbed from the gut in normal healthy horses without consequence. However, the sudden, massive increase in circulating endotoxin with colitis overwhelms the normal defences and can have devastating consequences including systemic inflammatory response and organ failure.

Endotoxin in circulation is highly inflammatory and immunogenic, and triggers multiple physiologic cascades in the body's attempt to isolate and eliminate it.

LPS directly damages the endothelium of vessels, creating widespread inflammation (vasculitis); as a consequence, activation of the coagulation and complement cascades ensues. Additionally, fibrinolysis is increased in an attempt to restore balance. The result is a hypercoagulable state in addition to the vasculitis. This hypercoagulability results in rapid consumption of clotting factors and platelets and a condition known as DIC, characterized by thrombocytopenia, increases in PT, PTT, and ACT times as well as increased D-dimers and fibrin degradation products. This hypercoagulable state creates conditions whereby thrombi may form in the small vessels of organs, resulting in organ failure (as in this case). Here, the ischemia and eventual necrosis of the infarcted bowel likely created additional bowel inflammation, with further increased endotoxin leakage, escalating what previously had probably been subclinical DIC to symptomatic, uncontrollable DIC.

The vascular damage, resulting inflammation, and DIC all contribute to a rapid systemic increase in vascular permeability. This increase in permeability can result in the transfer of large amounts of fluid from the circulation to body cavities, in this case the peritoneal cavity (i.e. third spacing). The transfer of fluids can cause rapid decreases in circulating volume, resulting in rapid decreases in blood pressure, ending in cardiovascular compromise (a.k.a. cardiovascular shock). It is very difficult to replace the fluid volume of an animal as large as a horse at the rate they are losing fluids when they develop severe DIC. This combination of hypercoagulability (DIC) and severe vasculitis is also known as the systemic inflammatory response syndrome.

Reference
Monreal L, Cesarini C (2009) Coagulopathies in horses with colic. *Vet Clin North Am Equine Pract* 25(2):247–258.

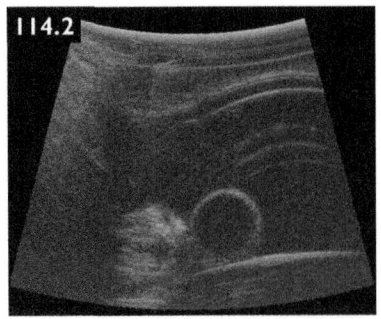

114.2

CASE 114
1 What are the differential diagnoses? Cystic corpora nigra, neoplasia (melanoma), and inflammatory infiltrate.
2 What diagnostic test is indicated to confirm the diagnosis? A diagnosis of cystic corpora nigra can usually be made based on clinical appearance, but it can be confirmed with ultrasonography (114.2).

CASE 115

1 What is suspected? Severe mucosal inflammation and necrosis, probably associated with instillation of iodine solution into the GPs.

2 What is the prognosis? The prognosis for recovery is poor. The necrosis of the mucosa will likely cause severe neurologic damage to the CNs within the pouches.

Discussion

It has been well established that povidone–iodine infusions into healthy GPs causes considerable inflammation and damage to the mucosa, even when diluted to a 10% solution. In this case, it is quite likely that the integrity of the GP mucosa was compromised to some degree prior to the infusion due to some unknown pathology (presumed mycosis). As visualization of the interior of the pouches was not possible, the extent of damage and inflammation present in the pouch prior to infusion could not be determined. This case illustrates the high risk associated with infusing an inflammatory substance into a potentially compromised pouch without a visual assessment of its integrity and the associated CNs and arteries. Extreme care must be taken to ensure appropriate dilution to prevent chemical mucosal injury. It is generally recommended that iodine solutions should not be used in the upper respiratory tract because of their known inflammatory properties.

References

Sherlock CE, Hawkins FL, Mair TS (2007) Severe upper airway damage caused by iodine administration into the guttural pouches of a pony. *Equine Vet Educ* **19**:515–529.

Wilson J (1985) Effects of indwelling catheters and povidone iodine flushes on the guttural pouches of horses. *Equine Vet J* **17**:242–244.

CASE 116

1 Describe the lesion and provide possible etiologies. There are peripapillary areas of hypopigmenation nasal and temporal to the optic disk ('butterfly lesion'). This lesion is described in horses with previous posterior uveitis; this horse had disinsertion of the iridocorneal angle laterally consistent with previous blunt trauma.

CASE 117

1 What is the presumptive diagnosis? The clinical signs and radiographic and ultrasonographic findings are compatible with a diagnosis of a pulmonary granular cell tumor.

2 How would this suspicion be confirmed? Transendoscopic biopsy may be possible. Alternatively, biopsy forceps may be introduced via a distal cervical tracheotomy.

3 What treatment options should be considered? Reported treatments for this tumor have included transendoscopic electrosurgery, laser ablation, and lung resection.

Discussion

Primary lower respiratory neoplasia is rare in the horse, less common than metastasis to the lung by other malignant neoplasms. Of these rare primary neoplasms, the pulmonary granular cell tumor (a.k.a. myoblastoma or putative Schwann cell tumor) is the most commonly diagnosed. Histologically, the tumors are comprised of loosely arranged round to polygonal cells divided by a fine fibrovascular stromal network. The cells have abundant eosinophilic cytoplasm and granules, but mitotic figures are rarely seen. Grossly, the masses are described as being firm, white to tan, round to multilobulated masses covered by normal respiratory epithelium. Immunohistochemical staining suggests a Schwann cell origin. Uni- and bilateral distribution as well as solitary and multinodular cases have been reported, but it appears that unilateral mulitinodular disease may be more common than either bilateral disease or solitary masses. No cases of metastasis to lymph nodes or other organs have been reported. This, along with the rare observance of mitotic figures, gives reason to believe that this is a slow growing neoplasm with limited metastatic potential.

While there is no observed breed predilection, among the limited number of cases reported there appears to be a sex predilection for females over males. As with many other neoplasms, older animals are more likely to be affected. Presenting clinical signs are likely to be similar to this case: chronic cough and exercise intolerance with or without a concurrent history of weight loss. Other rarer presentations, such as hypertrophic osteopathy and epistaxis, have also been reported. Physical examination is likely to be normal with the exception of quiet to absent breath sounds over one hemithorax and an increased RR. Hematology/biochemistry and ultrasound are typically normal, but radiography may reveal soft tissue opacities consistent with neoplasia. Bronchoscopy may show masses protruding into or occluding mainstem bronchi, but pinch mucosal biopsies are non-diagnostic, as only the normal overlying mucosa is collected. Diagnostic biopsies may be obtained as suggested above. Depending on the distribution and potential access, surgical treatment may be attempted, but overall the prognosis is guarded and most horses are euthanized due to worsening disease.

References

Facemire PR, Chilcoat CD, Sojka JE *et al.* (2000) Treatment of granular cell tumor via complete right lung resection in a horse. *J Am Vet Med Assoc* **217**:1522.

Heinola T, Heikkilä M, Ruohoniemi M *et al.* (2001) Hypertrophic pulmonary osteopathy associated with granular cell tumour in a mare. *Vet Rec* **149**(10): 307–308.

Ohnesorge B, Gehlen H, Wohlsein P (2002) Transendoscopic electrosurgery of an equine pulmonary granular cell tumor. *Vet Surg* **31**:375–378.
Pusterla N, Norris AJ, Stacy BA *et al.* (2003) Granular cell tumours in the lungs of three horses. *Vet Rec* **153**(17):530–532.

CASE 118

1 What is the likely cause of the swelling of the neck and the radiological findings? The swelling of the neck and the radiological findings (tracheal intraluminal soft tissue thickening, peritracheal and subcutaneous emphysema, and pneumomediastinum) are all suggestive of tracheal rupture.

2 What is the likely cause of the cobblestone appearance of the tracheobronchial mucosa? It could be caused by inflammatory or neoplastic disease. Biopsies of the lesions would be required to diagnose the precise cause. In this case, biopsies identified the presence of well-demarcated cartilaginous nodular masses compatible with tracheobronchopathia osteochondroplastica.

Discussion

Tracheobronchopathia osteochondroplastica is the development of cartilaginous or mineralized nodules in the submucosa or mucosa of the trachea and/or bronchi. The condition has been commonly reported in humans. Despite its known existence since the 19th century, the etiology and pathogenesis of this rare condition remain unknown. It is, however, considered a benign condition typically diagnosed incidentally via bronchoscopy or CT and confirmed by histopathology. In humans, the condition is often asymptomatic but can present with non-specific complaints (e.g. chronic cough). In recent years a condition consistent with tracheobronchopathia osteochondroplastica was reported in a dog and a horse (this case). In these cases, the distribution, visual appearance, and histopathology were consistent with what has been reported in the human literature. In both animals the condition was diagnosed via bronchoscopy and could potentially be considered incidental, although in this case no other cause for the tracheal rupture was identified. As in humans, no therapy is necessary, but the condition should be considered as a differential for nodular tracheal disease in horses and, if possible, biopsies evaluated.

References

Sellon RK, Johnson JL, Leathers CW *et al.* (2004) Tracheobronchopathia osteochondroplastica in a dog. *J Vet Int Med* **18**(3):359–62.
Spanton JA, Henderson ISF, Krudewig C *et al.* (2008) Tracheal rupture in a native pony mare associated with a condition resembling tracheobronchopathia osteochondroplastica. *Equine Vet Educ* **20**:582–586.

Zhang XB, Zeng HQ, Cai XY *et al.* (2013) Tracheobronchopathia osteochondro-plastica: a case report and literature review. *J Thorac Dis* 5(5):E182–184.

CASE 119

1 What other examinations should be performed? The tracheal aspirate or a nasal/nasopharyngeal swab should be tested by PCR for respiratory viruses. In this case, a nasal swab was positive for equine influenza virus (EIV) (H3N8).

2 What is the diagnosis? Bacterial bronchopneumonia secondary to EIV infection.

3 How should this case be managed? The mare should be kept quarantined owing to the contagious nature of equine influenza. Antimicrobial therapy based on the results of sensitivity testing of the bacterial isolate from the tracheal aspirate should be continued until clinical recovery. The mare should be kept in a 'dust-free' environment until clinical resolution in order to reduce the risk of developing chronic inflammatory airway disease.

Discussion

EIV is a highly contagious member of the *Orthomyxoviridae* family and accounts for the majority of viral upper respiratory tract disease in horses. *Orthomyxoviridae* are single-stranded RNA viruses that replicate in the nucleus and bud from the plasma membrane. This family includes influenzas A, B, C, and Thogotovirus, with influenza A virus being the most widespread and including EIV. The official name for each influenza virus includes the host, geographic origin, strain number, and year of isolation, as well as the hemagglutanin (H) and neurominidase (N) types, but most refer to them only by the species and H and N type (e.g. equine influenza H3N8) or by geographic origin and/or strain number (e.g. Florida clade 1 or just clade 1). There are two known subtypes of EIV, H7N7 and H3N8, with H3N8 being the strain isolated in all outbreaks since 1979. A major hallmark of the influenza A viruses is their ability to undergo antigenic drift and antigenic shift.

The virus is spread by aerosol then infects and replicates in the cells of the upper and lower respiratory tract. Inflammation and serous discharge begins within 24 hours of infection, progressing to a mucopurulent discharge and eventual impairment of the mucocilliary apparatus. Horses are febrile for 4–5 days, often have a cough, and may exhibit depression and/or inappetence. Damage to the respiratory epithelium and impairment of the mucociliary apparatus predisposes infected horses to secondary bacterial infection, most often with gram-negative aerobic bacteria, but in more severe cases gram-positive and/or anaerobic bacteria may be involved. Viral infection in healthy horses is self-limiting and the virus is typically cleared in 2–3 weeks. Viral shedding occurs for approximately 5 days after clinical signs begin. It is recommended to rest and quarantine affected horses for at least this length of time to limit complications and spread of disease. Swabs of nasal mucus taken early in the course of disease can be submitted for viral isolation or RT-PCR. Serologic diagnosis using paired sera is possible later in the disease course.

Inactivated vaccines containing both the H7N7 and H3N8 viruses are available. It is preferable for the vaccine to contain virus that is closest to the current strains. The most common viruses isolated by far in recent years are the Florida clade 1 and clade 2 viruses (both are H3N8). Most current vaccines are IM adjuvanted vaccines that induce antibody production but questionable cell-mediated immunity. The current recommended vaccine protocol for high-risk animals (e.g. young animals and those frequently traveling, showing, or racing) is an initial vaccine with boosters at roughly 3 weeks and 5 months, followed by yearly boosters. A modified live intranasal vaccine is available and may offer better cell-mediated protection. It is important to remember (and communicate to clients) that a vaccine may help reduce clinical signs, but does not necessarily prevent infection or transmission.

References

Cullinane A, Gildea S, Weldon E (2014) Comparison of primary vaccination regimes for equine influenza: working towards an evidence-based regime. *Equine Vet J* **46(6):**669–673.

Elton D, Cullinane A (2013) Equine influenza: antigenic drift and implications for vaccines. *Equine Vet J* **45(6):**768–769.

Marr C (2013) Influenza: are we protecting our horses effectively? *Equine Vet J* **45(6):**766–767.

Paillot R, Prowse L, Montesso F *et al.* (2013) Duration of equine influenza virus shedding and infectivity in immunised horses after experimental infection with EIV A/eq2/Richmond/1/07. *Vet Microbiol* **166(1-2):**22–34.

CASE 120

1 What abnormalities can be seen in the CT scan? There is soft tissue swelling adjacent to the left temporomandibular joint, destruction of the temporal bone, and absence of the articulation at the caudal aspect of the joint. These changes are compatible with an invasive tumor that has caused osseous destruction around the left temporomandibular joint.

2 What further diagnostic procedures should be undertaken? Biopsy of the subcutaneous mass; this confirmed the presence of malignant melanoma.

Discussion

Regardless of coat color, this case is likely to be anaplastic malignant melanoma. Both the location and apparent bone destruction/invasion indicate that surgical intervention is not possible and medical therapy is highly unlikely to be successful; together these create a very poor prognosis.

Melanoma is typically associated with older gray horses but can affect horses of any age or coat color. Melanocytic tumors may be classified as gray horse dermal melanoma, gray horse dermal melanomatosis, melanocytoma, or anaplastic malignant melanoma. Anaplastic malignant melanoma is the most aggressive and often results in death within a year of diagnosis for non-gray horses. While many melanocytic tumors are considered benign, any melanocytic tumor has the potential to become malignant, with up to 66% becoming so at some point. The most common dermal sites of melanocytic tumor development include the tail base, perineum, sheath, parotid area, subauricular lymph nodes, and commissures of the lips. Particularly in gray horses, melanoma may develop in other organs (e.g. spine, GI tract, guttural pouch, salivary gland) without any evidence of dermal melanoma. Rapidly growing tumors are very firm and may outgrow their blood supply and develop central necrosis, which may drain externally and develop secondary bacterial infections. Left untreated, these rapidly growing masses are likely to metastasize to other areas of the body within a relatively short period of time. A biopsy is necessary (fine needle aspirate is not adequate; excisional or Tru-cut biopsy is necessary) to determine the likelihood of malignancy and differentiate melanoma from other tumors. Biopsy should especially be performed in horses with a solitary mass, non-gray horses, or a mass in an uncommon location.

Several therapies have been reported including intralesional, topical, and systemic therapies, but these tend to be effective for a very small proportion of cases and typically only for horses with a low number of small (<2–4 cm) tumors. More aggressive, rapidly growing tumors are unlikely to respond to any of the available medical treatments and surgery is indicated whenever possible. Complete removal of large or aggressive tumors is difficult as the masses infiltrate deeper tissues, making it difficult to obtain clean margins on all surfaces.

References

Moore JS, Shaw C, Shaw E *et al.* (2013) Melanoma in horses: current perspectives. *Equine Vet Educ* **25**(3):144–150.

Phillips JC, Lembcke LM (2013) Equine melanocytic tumors. *Vet Clin North Am Equine Pract* **29**(3):673–687.

CASE 121

1 Other than continuing corticosteroid or other immunosuppressive therapy such as azathioprine, what additional treatments could be offered for the inflammatory bowel disorder (IBD)? Additional medical treatments could include metronidazole (generally doses lower than the systemic antimicrobial dose) or sulfasalazine

(10–20 mg/kg PO q24h). Administration of a probiotic, in the hope of changing the intestinal microbiome, could be attempted but seldom appears to be clinically successful. An effort should be made to decrease the horse's exposure to possible dietary or parasitic antigens (daily pyrantel tartate could be tried but has seldom been effective in horses with IBD) and to provide quality nutrients to the healthiest part of the bowel (e.g. small intestine or large intestine – whichever is less affected).

Follow up/discussion
In this case, numerous fecal flotation tests had been negative and anthelmintic administration for all of the horses on the farm appeared to be effective based on the low number of fecal parasites identified. Both the small and large bowel were affected based on the duodenal and rectal biopsy results. If only the small intestine had been involved, feeding a high-quality forage that could be properly digested in the large bowel would have been recommended. Skin testing for food allergens could have been performed, but results have rarely been helpful as many of the same food substances frequently cause equal skin reactions in both normal and IBD horses.

In this particular case, the mare was placed on a low gluten diet and the dexamethasone dose was increased (0.05 mg/kg) and administered IM daily for 10 days followed by a tapering dose over 10 weeks. On recheck examination 5 months later, the mare had gained 200 lb (90 kg). Mild colic episodes were still present during the following year, but the body weight remained excellent.

Chronic inflammatory and infiltrative bowel disease may be the result of lymphoplasmacytic, eosinophilic, or granulomatous enterocolitis. Lymphosarcoma may also cause IBD. Weight loss, often with a good appetite, and low plasma protein are characteristic findings. The glucose absorption test can be used to confirm malabsorption but intestinal biopsy is needed to confirm the infiltrate. The role of gluten diets and gluten-dependent antibodies has recently been investigated in horses with IBD. Although further investigation is needed, a 14-year-old Warmblood with IBD had increases in some gluten-dependent antibodies and responded clinically to diet changes (as did the mare in this case) with a decline in gluten-dependent antibodies. Low-gluten diets used in the case report and in this case have included alfalfa hay and crushed black oats. This 1,320 lb (600 kg) mare was fed 15 lb (6.75 kg) of alfalfa hay and 6 lb (2.75 kg) of oats daily. More research and case reports are needed to determine if gluten diets and gluten antibodies play a causative role in equine IBD.

Reference
van der Kolk JH, van Putten LA, Mulder CJ *et al.* (2012) Gluten-dependent antibodies in horses with inflammatory small bowel disease (ISBD). *Vet Q* **32(1)**:3–11.

CASE 122

1 Identify the linear corneal lesions present in this picture (122.1). What are the possible etiologies? Haab striae, which occur as a result of breaks in Descemet's membrane. They can be secondary to chronic glaucoma and buphthalmos or can be an incidental finding in otherwise normal eyes.

CASE 123

1 Can any abnormalities be identified in the skull radiograph? The suture line between the basioccipital and basisphenoid bones is open. This suture line does not fuse until 5 years of age.

2 What is the significance of the submucosal hemorrhages in the wall of the right guttural pouch? Horses that flip over backwards may rupture the muscles within the guttural pouch (longus capitis and rectus capitis ventralis muscles), possibly resulting in hemorrhage.

3 What abnormalities can be seen in the CT images? The dorsal CT image (**123.3**) shows a displaced fracture through the suture between the basioccipital bone and the basisphenoid bone. There is soft tissue swelling on both sides of the basilar bones (worse on the left) with bone fragments within the soft tissue. The transverse CT image (**123.4**) reveals a small, minimally-displaced fracture of the right temporal bone and a cortical fragment of the basioccipital bone to the right of the basioccipital bone.

Discussion

Basilar skull fractures are not uncommon in young, inexperienced horses and often occur when resistance to training causes the horse to rear up and fall over backwards. The basilar aspect of the skull may hit the ground with force great enough to fracture the skull, with a correlation between the hardness of the surface and degree of damage (i.e. soft pasture creates less damage than concrete). Less damage may be noted when the incident happens in a stall, where the head contacts the ceiling or stall wall instead of the floor. There may be concurrent rupture of, or avulsion fracture from, the muscles that insert at the base of the skull (e.g. longus capitis), as in this case. These potentially lead to hemorrhage into the guttural pouches and have been associated with poorer prognosis. Clinical signs are most often neurologic in nature and are widely variable depending on several factors including, but not limited to, location and number of fractures, presence and degree of fracture dislocation, evidence of intracranial hemorrhage, and cervical spinal involvement. Therapy is generally supportive, based on clinical signs (although in rare cases surgery may be attempted), should be instituted as soon as possible, and may require referral to a hospital setting capable of providing 'round the clock' intensive nursing care.

Prognosis is highly dependent on clinical signs, response to therapy (which is largely symptomatic), and imaging findings.

Imaging includes skull radiographs, upper airway endoscopy, and CT. Lateral and ventrodorsal radiographic views often give very good information, but oblique views are typically of little value owing to the overlap of structures creating difficulties in interpretation. Radiographic findings may include: displacement of basisphenoid-basioccipital suture line; soft tissue opacity or bony fragments in the guttural pouch; ventral deviation of the dorsal pharyngeal wall; attenuation of the nasopharynx; or the presence of gas in the cranial vault or cranial cervical spine. The lateral view is the most diagnostic for a basilar fracture as it is the view most likely to show the presence or absence of displacement, but interpretation is difficult if displacement is minimal. Radiographs also offer no insight into potential damage of intracranial structures or intracranial bleeding.

CT has proven far more effective at evaluating intracranial structures as well as determining the location, displacement, and fragmentation of basilar skull fractures. Much of this is due to the ability to view cross-sectional slices and the more recent ability to create 3-D images, allowing a more complete analysis of the skull. More subtle fractures or displacements, as well as soft tissue damage, can be visualized using CT, thereby aiding in prognostic and therapeutic determinations. The added risk associated with GA must be considered, but standing CT has been developed for horses and may be preferable if available. The additional cost of CT (standing or otherwise) is likely to also be a factor, and should be discussed in light of radiographic findings, clinical signs, and other factors. While prognosis is often poor, there are many horses that have recovered and returned to work.

References

Anderson JM, Hecht S, Kalck KA (2012) What is your diagnosis? Skull fracture in a foal. *J Am Vet Med Assoc* **241(2)**:181–183.

Feary DJ, Magdesian KG, Leman MA *et al.* (2007) Traumatic brain injury in horses: 34 cases (1994–2004). *J Am Vet Med Assoc* **231(2)**:259–266.

Lim CK, Saulez MN, Viljoen A *et al.* (2013) Basilar skull fracture in a Thoroughbred colt: radiography or computed tomography? *J S Afr Vet Assoc* **84(1)**:E1–5.

MacKay RJ (2004) Brain injury after head trauma: pathophysiology and treatment. *Vet Clin North Am Equine Pract* **20(1)**:199–216.

CASE 124

1 What abnormalities can be seen in this laparoscopic image (124.2)? Enlarged colonic mesenteric lymph nodes.

2 What abnormalities can be seen in the large intestinal biopsy (124.3)? The biopsy shows encysted cyathostomin larvae within the colonic mucosa.

Answers

Discussion

This is a case of chronic larval cyathotominosis. As with the acute form, it is an inflammatory condition caused by the larval stages of cyathostomins. Unlike the acute disease, chronic disease and clinical signs progress slowly over a greater period of time and may result in protein-losing enteropathy (as in this case) or failure to thrive and grow. In most normal horses, the process of cyathostomin larvae encysting and being encysted causes little to no inflammation. However, in some horses the inflammation is more significant, leading to thickening of the bowel wall and eventual clinical signs. Treatment with a larvicidal anthelmintic (fenbendazole or moxidectin) should be instituted alongside supportive/symptomatic therapies.

As with antibiotic resistance, anthelmintic resistance is becoming more common globally. In response, many practitioners and organizations are recommending deworming that is less frequent and, when possible, more targeted based on fecal egg counts. There are differing concerns between age groups, and therefore different programs for these groups have emerged rather than a single blanket program for all horses. Other aspects of management (i.e. adequate manure removal, pasture rotation) are just as essential, if not more so, to managing parasite burdens and should be given appropriate attention and emphasis.

References

Nielsen MK, Reinemeyer CR, Donecker JM *et al.* (2014) Anthelmintic resistance in equine parasites. Current evidence and knowledge gaps. *Vet Parasitol* 204(1-2): 55–63.

Traversa D (2008) The little-known scenario of anthelmintic resistance in equine cyathostomes in Italy. *Ann N Y Acad Sci* 1149:167–169.

CASE 125

1 What are the differentials for an aberrantly enlarged mammary gland? Include: abscess (mammary gland abscesses due to *Corynebacterium pseudotuberculosis* or *Streptococcus equi* are usually not associated with lactation); mastitis, most commonly caused by *Streptococcus* non-*equi* spp.; trauma; inappropriate lactation; avocado toxicity; placentitis; twins; and neoplasia. Abscess and mastitis would likely be painful on palpation. Inappropriate lactation, avocado toxicity, placentitis, and twins would likely cause bilaterally symmetrical swelling.

2 What does the hypercalcemia suggest? Since renal indices were normal, the hypercalcemia would be highly suggestive that the mare had a neoplastic disorder.

3 What other blood test could be performed to help confirm that the hypercalcemia was associated with, or caused by, a neoplasm? Parathormone-related protein (PTHrP) is increased in some neoplastic diseases, leading to hypercalcemia.

4 What would be the next diagnostic procedure(s)? Needle aspiration of the mammary lymph node or the mammary tissue should be performed.

Follow up/discussion

In this mare, the serum PTHrP was 13.00 pmol/l (normal <1.0). The parathormone was mildly elevated at 2.60 pmol/l (normal <2.0). Calcitonin was elevated at 164 pg/ml (normal <40). Plasma ionized calcium was elevated at 9.0 mg/dl (2.24 mmol/l). On complete and careful examination of the mare, a small mass was found by palpation in the mid-thorax involving the eleventh rib. Abdominal palpation per rectum revealed several 5–10 mm nodules in the left

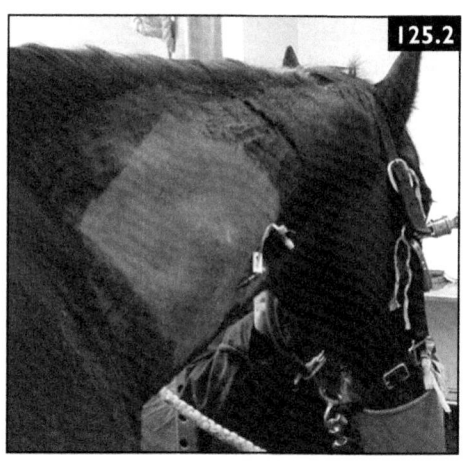

broad ligament. Ultrasound examination of the neck (attempting to rule out a parathyroid tumor) revealed a large soft tissue mass within the right cervical musculature (**125.2**).

Aspirates of the cervical and rib masses and the mammary tissue revealed intact and ruptured cells with round nucleoli having finely stippled chromatin and multiple large nucleoli. The diagnosis was an endocrine tumor. The mare was euthanized and the final diagnosis was a mammary gland papillary adenocarcinoma with local metastasis to the inguinal area and the medial left thigh. It was closely associated with the saphenous vein, which explained the limb edema. The masses on the left thorax and right cervical area were metastatic lesions.

Mammary gland disease in horses is rare compared with most domestic species, but is occasionally seen in mares >12 years of age. Mammary neoplasia should be considered when only one side of the udder is enlarged (often not painful), regional lymph nodes are enlarged, and the milk is bloody and cellular. Ulceration of the udder may also occur. Most equine mammary neoplasms are malignant and have a poor prognosis. They may be solid carcinomas, tubular adenocarcinomas, or papillary ductal adenocarcinomas, as in this case. Successful treatment requires early mastectomy and removal of regional lymph nodes.

References

Brendemuehl JP (2008) Clinical commentary: mammary gland enlargement in the mare. *Equine Vet Educ* **20**(1):8–9.

Shank M (2008) Mammary gland tumors in mares. *Equine Disease Quarterly* **17**(3).

Wynn S, Fougère B (2007) *Veterinary Herbal Medicine*. Elsevier, St. Louis.

CASE 126

1 What is this condition? Vitiligo.

2 What is its cause? This is an acquired depigmentation syndrome, possibly associated with an autoimmune disease.

3 What treatments are available? There is no remedy. Affected areas may be at increased risk of actinic dermatitis (sunburn) and squamous cell carcinoma. Protection of the affected areas from ultraviolet light is therefore prudent. Dyes, stains, or tattooing may be used to cover the affected areas.

Discussion

Vitiligo is an uncommon, likely heritable, depigmentation of annular areas of skin. A breed predilection for young Arabians, often gray, has been noted, but vitiligo can occur on any coat color at any age. Pregnant or recently foaled mares may be overrepresented. Affected skin is typically on the muzzle and around the eyes, and occasionally affects the external genitalia, hooves, or other areas of the body. There are multiple theories of pathogenesis, but autoimmune is the most favored. Antimelanocytic antibodies have been identified in the serum of some cases and a T-lymphocyte infiltration has been noted in early lesions. The skin is otherwise normal aside from the lack of pigmentation (melanocytes) and no treatment is necessary.

References

Montes LF, Wilborn WH, Hyde BM *et al.* (2008) Vitiligo in a quarter horse filly: clinicopathologic, ultrastructural, and nutritional study. *J Equine Vet Sci* 28(3):171–175.

Scott D, Miller W (2011) *Equine Dermatology*, 2nd edn. Elsevier, Maryland Heights, pp. 390–393.

CASE 127

1 What is hemochromatosis? Iron overload, or hemochromatosis, is associated with chronic hepatic cirrhosis in adult horses. Histologic lesions include disruption of hepatic architecture, bridging fibrosis, and bile duct hyperplasia. Iron accumulation is present within hepatocytes, macrophages, and Kupffer's cells, as indicated by Prussian blue staining. Iron accumulation is not noted in other tissues in these horses. Serum iron concentration may be high. A history of excessive dietary iron has not been a consistent feature in these horses, and it has been suggested that for unknown reasons excessive intestinal iron absorption occurs with resultant accumulation of iron in the liver. It is unknown whether the accumulation of iron in the liver is the result of liver failure or the cause of liver failure.

2 **What is the prognosis for this pony?** Poor due to liver failure and fibrosis.

3 **What treatments should be recommended?** There is no specific treatment for hemochromatosis. Horses with liver failure and hepatic encephalopathy should not be stressed. Sedation should be used only when necessary and diazepam should be avoided. IV fluid therapy using a balanced electrolyte solution, preferably without lactate, should be administered; this should be supplemented with potassium and dextrose. Supplemental vitamins can be administered but may not be necessary. Neomycin can be administered PO in order to decrease ammonia production in the bowel. Lactulose may also be used PO to decrease ammonia absorption from the bowel and can be used concurrently with neomycin. Vinegar (acetic acid) may be effective in decreasing blood ammonia when administered PO. The horse should be fed high-carbohydrate, high-branch chain amino acid feeds, with moderate to low TP content. The horse should be protected from sunlight in order to prevent photosensitization. Anti-oxidant, anti-inflammatory and anti-edema therapy is indicated in acute hepatic failure. These treatments include dimethylsulfoxide, S-adenosylmethionine, acetylcysteine, and mannitol given IV and vitamin E given IM. Anti-inflammatory therapy should include flunixin meglumine and pentoxifylline.

Discussion

A major part of iron homeostasis is control of GI absorption, as there is no efficient mechanism for excretion. Intestinal absorption increases or decreases depending on physiologic need and occurs mainly in the small intestine in two phases (a fast phase in the duodenum and a slow phase in the ileum), although some absorption occurs in the stomach and colon as well. Ferrous iron is more readily absorbed than ferric iron, but overall only 5–10% of iron is absorbed from ingesta. Iron that is taken up by enterocytes either remains in the enterocyte until the cell is sloughed or is transferred to the plasma. Once iron is in the plasma, it is bound to transferrin and delivered to the bone marrow for storage (as ferritin or hemosiderin) or incorporation into erythrocytes. Some additional storage occurs in the spleen and liver. Some excretion of iron is possible via bile, but up to 40% of iron excreted in this manner will be re-absorbed. Excessive iron accumulation in tissues may result in tissue peroxidation and organelle dysfunction mediated by free radical production. It is unclear whether liver damage precedes or is a consequence of hemochromatosis in horses, but the histologic lesions present in this case are typical of this condition.

References

Lavoie JP, Teuscher E (1993) Massive iron overload and liver fibrosis resembling haemochromatosis in a racing pony. *Equine Vet J* **25**(6):552–554.

Pearson EG, Andreasen CB (2001) Effect of oral administration of excessive iron in adult ponies. *J Am Vet Med Assoc* **218**(3):400–404.

Pearson EG, Hedstrom OR, Poppenga RH (1994) Hepatic cirrhosis and hemochromatosis in three horses. *J Am Vet Med Assoc* 204(7):1053–1056.

CASE 128

1 What conditions are suspected? Squamous cell carcinoma (SCC), habronemiasis, exuberant granulation tissue.

2 How could the diagnosis be confirmed? Biopsy of the ulcerated mass and of one of the enlarged submandibular lymph nodes.

3 What treatment options should be considered? Biopsy confirmed SCC. The oral cavity is a predilection site for this tumor (but rarer than the external genitalia and eyes). Treatment choices for confirmed SCC depend on evidence of spread locally or to distant sites. If the lesion is confined to the gingiva with no evidence of metastasis locally (based on histopathology of excised local lymph nodes, radiographs looking for osseous destruction, and a detailed oral examination to identify other sites of tumor development), surgical excision (partial maxillectomy) with adjunctive radiotherapy may be indicated.

Discussion

Oral masses and malignancy are uncommon in horses and differentials for soft tissue masses include SCC, fibrosarcoma, hemangiosarcoma, rhabdomyoma/sarcoma, lymphosarcoma, and myxomatous tumors. While it is one of the most commonly diagnosed malignancies in horses, SCC is rarely found in the oral cavity, although it is the most commonly diagnosed oral neoplasia. It may be locally invasive into bony and soft tissues and metastasis to regional lymph nodes is common in more advanced cases. Tumor location probably plays a large role in prognosis and treatment, as more rostral lesions are likely to be diagnosed earlier and are more accessible than more caudal lesions. Pathogenesis is unknown, but, unlike with penile SCC, there does not appear to be any connection to papillomavirus infection, although it has been established that papillomavirus can infect oral mucosa. Treatment is most often surgical, with medical therapies being largely anecdotal. There is one report of using a COX-2 inhibitor (piroxicam) to successfully control SCC at the mucocutaneous junction of a horse's lip.

References

Knight CG, Dunowska M, Munday JS *et al.* (2013) Comparison of the levels of *Equus caballus* papillomavirus type 2 (EcPV-2) DNA in equine squamous cell carcinomas and non-cancerous tissues using quantitative PCR. *Vet Microbiol* 166(1-2):257–262.

Moore AS, Beam SL, Rassnick KM *et al.* (2003) Long-term control of mucocutaneous squamous cell carcinoma and metastases in a horse using piroxicam. *Equine Vet J* 35(7):715–718.

CASE 129

1 What condition might be causing the behavioral abnormalities? Primary hyperammonemia. Horses with this condition commonly have a history of GI disease, including colic or diarrhea.

2 How can this be confirmed? By measuring blood ammonia concentration (blood ammonia >210 µg/dl [150 µmol/l] is supportive). Blood ammonia should be measured immediately after sample collection; alternatively, the sample should be kept on ice and centrifuged within 15 minutes. Concurrent liver failure (i.e. with hepatic encephalopathy) should be ruled out by evaluating liver enzymes and bile acid concentrations.

3 What treatments should be administered if the diagnosis is confirmed? Sedation with phenobarbitone (phenobarbitol) or low doses of xylazine may be required. Oral neomycin, lactulose, or magnesium sulfate may be used to decrease ammonia production in the intestines. IV fluid therapy should be maintained.

Discussion

Primary hyperammonemia is an uncommon to rare complication in horses with significant GI disease. It is commonly associated with large bowel disease, but has also been recorded in horses with small bowel disease or a combination of the two. It appears to be particularly common in horses with coronavirus enteritis. To date, no convincing etiologic agent has been found. However, it has been suggested that overgrowth of urease-producing bacteria may contribute, as this has been shown to be a cause in other species, including man. Culturing a urease-producing bacteria has not been shown to be predictive for a horse to develop hyperammonemia and related clinical signs. The pathogenesis is unknown, but is likely a combination of increased production in the GI tract, increased permeability and therefore absorption of ammonia by the inflamed gut, and decreased hepatic clearance as a result of either the GI disease itself or the pharmaceuticals used for its treatment.

It is important to note that while blood ammonia concentration is the main diagnostic test, clinical signs are not consistent across cases, and the severity of signs does not correlate with the degree of increase in ammonia. Treatment is largely based on clinical signs and attempts to bind or eliminate the ammonia. Unfortunately, the prognosis is guarded for horses that develop hyperammonemia, with many succumbing to their illness (either the hyperammonemia or the precipitating condition) or requiring euthanasia. Those that survive make a full recovery and return to normal function. There is a 40–50% survival rate, with surviving horses recovering neurologic function within 24–48 hours.

References

Dunkel B, Chaney KP, Dallap-Schaer BL *et al.* (2011) Putative intestinal hyperammonaemia in horses: 36 cases. *Equine Vet J* **43**:133–140.

Peek SF, Divers TJ, Jackson CJ (1997) Hyperammonaemia associated with encephalopathy and abdominal pain without evidence of liver disease in four mature horses. *Equine Vet J* 29:70–74.

CASE 130

1 What are the common causes of malabsorption in the adult horse? Common causes include: extensive small intestinal resection; chronic inflammatory bowel diseases (granulomatous enteritis, idiopathic eosinophilic enterocolitis, multisystemic eosinophilic epitheliotrophic disease, lymphocytic/plasmacytic enterocolitis); alimentary lymphoma (lymphosarcoma); idiopathic villous atrophy; parasitism (cyathostominosis). Less common causes include: amyloidosis; enteric infections (mycobacterial infection, *Rhodococcus equi* [rare in adult horses], *Lawsonia intracellularis* [rare in adult horses], enteric fungal infections); congestive heart failure; intestinal fibrosis.

2 What further diagnostic tests could be performed to elucidate the precise cause of this horse's intestinal malabsorption? Confirmation of the underlying pathologic process generally requires intestinal biopsies. Rectal biopsy and duodenal mucosal biopsies can be considered, but are rarely diagnostic in cases of small intestinal disease (unless the same disease process affects the ascending and descending colons). Full-thickness intestinal wall biopsies can be obtained from the small intestine through either a midline celiotomy performed under GA or a flank laparotomy performed in the standing sedated patient with local anesthesia.

Discussion

Chronic weight loss is a non-specific clinical sign associated with a lengthy list of differentials that can be categorized in a number of ways including: decreased dietary intake; decreased absorption or increased loss of protein/nutrients; increased rate of utilization of nutrients; and miscellaneous. Some of the underlying causes (e.g. feeding management, dental or other oral disease, dysphagia/esophageal obstruction, sources of pain, neurologic or infectious disease) are relatively easily determined with a good history, physical examination, and routine laboratory testing, while others (e.g. protein-losing enteropathies/nephropathies, neoplasia, hepatic disease, endocrinopathies) are much more difficult to determine without additional diagnostics. For cases in which the underlying cause is not as easily determined, additional diagnostics may be necessary, including parasitic fecal egg count, cytologic evaluation of abdominal, pleural, or pericardial fluid, imaging (ultrasound, radiography, CT), gastroscopy, absorption tests, and in some cases laparoscopy or laparotomy. Prognosis depends almost entirely on the underlying cause, although in many cases chronicity plays an important role as treatment earlier in the disease process generally yields better results. A low serum albumin concentration has been associated with a worse prognosis. Similarly, abnormal peritoneal fluid and complete malabsorption (determined by an absorption test) are also associated with a poorer prognosis.

References
Lecoq L, Lavoi J-P (2009) Diagnostic approach to chronic weight loss. In: *Current Therapy in Equine Medicine 6.* (eds. NE Robinson, KA Sprayberry) Saunders, St. Louis, pp. 898–901.
Metcalfe LV, More SJ, Duggan V *et al.* (2013) A retrospective study of horses investigated for weight loss despite a good appetite (2002–2011). *Equine Vet J* 45(3):340–345.

CASE 131

1 What diagnostic procedures should be performed next? Ultrasound examination of the abdomen. Peritoneal fluid collection and analysis are also recommended. Gastroscopic examination might be considered to rule out gastric ulcers, although the recent history (no work, turnout) would not be supportive of the diagnosis. Fecal flotation should also be performed.

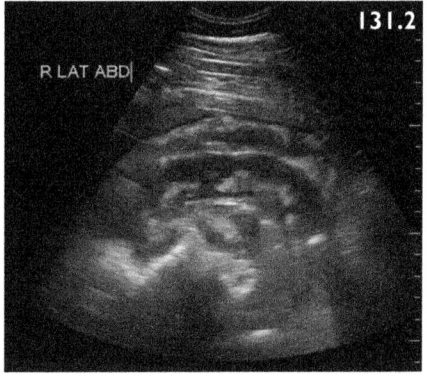

2 Interpret the result of the procedure performed. There is marked thickness and edema of the right dorsal colon (**131.2**).

Follow up/discussion
No parasite eggs were seen on fecal flotation. The only disorders known to cause the localized thickness seen on ultrasound are right dorsal colitis and lymphosarcoma. The history and normal plasma protein and albumin concentration were not compatible with right dorsal colitis. Likewise, one would expect a low plasma protein and albumin concentration and a more prolonged medical history of weight loss and colic if the thickness was due to lymphosarcoma. The horse underwent an exploratory laparotomy under GA and the marked thickness of the right dorsal colon was confirmed. The right dorsal colon appeared inflamed and felt edematous and the regional lymph nodes were enlarged. A full-thickness biopsy was taken and both frozen and formalin-fixed tissues were examined. Severe diffuse eosinophilia

and lymphoplasmacytic colitis with marked submucosal edema and lymphatic dilatation were observed. There were mucosal erosions with neutrophilic infiltrates and large amounts of fibrin. The horse was treated with penicillin and gentamicin IV and metronidazole PO. He was also administered misoprostol, pentoxifylline, and omeprazole PO. The 2 days following surgery did not result in any improvement and, in fact, the colic signs and diarrhea were progressively more pronounced. GI borborygmi were diminished, there was no change in the ultrasound findings of the right dorsal colon, and there was mild thickening (up to 0.6 cm) of the left ventral colon. The grain was discontinued, he was fed only Timothy pellets and grass hay, and he was started on azathioprine (0.3 mg/kg q24h) as a treatment for the colitis. Within 2 days, the horse had normal appetite and manure and the ultrasound appearance of the colon was normal. Antimicrobials and azathioprine were continued for 1 week. Three months after discontinuing all treatments and continuation of the grass hay and Timothy pellets, the horse appeared clinically normal and returned to successful athletic competition.

The cause of the marked inflammatory response in the right dorsal colon was unproven but an allergic or toxic reaction to the herbal 'liver flush' was suspected. A recommendation was made not to use the product in this horse again. The right dorsal colon is most severely affected with NSAID toxicity, but in this case the colic

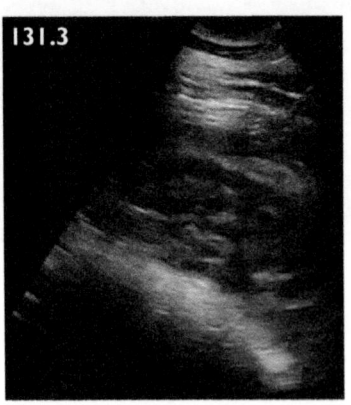

signs started before the flunixin meglumine was given. Furthermore, the normal plasma protein concentration made NSAID-induced right dorsal colitis unlikely. The ultrasound images of the right dorsal colon were very similar to ones that were seen later in the year in a Friesian horse with lymphosarcoma that showed weight loss, diarrhea, and severe infiltration of the right dorsal colon (**131.3**). In a recent systematic review of adverse effects of herbal medicines in human, the most common mild adverse reactions were allergic reactions, GI upsets, diarrhea, and nausea.

Reference
Posadzki P (2013) Adverse effects of herbal medicines: an overview of systemic reviews. *Clin Med* **13**(10):7–12.

CASE 132
1 **What is the diagnosis?** Verrucous pododermatitis ('grapes') with superficial pyoderma.

2 What is the cause of this lesion? The papillomatous or polypoid lesions are generally the result of chronic pastern dermatitis. Heavy horses and Cobs affected by chronic lymphedema appear to be predisposed.

3 What treatments should be recommended? Topical cleansing and topical antimicrobials may help to control the associated superficial infections. Surgical resection or cryotherapy may be required to debulk the proliferative tissue.

Discussion

Verrucous pododermatitis is a chronic, hyperplastic form of pastern dermatitis characterized by clearly demarcated calluses and wart-like projections often accompanied by a 'greasy' malodorous exudate. As with milder presentations of pastern dermatitis, feathered Draft breeds are overrepresented, non-pigmented skin may be more commonly affected, and secondary bacterial infections are common (likely the source of the malodorous exudate). Often the chronicity of the lesions prevents identification of an etiologic agent and the lesions are typically refractory to attempted therapies for routine pastern dermatitis. Diagnostics include skin scrapings and biopsies for both histopathology and culture/sensitivity. All pastern dermatitis, however, is associated with wet or unclean conditions, which allows maceration of the skin, with or without concurrent physical or chemical damage. Management is similar to other forms of pastern dermatitis (i.e. clean, dry environment and cleansing with antimicrobial shampoo for secondary bacterial infection) (see **case 153**). However, systemic therapy or surgical removal is more likely indicated for chronic lesions, particularly if they are causing lameness.

References

Poore LA, Else RW, Licka TL (2012) The clinical presentation and surgical treatment of verrucous dermatitis lesions in a draught horse. *Vet Dermatol* **23**(1):71–75, e17.

Rashmir-Raven AM, Black SS, Rickard LG *et al.* (2000) Papillomatous pastern dermatitis with spirochetes and *Pelodera strongyloides* in a Tennessee Walking Horse. *J Vet Diagn Invest* **12**(3):287–291.

Scott D, Miller W (2011) *Equine Dermatology*, 2nd edn. Elsevier, Maryland Heights, pp. 460–461.

CASE 133

1 What is the diagnosis? The persistent unilateral nasal discharge with purulent material draining into the caudal left middle nasal meatus is consistent with chronic sinusitis. The radiograph shows diffusely increased radiopacity in the maxillary sinuses consistent with fluid accumulation; no discrete fluid lines are visible, suggesting that the sinuses are completely filled with fluid (or soft tissue).

The discrete radiopacity in the area of the rostral maxillary sinus is likely to be an osteoma. Osteomas are smooth, solitary osseous benign growths protruding from the surface of a bone.

2 What is the significance of the profuse growth of *S. equi*? Horses with a *S. equi* infection (strangles) shed the bacteria continuously or intermittently from the nasal discharge during the acute clinical disease, and for variable periods thereafter. In a small number of recovered horses, a carrier state is established with intermittent shedding of bacteria in nasal secretions. This bacterial carriage may persist for several months or even years. These horses often show no clinical

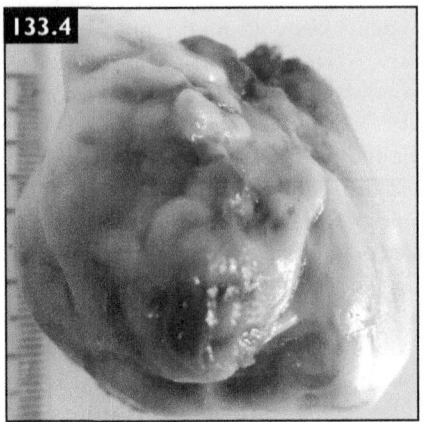

signs of infection (and rarely have a nasal discharge, unlike this horse). The site of carriage of persistent infection is most commonly in the guttural pouches.

3 How should this horse be treated? Treatment should include lavage of the maxillary sinuses via trephine holes or a bone flap. Surgical removal of the osteoma from the rostral maxillary sinus was recommended in this case (**133.4**), although its precise role in the pathogenesis of sinusitis or carriage of *S. equi* remains uncertain.

Follow up/discussion

In this horse, a persistent maxillary sinusitis developed following the strangles; culture of material taken directly from the maxillary sinus yielded a profuse growth of *S. equi*, suggesting that persistent infection had developed within the sinuses.

Paranasal sinus tumors are uncommon and the list of differentials for calcified tumors in the paranasal sinus is short and includes osteoma, osteosarcoma, ameloblastic odontoma, compound and complex odontoma, cementoma, and osseous fibroma. Osteomas are slow-growing, benign, space-occupying masses that typically are found unilaterally in the maxillary or conchofrontal sinus. They may be present at birth or very young ages, but are so slow-growing that horses are often several years old before any clinical signs are noted. The most common clinical signs are deformation of the facial bones overlying the affected sinus, unilateral nasal obstruction, nasal discharge (which may be mucopurulent), and unilateral epiphora. Exophthalmos and blepharospasm may also occur. Diagnosis is by radiography, CT, and/or biopsy. Surgical removal is recommended and masses do not usually return after resection, thus offering a very good prognosis. Resected tissue or biopsy material should be submitted for histopathology for definitive diagnosis.

References

Cilliers I, Williams J, Carstens A *et al.* (2008) Three cases of osteoma and an osseous fibroma of the paranasal sinuses of horses in South Africa. *J S Afr Vet Assoc* **79(4)**:185–193.

Schaaf KL, Kannegieter NJ, Lovell DK (2007) Calcified tumours of the paranasal sinuses in three horses. *Aust Vet J* **85(11)**:454–458.

CASE 134

1 What is the most likely diagnosis? Endothelial lesions are located ventrally, with pigmented cells associated with corneal endothelium and diffuse edema. These findings are most consistent with a diagnosis of endothelial immune-mediated keratitis or endotheliitis.

CASE 135

1 What abnormalities can be identified in the radiograph and ultrasonogram? There is diffuse increased radiopacity in the cranial thorax/cranial mediastinum. The ultrasonogram shows diffusely increased echogenicity of the retroperitoneal fat.

2 What is the provisional diagnosis? Generalized (pan)steatitis/yellow fat disease.

3 What gross pathologic findings would be expected on necropsy with this disease? The fat throughout the body is likely to be discolored yellow–brown, hardened, multinodular, and focally hemorrhagic (**135.4–135.6**). Subcutaneous edema of the

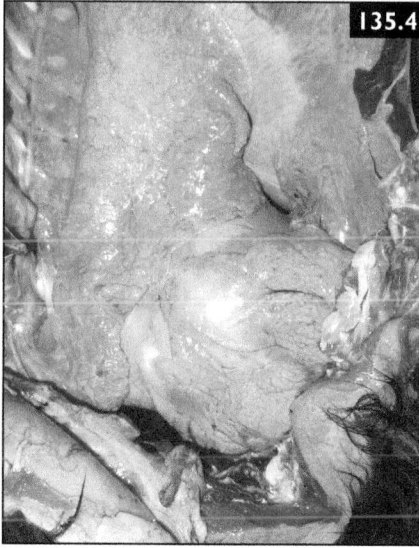

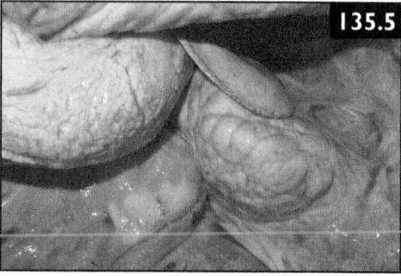

ventral abdominal wall is common. Edema of the lower parts of the hindlimbs, hydrothorax, and hydroabdomen are seen less frequently.

Discussion

Yellow fat disease or generalized (pan)steatitis is an uncommon condition that generally affects foals and young adults (<4 years), with rare cases in older animals. Ponies and 'cold blooded' (a.k.a. Draft-type) horses appear to be overrepresented, with one reported case in a donkey foal. Clinical presentation is often similar to the case above. The most common presenting signs include significant tachycardia, pyrexia, and ventral edema with hardened painful areas around the nuchal ligament, axillary, and/or groin regions. While the subcutaneous areas are commonly affected, in some cases the internal fat deposits are affected, making diagnosis more difficult. These signs are typically accompanied by a history of anorexia and abnormal fecal consistency: either diarrhea or firm feces (as in this case). A presumptive diagnosis can often be made based on physical examination findings, case signalment, and history. The most consistent hematology finding is anemia accompanied by either leukocytosis or leukopenia, while serum biochemistry often reveals large increases (up to 4 times or more above reference range) in LDH and AST with possibly a moderate increase in CK. If tested, affected individuals may have low serum levels of vitamin E and/ or selenium.

Definitive diagnosis is typically made either by biopsy of affected adipose tissue or on necropsy, with gross findings similar to those described above. Histopathology shows fat necrosis and an infiltrate of macrophages with varying amounts of other cell types. Skeletal or cardiac muscle may also be affected with a similar cellular infiltrate and white striations grossly. There is no known etiologic agent and no clear pathophysiology for yellow fat disease in horses. Histopathologic changes consistent with yellow fat disease have been reported in the late-term fetus. This, in conjunction with the apparent breed or type predisposition, suggests a possible genetic component. While not consistent, the low plasma vitamin E levels noted in many cases suggest a possible role in pathogenesis. Regardless, prognosis for yellow fat disease is poor. Supportive therapy in the early stages may be successful, but the majority of cases are advanced at the time of diagnosis and either succumb to the disease or are euthanized.

Reference

de Bruijn CM, Veldhuis Kroeze EJB, Sloet van Oldruitenborgh-Oosterbaan MM (2006) Yellow fat disease in equids. *Equine Vet Educ* **18**:38–44.

CASE 136

1 What is this condition? The abnormal gait is characteristic of stringhalt.
2 What is its cause? Three forms of the disease are recognized: sporadic, epidemic (Australian stringhalt refers to a syndrome that occurs in outbreaks with involvement, often profound, of both hindlimbs and often the forelimbs; there is an unproven association with dandelion and flatweed intoxication), and intoxication with sweet pea plants (lathyrism). These syndromes may be the result of a sensory neuropathy, a myopathy, or spinal cord disease. In lathyrism and Australian stringhalt, a Wallerian degeneration of nerve fibers is seen with evidence of neurogenic muscle atrophy in severe cases.
3 How should this horse be treated? Treatment should include removal of the horse from any toxic plant or area. Many cases improve slowly with time (days to weeks). Tenotomy or tenectomy of the lateral digital extensor tendon may help, even in cases of lathyrism, although this does not cure severely affected horses. Treatment with oral phenytoin (15 mg/kg PO for 14 days) has been helpful for Australian stringhalt in improving clinical signs and getting these horses back into exercise.

Discussion

Stringhalt (a.k.a. equine reflex hypertonia) is not a disease but rather a clinical sign characterized by sudden, involuntary hyperflexion of one or both hindlimbs. Severity is graded on a I–V scale. In mild to moderate cases (grades I–III), the flexion may only be observed during backing of the horse (grade I), stress, or during forward movement, while severe cases may exhibit severe flexion at rest (grade IV), be reluctant to move, and/or exhibit 'bunny hopping' (grade V). There are two clinical presentations of stringhalt: idiopathic (a.k.a classic or spontaneous) and acquired bilateral stringhalt, which includes lathyrism and Australian types. The idiopathic form typically presents as unilaterally affected single animals, while the acquired forms typically present as bilaterally affected animals, and often multiple animals are affected as an 'outbreak', typically in late summer/early autumn. Treatment is symptomatic and may include the use of muscle relaxants or sedatives in addition to phenytoin. Severe or chronic cases may benefit from myotenectomy of the lateral digital extensor tendon. Prognosis is variable and determined by chronicity, suspected cause, and severity. Recovery may take months, but surgical therapy, while palliative in nature, often provides rapid symptomatic relief for severe cases.

Little is known about the etiopathogenesis of the idiopathic presentation, but the acquired bilateral presentation is known to be the result of distal axonopathy resulting in axonal demyelination and degeneration with secondary muscle atrophy.

The most commonly affected nerves are the superficial and deep peroneal and the recurrent laryngeal nerves, though laryngeal nerve dysfunction is not always apparent on laryngoscopy. In these cases, presumptive diagnosis is by history and clinical signs, but definitive diagnosis is by histopathology of appropriate muscle and nerve biopsies. Recovery of horses with acquired forms of stringhalt is dependent on axonal regeneration, therefore early intervention and severity are major factors affecting prognosis.

Differentials, particularly for mild cases of stringhalt, include shivers and fibrotic myopathy. The convention is that shivers horses only exhibit hyperflexion when asked to back up, but mild cases of stringhalt may only exhibit hyperflexion with backing as well. A recent study attempting to determine clinical characteristics to differentiate the two conditions noted that shivers horses are more likely to abduct the hyperflexed limb during forward motion, while stringhalt horses typically follow a straight flightpath. Horses with stringhalt also appear to hyperflex the limb(s) in a greater percentage of strides, and hyperflexion persists at the trot, whereas shivers horses have a normal gait at the trot. Hyperflexion of the limb on manual elevation was not found to be a characteristic of only shivers horses. Additional historical information is likely to be very helpful in differentiating between these two diseases.

References

Armengou L, Añor S, Climent F *et al.* (2010) Antemortem diagnosis of a distal axonopathy causing severe stringhalt in a horse. *J Vet Intern Med* **24(1)**:220–223.

Draper AC, Trumble TN, Firshman AM *et al.* (2014) Posture and movement characteristics of forward and backward walking in horses with shivering and acquired bilateral stringhalt. *Equine Vet J* **47(2)**:175–181.

Torre F (2005) Clinical diagnosis and results of surgical treatment of 13 cases of acquired bilateral stringhalt (1991–2003). *Equine Vet J* **37(2)**:181–183.

CASE 137

1 What is the provisional diagnosis? Neoplasia, most likely squamous cell carcinoma (SCC) or transitional cell carcinoma (TCC).

2 How could the diagnosis be confirmed? Transendoscopic pinch biopsies. In this case the histopathologic diagnosis was TCC.

3 What treatment options should be considered? Surgical debulking may be undertaken via a urethrotomy incision, followed by laser ablation and/or treatment with piroxicam or a selective cyclooxygenase (COX)-2 inhibitor.

Discussion

TCC is the second most common primary neoplasm in the equine bladder following SCC. As with all neoplasia in the horse, bladder neoplasia is uncommon to rare, but should be on the list of differentials for hematuria (which occurs due to ulceration of the overlying mucosal epithelium). TCCs tend to be both locally invasive and metastatic (via lymphatics) in nature, leading to a poor prognosis in most cases. Depending on the location and size of the mass, pyelonephritis and/or hydronephrosis may develop as secondary complications. Histologically, TCC tends to be poorly demarcated, containing cuboidal to polygonal cells with eosinophilic cytoplasm arranged in lobules and cords. Tumor cells are likely to invade deeply into the submucosa and muscularis and areas of necrosis may be present. They are classified as papillary or non-papillary and infiltrating or non-infiltrating, with non-papillary, non-infiltrating masses carrying the best prognosis. Treatment, when attempted, is usually surgical.

COX enzymes produce prostaglandins (PG) from arachidonic acid. PGs (particularly PGE 2) play a role in tumor growth, angiogenesis, and metastasis. In humans and dogs, COX-2 has been shown to be expressed in high concentrations in epithelial carcinomas. Subsequently, COX-2 inhibitors have been shown to have some antitumor activity in humans and dogs, either alone or as an adjunct to chemotherapy. The use of COX-2 inhibitors in horses is extrapolated from these studies, but very little investigation of the expression of COX-2 or the use of COX-2 inhibitors in horses has occurred. There is one report of treatment with the COX-2 inhibitor piroxicam producing tumor shrinkage and long-term control of an SCC on the upper lip of a horse. More recently, attempts have been made to determine COX-2 expression in ocular and preputial SCC. Ocular SCC in the horse has been shown to have significantly increased COX-2 expression when compared with normal tissue, but the results for preputial SCC expression of COX-2 are more mixed, with newer masses expressing COX-2 at higher concentrations than more chronic masses. No studies have been conducted on either COX-2 expression or treatment with a COX-2 inhibitor on equine bladder SCC or TCC.

References

Fischer AT Jr, Spier S, Carlson GP *et al.* (1985) Neoplasia of the equine urinary bladder as a cause of hematuria. *J Am Vet Med Assoc* **186**(12):1294–6.

Moore AS, Beam SL, Rassnick KM *et al.* (2003) Long-term control of mucocutaneous squamous cell carcinoma and metastases in a horse using piroxicam. *Equine Vet J* **35**(7):715–718.

Patterson-Kane JC, Tramontin RR, Giles RC Jr *et al.* (2000) Transitional cell carcinoma of the urinary bladder in a Thoroughbred, with intra-abdominal dissemination. *Vet Pathol* **37**(6):692–5.

Van Den Top JGB, Harkema L, Ensink JM *et al.* (2013) Expression of cyclo-oxygenase-1 and -2, and microsomal prostoglandin E synthase-1 in penile and preputial papillomas and squamous cell carcinomas in the horse. *Equine Vet J* 46(5):618–624.

CASE 138

1 What abnormalities can be seen on the gastroscopic image? Ulceration/erosion of the squamous mucosa of the non-glandular region of the stomach. These findings are compatible with equine gastric ulcer syndrome (EGUS).

2 What clinical signs can be associated with this disease? Clinical signs of EGUS are wide and varied but include recurrent colic, poor appetite, weight loss, hair coat changes, poor performance, behavioral changes, and pain on tightening of the girth.

3 What treatment should be recommended? Standard treatment for horses with clinical evidence of EGUS is administration of omeprazole.

Discussion

Gastric ulcers in adult horses may affect the glandular mucosa, but they more commonly affect the squamous mucosa, with the majority being located at the margo plicatus. Identification is by gastroscopy following a period of fasting (at least 6 hours) and the ulcers are graded based primarily on depth, which has been determined to be the most important factor in ulcer healing. The incidence varies by population, with some reports indicating that over 90% of Thoroughbred racehorses have gastric ulcers. Other reports have indicated a lower prevalence among other populations (event horses, competition horses) with horses not in work and living on pasture having the lowest incidence. The etiology is not fully known and is most certainly multifactorial, but the pathogenesis is erosion of the squamous epithelium due to acid exposure with ulcer severity related to duration of exposure. Reported risk factors for ulcer development include stress, high-energy feed, intermittent feeding, intense exercise/training/racing, stall confinement, and type of roughage fed. It is also well known that horses may develop gastric ulcers with as little as 48 hours of feed deprivation.

Clinical signs are varied and generally non-specific (see above), but many horses with gastric ulcers will be asymptomatic. Very often, especially with racehorses, the horse will present with the owner or trainer complaint of "poor performance", and it is important to remember that there are a great many variables that can contribute to poor performance with ulcers being only one of them. EGUS is diagnosed when a horse exhibits clinical signs potentially attributable to ulcers noted on gastroscopy. Often this diagnosis is given presumptively until improvement with treatment is demonstrated due to the knowledge that many horses with gastric ulcers do not show clinical signs. Treatment for ulcers in the squamous portion of the stomach ideally is with the proton pump inhibitor omeprazole, which irreversibly binds the H^+/K^+ pump to block acid secretion into the stomach. This irreversible binding allows for a prolonged effect, less frequent dosing, and is the only therapy with research demonstrating ulcer healing. Lower doses of omeprazole have been reported to prevent ulcer formation/recurrence either following treatment or prophylactically. H2 receptor antagonists (e.g. ranitidine, famotidine, cimetidine) must be dosed multiple times a day and have not shown efficacy in treating existing ulcers. Similarly, sucralfate may ease clinical signs for some horses but is not an effective therapy on its own, must be given multiple times a day, and may interfere with the absorption of other medications. Given the important role that management factors play as risk factors for EGUS, attention to the management and diet of horses affected by the condition is important. Access to hay or other sources of roughage and reduction in the soluble carbohydrate content of meals should help reduce acid exposure to the squamous mucosa. The addition of corn oil to the diet can also help decrease gastric acid production. A variety of other drugs and supplements have been used to treat EGUS, including various antacids and mucosal protectants.

References
Bell RJW, Mogg TD, Kingston JK (2007) Equine gastric ulcer syndrome in adult horses: a review. *N Z Vet J* 55(1):1–12.
Sykes BW, Sykes KM, Hallowell GD (2014) A comparison of two doses of omeprazole in the treatment of equine gastric ulcer syndrome: a blinded, randomised, clinical trial. *Equine Vet J* 46(4):416–421.

CASE 139
1 What is the diagnosis? Suture line periostitis (inflammation of the facial bone sutures). The nasofrontal and nasolacrimal sutures are likely involved in this case. Periostitis of the nasofrontal suture causes the development of an irregular firm swelling between and just rostral to the eyes. Suture periostitis of the nasolacrimal, maxillolacrimal and/or frontolacrimal sutures causes a diffuse swelling rostral to the medial canthus of the eye, with accompanying epiphora.
2 What is the likely cause of this condition? The condition is caused by inflammation of the craniofacial sutures, most commonly affecting the nasofrontal sutures, but periostitis of the nasolacrimal, maxillolacrimal, and frontolacrimal sutures have also been described. In most cases the cause of the suture periostitis is unknown, although some cases have a known history of head trauma (including facial bone/paranasal

sinus surgery) and others have a history of epistaxis of unknown origin. It has also been postulated that other forces, such as masticatory forces, could contribute to the development or persistence of nasofrontal suture line periostitis. Initially, the swellings are non-calcified, but within a few weeks, bony exostoses develop at both edges of the suture that surround a central radiolucent area and then enlarge both externally (visible as a swelling) and internally (identifiable radiographically). In time, the bony swellings remodel.

3 How should this case be managed? Generally, no treatment is required. The bony swellings usually remodel over time and may ultimately resume a normal contour. Nasolacrimal suture line periostitis, and the accompanying swelling and epiphora, often resolve in a few months. Occasionally, exudation through the overlying skin may occur with severe suture line periostitis. Rarely, surgical stabilization is required if the condition does not resolve after approximately 12 months.

Discussion

The nasofrontal suture in horses is typically fused in the first 6 months and obliterated by approximately 1 year of age. Suture periostitis (a.k.a. suture exostosis of the nasofrontal suture) is a relatively uncommon condition in adult horses resulting in non-painful, bilaterally symmetrical, bony proliferation with variable amounts of soft tissue swelling along the nasofrontal suture of the skull. In most cases the swelling is only cosmetic in nature, but in rare cases the proliferation may affect the nasolacrimal duct, causing epiphora. The etiology is unknown, but theories include trauma, previous surgery, and stresses to the suture line from muscle attachments. In cases where histopathology was performed the results were consistent with callus formation, which may support a theory of chronic fracture instability. In the few cases reported in the literature, Thoroughbreds appear to be overrepresented, although other breeds have been reported.

Most cases resolve without treatment or with NSAIDs (systemic or topical) and, if epiphora is present, cleaning of the skin below the eye and application of barrier creams to prevent secondary dermatitis. If epiphora is severe, or if the proliferation continues to grow, surgical treatment may be considered. Radiographs typically show periosteal reaction along the suture line, very often with a radiolucent defect at the suture line. CT may be of use in refractory cases or those presenting with severe epiphora, for whom surgery may be considered.

References

Carslake HB (2009) Suture exostosis causing obstruction in the nasolacrimal duct in three horses. *N Z Vet J* **57**(4):229–234.

Klein L, Sacks M, Furst AE *et al.* (2014) Fixation of chronic suture exostosis in a mature horse. *Equine Vet Educ* **26**(4):171–175.

CASE 140

1 What is the likely diagnosis? Arytenoid chondropathy (chondritis).

2 What is the cause of this condition, and what are the typical diagnostic features? Arytenoid chondropathy consists of the development of suppuration within the matrix of one or both arytenoid cartilages. The mechanism of development of microabscessation with discharging tracts is not clear. Horses of any breed and age can be affected. The condition appears to be more prevalent in the USA than in Europe. The diagnosis of arytenoid chondropathy is primarily made by endoscopy, which shows distortion of the affected cartilage(s). As the microabscesses develop, the cartilage thickens and displaces axially towards the midline with reduced motility. As the condition advances, the distortion of the cartilage becomes more severe and granulomatous eruptions appear on the corniculate process. Contact lesions may develop on the contralateral arytenoid cartilage. Lateral radiographs of the larynx may show focal mineralization, even in early cases.

3 What treatment options are available? The progress of arytenoid chondritis may be arrested in the early stages by prolonged (i.e. 6 weeks) use of potentiated sulfonamide medication and antimicrobial/anti-inflammatory throat sprays, but the only treatment option is arytenoidectomy once the chronic stage has been reached.

Discussion

The etiopathogenesis of arytenoid chondritis is unknown. The highest incidence occurs in racing Thoroughbreds, but Thoroughbred broodmares and yearlings, as well as horses of other breeds, may be affected. Young athletic horses often present for evaluation of a respiratory noise during exercise, while broodmares and older animals are more likely to present with more severe respiratory signs, including respiratory distress. The higher exercise demands of younger horses, leading to the recognition of abnormal respiratory noise, probably allows a diagnosis earlier in the course of disease compared with broodmares and older horses.

The most widely accepted theory of pathogenesis is that arytenoid chondritis is related to laryngeal mucosal trauma (e.g. from nasogastric tubing or endoscopy), which allows bacterial or viral infection to occur. While this certainly may be true for older animals, it does not always explain the disease in very young animals. It is common practice to perform endoscopy on yearling Thoroughbreds at auction, but studies have shown that some yearlings have laryngeal mucosal lesions at the time of endoscopy (therefore not caused by endoscopy). Furthermore, very few of those yearlings with documented lesions will go on to develop chondritis. The same is true for adult horses; endoscopy and nasogastric tube passage are very common procedures, yet very few cases of chondritis develop.

Endoscopic examination of affected horses reveals an enlarged, thickened, irregularly shaped arytenoid cartilage potentially leading to incomplete abduction. Histologic evaluation reveals fibrous tissue in the center of the cartilage surrounded by hyaline-like cartilage to create a layered appearance. Mineralization may be present, but has also been noted in unaffected cartilage, so the clinical significance of this is unclear. Recently, the ultrasonographic appearance of arytenoid chondritis was described. The cross-sectional area of affected cartilage was found to be roughly double that of normal cartilage, as well as to have a heterogeneous increase in echogenicity. Mineralization could be seen in some affected cartilages, but was also observed in normal cartilages at a similar frequency. It was also possible to trace tracts leading to intracartilage abscesses. These findings suggest a role for ultrasound in diagnosing this condition, particularly in its earlier stages.

References
Garrett KS, Embertson RM, Woodie JB *et al.* (2013) Ultrasound features of arytenoid chondritis in Thoroughbred horses. *Equine Vet J* 45(5):598–603.
Kelly G, Lumsden JM, Dunkerly G *et al.* (2003) Idiopathic mucosal lesions of the arytenoid cartilages of 21 Thoroughbred yearlings: 1997–2001. *Equine Vet J* 35:276–281.
Smith RL, Perkins NR, Firth EC *et al.* (2006) Arytenoid mucosal injury in young Thoroughbred horses: investigation of a proposed aetiology and clinical significance. *N Z Vet J* 54(4):173–177.

CASE 141

1 What is the most likely diagnosis? Eosinophilic granulomas (a.k.a. eosinophilic granuloma with collagen degeneration, nodular collagenolytic granuloma, acute collagen necrosis, nodular necrobiosis).
2 How can the diagnosis be confirmed? With a skin biopsy.
3 What is the etiology of these lesions? The etiology is unknown but it may be multifactorial. Insect hypersensitivity has been suspected; this may explain the chains of lesions seen in some cases (as in this case). Many lesions occur in the saddle region, suggesting that trauma may be a predisposing factor.
4 How should this disease be treated? Treatment may not be required in most cases. In many cases, the lesions gradually reduce and may even resolve completely over several months. Systemic or intralesional corticosteroids may be effective. Chronic lesions often become calcified, in which case surgical excision may be the only effective treatment.

Discussion

Eosinophilic granuloma presents as single or multiple non-painful, non-pruritic, firm subcutaneous masses. It is typically considered to be a hypersensitivity reaction, although is likely multifactorial. Seasonality has been noted, with differences between regions (spring/summer in the US northeast, autumn/winter in the Pacific northwest), but there does not appear to be any sex or age predilection. Mules may also be affected. Diagnosis is by excisional biopsy, as these nodules do not tend to exfoliate well. Central areas of necrosis are not common and histopathology is needed to differentiate eosinophilic granuloma from other nodular skin disease. Treatment is not necessary in most cases, but if instituted can be systemic or intralesional. A change in hair color is possible following intralesional therapy and may be temporary. (**Note:** Chronic, mineralized lesions will not respond to medical therapy, so surgical excision may be necessary in those cases.)

References

Scott D, Miller W (2011) *Equine Dermatology*, 2nd edn. Elsevier, Maryland Heights, pp. 436–440.

Valentine BA, Plant JD (2009) Non-neoplastic nodular and proliferative lesions. In: *Current Therapy in Equine Medicine 6*. (eds. NE Robinson, KA Sprayberry) Saunders, St. Louis, pp. 681–686.

CASE 142

1 What condition is suspected, and what differentials should be considered? The most likely diagnosis based on the history (chronic cough and a previous episode of respiratory distress) and the clinical features (expiratory dyspnea and widespread wheezes and crackles over the lung fields) is recurrent airway obstruction (RAO) or summer pasture-associated RAO. Differential diagnoses include lower airway infection (bronchopneumonia) and interstitial lung disease (e.g. equine multinodular pulmonary fibrosis).

2 What diagnostic tests can be performed to confirm the diagnosis? Endoscopy to exclude other obstructive lesions of the trachea and bronchi, demonstrate exudate in the trachea and bronchi, and assess inflammation of the bronchi or 'blunting' of the carina (due to mucosal swelling). Tracheal aspirates and bronchoalveolar lavage samples generally have a marked neutrophilic inflammatory response; bronchoalveolar lavage cytology is more specific and sensitive for accurate diagnosis.

3 How should this pony be treated? Treatment options include: environmental control (i.e. 'dust-free' management); corticosteroids; bronchodilators; mucolytics, expectorants, and mucokinetic agents; and antioxidants.

Answers

Discussion

RAO, or heaves, is a disease of the lower airways characterized by bronchospasm/ bronchial hyperreactivity, increased mucus production and accumulation, and neutrophilic airway inflammation, with subsequent airway remodeling. It is the most common cause of chronic cough in horses, with other possible clinical signs including increased expiratory effort with or without an abdominal component, flared nostrils, and exercise intolerance, with nasal discharge being an inconsistent sign. Thoracic auscultation typically reveals expiratory wheezes and early inspiratory crackles. Clinical signs and auscultation vary with the severity of disease. Diagnostics and therapy are described above. Risk factors include age (>4 years), stabling, and season (typically clinical signs are worse during the winter).

Pathogenesis is incompletely understood, but generally thought to be the result of a hypersensitivity to inhaled antigens. A complex genetic component has also been identified. Clinical signs in affected horses are exacerbated in dusty, poorly ventilated environments and improve when kept on pasture. There is evidence that even when clinical signs are mild and clinical remission is achieved, there is residual inflammation in the lower airway and repeated exacerbations and chronicity induce irreversible ultrastructural changes. Environmental management is the cornerstone to managing clinical signs with the major goal being reduction of dust in the environment and feed. When possible, keeping the horse on pasture is best, but using low-dust bedding (e.g. wood shavings or shredded paper) and soaking hay, or feeding cubed or other low-dust forages, can also make a big difference. For acute exacerbations and severe cases, medical therapy may be necessary and is described above.

Differentials for RAO include summer pasture associated-RAO (SPA-RAO) and inflammatory airway disease (IAD), both of which share clinical attributes with RAO, making differentiation between the conditions challenging in some cases. The former presents very similarly to RAO and is also the result of neutrophilic inflammation and lower airway obstruction (which may have an acute onset), but typically presents during the summer in horses on pasture rather than in stabled horses during the winter. It is exacerbated by heat and humidity and it has been suggested that the major difference between SPA-RAO and RAO is simply in the inhaled antigen, with pollen and mold spores (in humid environments) in the former and dust in the latter. Environmental management of SPA-RAO is less rewarding than for RAO. IAD shares the clinical signs of cough and exercise intolerance as well as mucus accumulation in the trachea, with often a neutrophilic inflammation on bronchoalveolar lavage cytology. However, clinical signs tend to be subtle or absent at rest and horses of any age may be affected (RAO generally is only seen in mature horses). Lower ratios of neutrophils in bronchoalveolar lavage fluid of IAD horses may help differentiate the two conditions, although there is considerable overlap. The etiology of IAD is unknown.

Reference
Pirie RS (2014) Recurrent airway obstruction: a review. *Equine Vet J* **46**(3): 276–288.

CASE 143

1 What is the most likely diagnosis? Squamous cell carcinoma (SCC). The penis and prepuce are the commonest sites for this tumor in male horses. Other tumors at this site are rare, but include sarcoid, papilloma, melanoma, and hemangioma/hemangiosarcoma. Habronemiasis can also cause granulomatous growths on the penis, especially around the urethral orifice.

2 What treatment options should be discussed with the owner? Treatment is likely to include surgery (e.g. phallectomy or *en bloc* resection of the penis, prepuce, and superficial inguinal lymph nodes).

3 What will likely happen if the owner decides not to treat this lesion? Widespread metastatic dissemination to the abdomen and thorax is likely.

Discussion
SCC is the most common neoplasm diagnosed on the external genitalia in horses, and occurs on the penis or prepuce of stallions and geldings with equal frequency. It is also commonly associated with the eye and ocular adnexa, particularly on non-pigmented skin, and is less frequently diagnosed in the GI or urinary tract. Penile tumors are typically located on the glans, often infiltrate the corpus cavernosum, and have the potential for regional lymph node metastasis. Recurrence after treatment is possible and prognosis may be poor, depending on chronicity and severity. Medical therapies, including the use of COX-2 inhibitors, are largely anecdotal, but recently attempts have been made to determine COX-2 expression in ocular and preputial SCC. Ocular SCC in horses has been shown to have significantly increased COX-2 expression compared with normal tissue, but the results for preputial SCC expression of COX-2 are more mixed, with newer masses expressing COX-2 at higher concentrations than more chronic masses. This has made the potential usefulness of COX-2 inhibitors difficult to determine in penile cases.

The pathogenesis is unknown, but given the increased incidence on non-pigmented skin, ultraviolet radiation exposure has been suggested, as has chronic inflammation, and in recent years progression of papillomas (the result of

Equus caballus papillomavirus type-2). Reports have shown a significant increase in both the probability of isolating papillomavirus DNA and in the quantity of DNA present in penile SCC versus other penile lesions or normal penile tissue. Papillomavirus DNA so far has not been consistently found in SCC elsewhere on the horse, indicating that a multifactorial pathogenesis is likely. This lends some credibility to the theory that penile SCC may be a progression of penile papilloma caused by papillomavirus. However, the relative lack of detectible DNA in non-penile SCC suggests that while penile SCC may progress from papillomavirus lesions, the virus may not be directly responsible for the progression to malignancy. Further investigation is necessary to elucidate the role of papillomavirus in penile SCC in horses.

References
Knight CG, Dunowska M, Munday JS *et al.* (2013) Comparison of the levels of *Equus caballus* papillomavirus type 2 (EcPV-2) DNA in equine squamous cell carcinomas and non-cancerous tissues using quantitative PCR. *Vet Microbiol* 166(1-2):257–262.

Newkirk KM, Hendrix DV, Anis EA *et al.* (2014) Detection of papillomavirus in equine periocular and penile squamous cell carcinoma. *J Vet Diagn Invest* 26(1):131–135.

van den Top JGB, Harkema L, Ensink JM *et al.* (2013) Expression of cyclo-oxygenase-1 and -2, and microsomal prostaglandin E synthase-1 in penile and preputial papillomas and squamous cell carcinomas in the horse. *Equine Vet J* 46(5):618–624.

CASE 144
1 **What clinical signs support this diagnosis?** Anterior uveitis: blepharospasm, epiphora, conjunctival hyperemia, diffuse corneal edema, keratic precipitates, aqueous flare, fibrin in anterior chamber, hypopyon, hyphema, rubeosis iridis, miosis, and changes to iris color.
2 **What are the differential diagnoses?** Include: Lyme disease, *Leptospira*, *Rhodococcus equi*, *Brucella*, EHV-1, EVA, *Onchocerca*, *Strongylus*, *Toxoplasma*, recurrent uveitis, other immune-mediated diseases, and trauma.

CASE 145
1 **What is the diagnosis?** Epiglottal entrapment (EE) and a subepiglottic cyst.
2 **What is the etiology of this condition?** The cartilage of the epiglottis becomes enveloped by a fold of glossoepiglottic mucosa arising between the epiglottis and the base of the tongue and extending laterally as the aryepiglottic folds.

The etiology of EE is usually not known, but involves a degree of stretching of the mucosa. In some cases the epiglottis is congenitally hypoplastic or there may be an associated subepiglottic cyst, as in this case. These possibilities should be checked by endoscopy and/or radiography/CT before surgical correction is attempted. Subepiglottic cysts are thought to be derived from the embryologic remnants of the thyroglossal duct, a structure that runs from the level of the epiglottis to the anterior mediastinum. It is believed that subepiglottic cysts are present from birth, although they may not be discovered until the horse is mature and commences training.

3 How can this condition be managed? The treatment options for epiglottal entrapment are: resection via ventral laryngofissure; axial division *per os;* axial division *per nasum;* transendoscopic diode laser resection. The subepiglottic cyst should also be surgically removed.

Discussion

Epiglottic fold entrapment is the most common epiglottic abnormality reported in horses and has a reported prevalence in Thoroughbred racehorses of approximately 0.9%. In this population, the presenting complaints are typically abnormal respiratory noise and exercise intolerance. In contrast, the presenting complaint in a non-racehorse is nearly twice as likely to be coughing and/or nasal discharge as it is to be exercise intolerance or abnormal respiratory noise. The athletic demands placed on racehorses probably lead to more frequent recognition compared with non-racehorses where the horse's performance is less likely to be affected by the entrapment. Furthermore, the exercise demands are lower in these horses, reducing the chances of an owner noting abnormal respiratory noise. Additionally, a presenting complaint of chronic cough generally lends itself more to suspicion of lower respiratory tract disease, illustrating the importance of a thorough examination of both the upper and lower respiratory systems. The chronic or recurrent entrapment of the epiglottis results in inflammation and eventual thickening of the subepiglottic tissue, exacerbating the airway obstruction. In the case above, the subepiglottic cyst likely created additional airway obstruction, leading to the complaints of exercise intolerance and abnormal respiratory noise. Both aryepiglottic entrapment and subepiglottic cysts are treated surgically.

References

Aitken MR, Parente EJ (2011) Epiglottic abnormalities in mature nonracehorses: 23 cases (1990–2009). *J Am Vet Med Assoc* **238**(12):1634–1638.

Lacourt M, Marcoux M (2011) Treatment of epiglottic entrapment by transnasal axial division in standing sedated horses using a shielded hook bistoury. *Vet Surg* **40**(3):299–304.

CASE 146

1 What is the likely diagnosis? Dermoid cyst.

2 What is the cause of this lesion? Dermoid cysts are developmental anomalies and are often congenital. Most dermoid cysts occur on the midline. They are thought to develop as a result of embryonic displacement of ectoderm into the subcutis. Lesions may be single or multiple.

3 What treatment should be recommended? Surgical removal.

Discussion

Dermoid cysts are considered congenital, can be single or multiple, often occur on the dorsal midline, and are non-painful and non-pruritic. The cysts are covered with normal, haired skin and histologically exhibit follicles and glands in their lining. Dermoid cysts are uncommon, and no treatment is needed unless the cyst is in an area that will be contacted by the horse's tack, causing rubbing and irritation. Surgical removal is curative, but care should be taken to ensure there is no communication to deeper tissues (i.e. the subarachnoid space, making the mass a dermoid sinus rather than a cyst) prior to removal.

References

Hillyer LL, Jackson AP, Quinn GC *et al.* (2003) Epidermal (infundibular) and dermoid cysts in the dorsal midline of a three-year-old thoroughbred-cross gelding. *Vet Dermatol* **14**(4):205–209.

Scott D, Miller W (2011) *Equine Dermatology*, 2nd edn. Elsevier, Maryland Heights, pp. 508–509.

Valentine BA, Plant JD (2009) Non-neoplastic nodular and proliferative lesions. In: *Current Therapy in Equine Medicine 6*. (eds. NE Robinson, KA Sprayberry) Saunders, St. Louis, pp. 681–686.

CASE 147

1 What is the diagnosis? Septic jugular thrombophlebitis with perivenous swelling. Thrombophlebitis includes thrombosis and mural inflammation, and is a common complication of IV catheterization in horses with GI disease.

2 What further investigations should be performed? Diagnostic ultrasound is useful to characterize the nature of the thrombophlebitis. In this case, the thrombus appears heterogeneous with anechoic regions representing fluid accumulation or necrosis and hyperechoic foci with reverberation artefacts representing gas formation. Ultrasound-guided aspiration of hypoechoic/anechoic areas should be performed for a sample for bacterial culture; both aerobic and anaerobic cultures should be performed. Blood culture may also be undertaken.

3 What treatments should be considered? Antibiotic therapy, ideally based on the results of bacterial culture and sensitivity testing, may be required for several weeks. Oral aspirin and topical hot-packing may be helpful. Low molecular weight heparin or regular heparin may be useful to decrease the risk of progressive thrombosis. In acute thrombosis, tissue plasminogen activator (tPA) may be injected just proximal to the thrombus to lyse the clot. In chronic non-responsive cases, surgical thrombectomy may be considered.

Discussion

In normal, healthy horses a balance exists between fibrin deposition and fibrinolysis. During illness this balance is often disrupted, allowing for thrombus formation. The most common illnesses that result in increased risk of thrombosis are GI disease and sepsis, both of which frequently involve endotoxemia or systemic inflammatory response syndrome and the resulting vascular inflammation/damage. Neoplasia and protein-losing nephropathy or enteropathy may also result in coagulopathies and predispose to venous thrombi formation. Iatrogenic damage to the venous endothelium either mechanically (with a catheter or needle and syringe) or chemically (with vascular or accidental perivascular injection of irritating pharmaceuticals) also increases the likelihood of thrombus formation.

Horses with any of the above conditions often require IV fluids, antimicrobials and/or anti-inflammatory therapy, which generally is provided through an indwelling venous catheter. Therefore, careful catheter placement, use, and maintenance are important to minimize the risk of thrombosis in these predisposed horses. Maintenance should include checking for thrombus formation at least daily and if any swelling is noted, the catheter should be carefully removed and the tip cultured for bacteria because thrombi are easily colonized, becoming septic as in this case. A thrombus may completely occlude blood flow of the affected jugular vein if not promptly treated, and septic thrombi may result in complications including infective endocarditis, pleuropneumonia, pulmonary thromboembolism, and sepsis.

Appropriate therapy should be instituted as soon as the thrombus is identified in an effort to prevent complete and permanent occlusion of the vein or other potential complications. When the thrombus is septic, antibiotic use should be prolonged, as drug penetration into the thrombus is likely poor. Prognosis for thrombus resolution is unpredictable and largely depends on chronicity. Partial to full re-canulization is possible in some cases, but some will remain occluded and may develop collateral circulation (generally within a year), assuming there is no surgical intervention.

References

Bäumer W, Herrling GM, Feige K (2013) Pharmacokinetics and thrombolytic effects of the recombinant tissue-type plasminogen activator in horses. *BMC Vet Res* **9**(1):158.

Dias DP, de Lacerda Neto JC (2013) Jugular thrombophlebitis in horses: a review of fibrinolysis, thrombus formation, and clinical management. *Can Vet J* **54(1):**65–71.

Divers TJ (2003) Prevention and treatment of thrombosis, phlebitis and laminitis in horses with gastrointestinal disease. *Vet Clin North Am Equine Pract* **19:** 779–790

Russell TM, Kearney C, Pollock PJ (2010) Surgical treatment of septic jugular thrombophlebitis in nine horses. *Vet Surg* **39(5):**627–630.

Traub-Dargatz JL, Dargatz DA (1994) A retrospective study of vein thrombosis in horses treated with intravenous fluids in a veterinary teaching hospital. *J Vet Intern Med* **8:**264–266.

CASE 148

1 What is the diagnosis? The condition has several different names, including pinnal acanthosis, hyperplastic aural dermatitis, and aural plaques.

2 What is the likely etiology? The condition is thought likely to be a form of papilloma caused by an as yet unidentified papillomavirus. The lesions are usually confined to the pinnae, although they are sometimes found in the inguinal region and thighs. The papillomavirus may be transmitted to these areas by black flies (*Simulium* spp.).

3 What treatment should be recommended? There is no well-ducumented effective treatment, although topical imiquimod cream has efficacy. Attempted treatments run the risk of making the condition worse and creating further behavioral problems by handling the affected horse's ears.

Discussion

Aural plaques begin as in the case above: small depigmented or erythematous papules that enlarge and may coalesce and become covered with a white/gray flaky crust. There is no breed, sex, or age predilection. Most horses are asymptomatic, but some will become sensitive, resistant to having their ears handled, and head shy. The plaques will not resolve spontaneously and visual inspection is usually enough for diagnosis in most cases. Sometimes, a biopsy may be necessary to rule out sarcoid or other skin abnormalities. As most cases are asymptomatic, no treatment is required, but some owners may request it for aesthetic reasons or because of the horse becoming head shy. There are many proposed topical therapies but only anecdotal evidence of their efficacy. A study of imiquimod showed that topical administration was effective at eliminating the plaques. Unfortunately, the therapy caused a great amount of discomfort to the horses, necessitating sedation prior to each application and exacerbating or creating head shy behaviour.

While aural plaques have long been thought to result from papillomavirus virus, isolation and DNA amplification for *Equus caballus* papillomavirus (EcPV)-1 and -2 have been unsuccessful. It is only recently that viral DNA has been amplified from aural plaque lesions, revealing four new equine papillomaviruses: EcPV-3, -4, -5, and -6. The papillomavirus is a double-stranded DNA virus. As a family, papillomaviruses typically exhibit very specific species and site preferences with an incubation period anywhere from 10 to 60+ days. Studies have shown that damaged epithelium is necessary for infection and the virus can persist in the environment for up to 3 weeks. Black flies have long been suspected to be vectors for the etiologic agent of aural plaques. This is probably true as they may not only act as a fomite, but also create damage to the epithelium with their bite. Fly control is therefore likely an important part of controlling spread.

References

Gorino AC, Oliveira-Filho JP, Taniwaki SA *et al.* (2013) Use of PCR to estimate the prevalence of *Equus caballus* papillomavirus in aural plaques in horses. *Vet J* **197(3):**903–904.

Scott D, Miller W (2011) *Equine Dermatology*, 2nd edn. Elsevier, Maryland Heights, pp. 468–473.

Torres SM, Koch SN (2013) Papillomavirus-associated diseases. *Vet Clin North Am Equine Pract* **29(3):**643–655.

Torres SM, Malone ED, White SD *et al.* (2009) The efficacy of imiquimod (Aldera) in the treatment of equine aural plaque: an open label pilot study. *Vet Dermatol.* **20(2):**216.

CASE 149

1 What condition is suspected? The clinical and postmortem features are compatible with acorn (oak) toxicity.

2 What is the pathogenesis of this disease? Oak tannins and metabolites bind to peptide linkages and precipitate proteins, acting like an astringent on affected surfaces and resulting in mucosal necrosis. Concurrent damage to vascular endothelium and intravascular edema result in ischemia. DIC may also occur. Complete necrosis of groups of proximal tubules with intratubular hemorrhage also occurs in the kidneys.

Discussion

Horses typically are fastidious eaters, assisted by their prehensile lips and keen senses of smell, taste, and texture. Thus, they generally choose to eat only the most nutritious plants/parts of plants while avoiding many harmful species of vegetation.

315

While this is true much of the time, when availability of their preferred feedstuff is lessened, horses often turn to less preferred sources, which may be harmful or even deadly. Oak toxicity is an example of this, as horses will typically seek out grasses on pasture, but in conditions of overgrazing or drought they will eat the leaves and acorns of oak trees. Relatively large amounts must be ingested for severe disease, but toxicity may be fatal. The mainstay of treatment for toxicities is supportive therapy, but if the underlying cause can be rapidly determined by history, physical examination, or clinical/laboratory testing, therapy may be better tailored. Toxicology testing on feces or gut contents may be performed, but may take several days and is sometimes unrewarding as the toxin may have been cleared by the time clinical signs are evident. Ensuring an adequate source of good quality feed is the best way to prevent toxicosis. Researching the common toxic plants for the region and educating owners is also essential in preventing toxicosis.

This case demonstrates oak/acorn toxicity, but other trees, shrubs, grasses, or pesticides may also be toxic, and some require ingestion of very little to cause disease or fatality. Additionally, many plants are not themselves toxic, but can become infected with fungi that produce potent mycotoxins.

References

Burrows G, Tyrl RJ, Knight AP *et al.* (2004) Plants. In; *Clinical Veterinary Toxicology.* (ed. KH Plumlee) Mosby, St. Louis, pp. 337–442.

Panter, KE, Gardener DR, Lee ST *et al.* (2007) Important poisonous plants of the United States. In: *Veterinary Toxicology: Basic and Clinical Principles.* (ed. RC Gupta) Academic Press, New York, pp. 825–872.

Poppenga RH, Puschner B (2008). Toxicology. In: *Equine Emergencies: Treatment and Procedures,* 3rd edn. (eds. Orsine JA, Divers TJ) Saunders, St. Louis, pp. 593–623.

CASE 150

1 What are the differentials for this horse? Enophthalmos in older horses is most likely related to reabsorbing orbital fat; in foals it is most likely due to severe dehydration. Other causes of enophthalmos include orbital fractures and sympathetic denervation of orbital smooth muscle. Enophthalmos must be differentiated from phthisis bulbi.

CASE 151

1 What abnormalities do the radiographs show? The lateromedial radiographs of both front feet show mild capsular rotation of the distal phalanx, consistent with chronic laminitis.

2 What is the significance, if any, of the body condition score and fat deposits? What condition is suspected, and how does this relate to the lameness? The history of previous episodes of lameness and the radiographic evidence of chronic laminitis are highly suggestive of repeated episodes of pasture-associated laminitis. This, plus the presence of fat deposits, is suggestive of equine metabolic syndrome (EMS) (insulin dysfunction). Physical characteristics of EMS include generalized obesity and/or regional adiposity. Regional adiposity takes the form of a cresty neck and abnormal adipose tissue deposits close to the tailhead, within the prepuce, or randomly distributed beneath the skin of the trunk region. Non-structural carbohydrates within pasture grasses play an important role in this disease by potentially causing hyperinsulinemic episodes. Episodes of laminitis usually occur when affected horses or ponies are grazing on pasture. Laminitis due to insulin dysfunction and EMS are potentially related through four mechanisms: (1) impaired glucose delivery to hoof keratinocytes, (2) altered blood flow or endothelial cell function within the hoof vessels, (3) insulin-like growth factor receptor response in the lamella leading to cellular proliferation, elongation, and stretching of the lamellae, and (4) pro-inflammatory or pro-oxidative states associated with chronic insulin dysfunction and/or obesity.

3 How can the suspicion be confirmed? Confirmation of EMS can be attempted by:

- Resting serum insulin. Resting fasting serum insulin concentrations are elevated in horses with moderate or severe insulin dysfunction and EMS. Horses should be brought in from pasture the night before testing and only one flake of grass hay should be available after 10 PM in order to perform the test under short-term fasting conditions. Blood samples should be collected in the morning and an insulin concentration ≥20 µU/ml supports the diagnosis of insulin dysfunction/EMS.
- The oral sugar test. This is a more sensitive way to detect horses with mild/ early insulin dysfunction. Blood is collected 60 and 90 minutes after administration of corn syrup, and glucose and insulin concentrations are measured. Hyperinsulinemia is diagnosed if the insulin concentration is >60 µU/ml at 60 or 90 minutes.
- Combined glucose-insulin test. This dynamic test should be considered when the fasting serum insulin concentration is <20 µU/ml and the oral sugar test does not confirm a diagnosis, but insulin resistance is still suspected. The test should be performed under short-term fasting conditions. A pre-infusion (baseline) blood sample is collected and then 150 mg/kg body weight 50% dextrose solution is quickly infused, immediately followed by 0.10 units/kg regular insulin (100 units/ml; 0.50 ml for a 500-kg horse). Blood samples are collected at 5, 15, 25, 35, 45, 60, 75, 90, 105, 120, 135, and 150 minutes

post infusion, but the test can be halted once the blood glucose level falls below baseline. Blood glucose concentrations are measured with a hand-held glucometer. Insulin concentrations should be measured prior to injection and at the 45 minutes post-injection time point. Insulin resistance is diagnosed if the blood glucose concentration remains above baseline for ≥45 minutes and/or the insulin concentration exceeds 100 µU/ml at 45 minutes.

- Insulin response to the dexamethasone suppression test (DST). Horses and ponies that suffer from insulin resistance as part of EMS show exaggerated insulin responses to dexamethasone. If a DST is performed to diagnose equine Cushing's disease, insulin concentrations should also be measured to detect insulin resistance. When the test is performed, insulin concentrations should be measured at 0 and 19 hours (24 hours is also acceptable) in addition to cortisol. Detection of hyperinsulinemia prior to dexamethasone injection indicates that the animal is insulin resistant and an insulin concentration >75 µU/ml at 19 hours post dexamethasone provides additional support for this diagnosis.

4 How should this case be managed? Medical treatment for insulin resistance should not be used as a substitute for diet and exercise strategies, but there are three indications for treatment: (1) medical management of insulin resistance during the time it will take for diet and exercise approaches to take effect (3–6 months), (2) for patients that do not respond to changes in diet and exercise, and (3) for horses in laminitic episodes to help mitigate the underlying cause of the laminitis. Metformin and levothyroxine sodium are currently used for the medical management of insulin resistance in horses and ponies.

Management, in particular dietary management in conjunction with exercise, is the cornerstone of treatment for horses with EMS. Many horses and ponies are overfed and reduction, or better yet removal, of concentrates (grains) from their diet along with increased exercise (assuming they are not currently experiencing clinical signs of laminitis) may be sufficient to reduce overall body weight and improve insulin resistance. Many, however, will have to undergo further energy restrictions, which may involve limiting the hay intake to no more than 1.5% (but no less than 1.0%) on a dry matter basis of ideal body weight, soaking the hay for 60 minutes prior to feeding to help reduce water soluble carbohydrates, and limiting or eliminating pasture access.

References

Frank N, Geor RJ, Bailey SR *et al.* (2010) Equine metabolic syndrome. *J Vet Intern Med* 24(3):467–75.

Frank N, Tadros EM (2014) Insulin dysregulation. *Equine Vet J* 46(1):103–112.

CASE 152

1 What abnormalities can be identified on the CT image, and what is the diagnosis?
There is soft tissue/fluid attenuation in the right tympanic cavity, mild irregular thickening of the right tympanic bulla, thickening of the soft tissues of the dorsal aspect of the medial compartment of the right guttural pouch, and mild thickening of the proximal extent of the right stylohyoid bone with new bone formation at the axial aspect of the proximal extent of this bone. These changes are compatible with otitis media. There is currently minimal evidence of temporohyoid osteopathy.

2 What treatment options should be considered? Treatment with broad-spectrum antibiotics for a prolonged period (30 days) and NSAID therapy is indicated. If the condition does not resolve, bacterial culture and sensitivity testing of an aspirate from the middle ear should be undertaken if feasible. The owner should be warned that it is possible that temporohyoid osteopathy will develop in the future and may require further treatment.

Discussion

Otitis media typically results from bacterial causes, although fungal otitis has been reported. While the pathogenesis is unknown, possibilities include the spread of infective agents through hematogenous routes or by extension of otitis externa. Recent or concurrent sinus or guttural pouch infections should be considered and investigated as sources for hematogenous spread. Trauma and ectoparasites may be a cause of otitis externa. The history in this case disclosed a previous recent mucopurulent aural discharge, making an extension of otitis externa more likely. The course of sulfonamide therapy was probably not long enough to treat the residual infection in the tympanic bulla, as a 30-day course is necessary. This residual infection created inflammation in the bulla and surrounding soft and osseous tissues, including the temporohyoid bone and joint. Left untreated, this inflammation could potentially lead to temporohyoid osteoarthropathy and eventual fusion of the temporohyoid joint. If fusion occurs, movement of the tongue and larynx may result in fracture of the petrous temporal bone and associated neurologic deficits.

Headshaking is a common and non-specific clinical sign and other causes (including EPM, trauma, polyneuritis equi, and headshaking syndrome) should be considered. However, the intermittent nature of the shaking in this case, along with the history, were more suggestive of otitis. Imaging confirmed the diagnosis.

Imaging, including upper airway/guttural pouch endoscopy, radiography, and/or CT, is the most valuable diagnostic modality in these cases. CT is best if available. The standard recommendation is to obtain a sample for culture and sensitivity, but with otitis this is difficult to impossible unless there is a concurrent, easily accessible possible source of infection (i.e. guttural pouch or sinus disease). In these cases, a culture from the concurrent site may help guide antibiotic choice, although it is important to remember that the two conditions may be unrelated even though they are concurrent. Thus, barring direct culture of the otitis media, preference should be given to broad-spectrum antibiotics.

References

Hassel DM, Schott HC, Tucker RL *et al.* (1995) Endoscopy of the auditory tube diverticula in four horses with otitis media/interna. *J Am Vet Med Assoc* **207**:1081–1084.

Hilton H, Puchalski SM, Aleman M (2009) The computed tomographic appearance of equine temporohyoid osteoarthropathy. *Vet Radiol Ultrasound* **50**:151–156.

Katz L (2006) Left otitis media/interna and right maxillary sinusitis in a Percheron mare. *Vet Clin North Am Equine Pract* **22(1)**:163–175.

Newton SA, Knottenbelt DC (1999) Vestibular disease in two horses: a case of mycotic otitis media and a case of temporohyoid osteoarthropathy. *Vet Rec* **145(5)**:142–144.

CASE 153

1 What conditions are suspected? This may be described as 'pastern dermatitis'. A variety of different conditions may cause these signs, including stapylococcal folliculitis, dermatophilosis, dermatophytosis, vasculitis, chorioptic mange, and pemphigus foliaceus.

2 What further diagnostic tests should be performed? Swabbing for bacterial culture and skin biopsy.

3 What treatments should be considered? In addition to the therapies that the owner was already undertaking, topical antimicrobial shampoos and emollients may help. Systemic antibiotics (based on the results of bacterial culture and sensitivity testing) may be needed in cases of bacterial infection. If the skin biopsy suggests vasculitis, systemic corticosteroids may be required.

Discusssion

Pastern dermatitis is a cutaneous reaction pattern with many possible underlying etiologies. Elucidating the underlying cause early in the course of disease yields the best outcome, as therapy can then be targeted. There is no sex predilection, but Draft breeds with feathering are overrepresented and the condition overwhelmingly affects adult animals. Dermatitis is typically bilaterally symmetrical, affecting the medial, lateral, and plantar aspect of the hindlimbs, although it can affect the palmar aspect of the forelimbs and in severe cases the dorsal aspect of a limb or limbs. Lesions are more commonly noted on non-pigmented skin. Early signs (mild presentation) include erythema and scaling, developing into an exudative condition/presentation causing matting of the hair, crust formation, alopecia, and erosion with or without pain or pruritis. Chronic lesions often become proliferative, hyperkeratotic, and lichenified and often fissure, leading to limb edema, draining tracts and possibly lameness. Secondary bacterial infections are common in chronic cases.

A thorough clinical history, including possible seasonality, previous therapy/response, and environmental factors, is imperative as is early intervention to avoid chronic changes. Diagnostic tests may include superficial skin scrape, adhesive tape impression, examination of hairs, culture (bacterial and/or fungal), and biopsy, but the underlying cause may remain elusive. Contact hypersensitivity, immune-mediated processes, and parasitic causes (especially in feathered Draft breeds) should be included on the list of potential etiologies.

Regardless of the inciting cause, a major part of therapy is careful management including: keeping the horse in clean, dry conditions; minimizing contact with potentially irritating substances; gentle clipping; cleansing, typically with antimicrobial shampoo or solutions with astringent and/or hypertonic properties; drying of affected areas; with or without topical application or systemic administration of antimicrobial or antifungal agents. If the cause is suspected to be autoimmune, systemic corticosteroid therapy may be necessary. In chronic, refractory cases, surgical debulking of the proliferative tissue may be necessary. Owners should be advised that this condition is likely to recur.

References
Scott D, Miller W (2011) *Equine Dermatology*, 2nd edn. Elsevier, Maryland Heights, pp. 460–461.
Yu AA (2013) Equine pastern dermatitis. *Vet Clin North Am Equine Pract* 29(3):577–588.

CASE 154
1 What is the diagnosis? Bilateral guttural pouch (GP) empyema with nasopharyngeal compression and airway obstruction.
2 What is the etiology of this condition? Empyema of the GP occurs when mucus and/or pus accumulates within the pouch because it is failing to drain satisfactorily. The primary etiologic factor in GP empyema is a dysfunction of mucociliary clearance followed by stagnation of mucus, opportunist bacterial infection, and finally purulent exudation. Stagnant pus within the pouch may become inspissated and lead to the formation of solid concretions (chondroids). Horses with chondroids are often found to be carriers for *Streptococcus equi* subsp. *equi* infection.

3 How should this disease be treated? An indwelling self-retaining Foley balloon catheter may be used for drainage of the GP and for long-term irrigation in the management of chronic cases. Inspissated caseous pus/chondroids may be liquefied by a process of repeated lavage via the pharyngeal ostium aided by the instillation of acetyl cysteine. Chondroids that do not respond to conservative management may be removed individually using transendoscopic grasping forceps (if they are small in number), otherwise surgical removal is required.

Discussion

GP empyema is the term used to describe purulent material in the GPs resulting from upper respiratory tract infection. While empyema is not pathognomonic for any specific pathogen, it is very commonly associated with infection with *S. equi* (a.k.a. strangles), with roughly 10% of infected horses developing chronic empyema and, potentially, chondroids. Unilateral or bilateral disease may occur. Horses with chronic empyema or chondroids are likely to be persistent, subclinical carriers/shedders of the bacteria and serve as sources of infection to other horses. Regardless of the etiologic agent, clinical signs of empyema are similar to the case above. The most consistent sign is a uni- or bilateral mucopurulent to purulent nasal discharge, which may be intermittent. Other clinical signs may include fever, decreased appetite, retropharyngeal pain, and/ or swelling and difficulty swallowing or breathing in some cases. Prognosis for GP empyema is typically very good for resolution of clinical signs and return to previous athletic performance.

Endoscopic examination of the upper airway may reveal a collapsed pharynx, dorsal displacement of the soft palate, and/or laryngeal paralysis, along with drainage of purulent material from one or both openings to the GPs. Depending on the amount of purulent material inside the GP and chronicity, enlarged, potentially abscessed retropharyngeal lymph nodes distorting the floor of the pouch and/or draining into the pouch may be visible. Treatment for empyema is as above or by lavage via the biopsy port of an endoscope, which allows visualization during the lavage. Lavage fluid should be physiologic saline or a balanced electrolyte solution. Chlorhexidine and other potentially irritating substances should not be infused into the pouch. Samples should be taken for culture and sensitivity prior to initiation of therapy. Use of systemic antimicrobials is controversial, but may be warranted in severe cases.

References

Newton JR, Wood JL, Dunn KA *et al.* (1997) Naturally occurring persistent and asymptomatic infection of the guttural pouches of horses with *Streptococcus equi. Vet Rec* **140**(4):84–90.

Trostle SS, Rantanen NW, Nilsson SL *et al.* (2004) What is your diagnosis? Guttural pouch empyema. *J Am Vet Med Assoc* **224**(6):837–838.
Walshe N, Duggan V (2011) Equine strangles: a review. *Vet Ire J* **1**(8):459.

CASE 155

1 What diagnostic tests are indicated? Sedation and (palpebral branch of) auriculopalpebral nerve block are required to facilitate ophthalmic examination. Cytology, aerobic/anaerobic bacterial cultures, and fungal cultures are warranted to direct treatment.

2 What are the most common etiologic agents causing melting ulcers in horses? Bacterial (*Staphylococcus*, *Streptococcus*, *Pseudomonas*); fungal (*Aspergillus*, *Fusarium*, *Alternaria*, *Candida*, *Mucor*).

CASE 156

1 Describe the abnormalities seen in the ultrasound images. There is muscle atrophy of the infraspinatus muscle of the right side. There is a multilobular soft tissue mass deep to the superficial muscle layers in the right caudal cervical region.

2 What disease process is suspected, and how can the diagnosis be confirmed? Neoplasia. This could be confirmed by performing an ultrasound-guided biopsy of the mass; this was performed and confirmed the presence of hemangiosarcoma.

Discussion

Atrophy of the supraspinatus and infraspinatus muscles (a.k.a. Sweeney) is well known but uncommon and is typically the result of injury to the suprascapular nerve as it courses over the cranial aspect of the neck of the scapula from the brachial plexus. Comparatively little tissue overlies and protects the nerve here, leaving it more vulnerable to injury. Shoulder joint instability and/or pain that results in lameness often accompanies damage to this nerve. When horses were the main mode of transportation and were used heavily for farming, logging, mining, etc, damage to the suprascapular nerve caused by ill-fitting harness collars was common, but has become uncommon in today's world. Other trauma (e.g. kick to the shoulder, brachial plexus injury), caudal cervical disease, or disuse atrophy may also contribute to muscle loss over the scapular region. Therapy is typically based on stall rest and physiotherapy, and the prognosis is determined by cause (when known), chronicity, and severity of the muscle loss. In most cases, some fibrosis and resultant permanent muscle loss occurs, but many horses are able to return to a level of work similar to that prior to injury, although recovery can take months.

The earlier the condition is recognized and the more mild the muscle atrophy, the better the prognosis for return to work.

In this case, the cause was a caudal cervical mass that likely produced a radiculopathy affecting the spinal nerves in the caudal cervical area. Given the size of the mass and progression of clinical signs, it was also probably compressing the suprascapular nerve against the scapula, creating neuropathy and muscle atrophy and contributing to lameness and pain. Prognosis was poor due to the underlying cause being a highly malignant neoplasia (hemangiosarcoma).

References

Devine DV, Jann HW, Payton ME (2006) Gait abnormalities caused by selective anesthesia of the suprascapular nerve in horses. *Am J Vet Res* **67**(5):834–836.

Dutton DM, Honnas CM, Watkins JP (1999) Nonsurgical treatment of suprascapular nerve injury in horses: 8 cases (1988–1998). *J Am Vet Med Assoc* **214**(11):1657–1659.

CASE 157

1 Describe the abnormalities and recommend treatment. Ophthalmic abnormalities include conjunctival hyperemia, blepharospasm, a fibrin clot within the anterior chamber, and a linear foreign body adherent to the posterior cornea. Surgical extraction of the foreign body, topical and systemic antibiotics, systemic NSAIDs, and topical atropine are indicated.

CASE 158

1 What are these structures? Chondroids, caused by inspissation of pus to form solid concretions that become smooth and ovoid because of compression within the GP by head movement.

2 What is the most likely cause? Chronic GP empyema and chondroid formation usually arise following upper respiratory infection, in particular infection by *Streptococcus equi* subsp. *equi* (i.e. strangles), with rupture of abscessed retropharyngeal lymph nodes into the pouch.

3 How should this case be treated? Transendoscopic removal of chondroids using a snare or basket forceps can be successful if only small numbers of chondroids are present. For larger numbers of chondroids, daily lavage of the affected GP with saline via an indwelling catheter could allow dissolution of the chondroids and eventual drainage. The addition of dilute acetyl cysteine solution has been recommended to speed up dissolution of the masses. Strict hygiene measures should be imposed because of the risk of a chronic carrier status for *S. equi* in these cases.

Discussion

Chondroids are a potential sequela of GP empyema and are often associated with the chronic carrier state of *S. equi*. Clinical signs are similar to those noted above, with a chronic nasal discharge and pain on pharyngeal palpation. Diagnosis of chondroids is typically accomplished via endoscopic visualization of the inspissated material within the pouch, although they can usually also be seen on radiographs of the region. Treatment is as described above and surgical removal is reserved for cases refractory to medical therapy. Historically, surgical removal of chondroids was performed under GA, but in recent years removal has been successful in the standing animal. Prior to initiation of treatment/removal, swabs or fluid samples should be taken for bacterial culture/sensitivity and sequence encoding M-like protein (SeM) PCR. Culture and sensitivity will take a few days and may be unsuccessful in chronic carriers, but are the only way to confirm the presence of live bacteria. While PCR is faster and very specific, it cannot distinguish between live and dead bacterial DNA. A recent study discussed the development of a real-time PCR for *S. equi* genes (eqbE) (as opposed to SeM). The benefit of real-time PCR for eqbE over PCR for SeM would be a faster turnaround time, but as with SeM PCR the distinction cannot be made between live and dead bacteria.

References

North SE, Wakeley PR, Mayo N *et al.* (2014) Development of a real-time PCR to detect *Streptococcus equi* subspecies *equi*. *Equine Vet J* **46**(1):56–59.

Perkins JD, Schumacher J, Kelly G *et al.* (2006) Standing surgical removal of inspissated guttural pouch exudate (chondroids) in ten horses. *Vet Surg* **35**(7):658–662.

Verheyen K, Newton JR, Talbot NC *et al.* (2000) Elimination of guttural pouch infection and inflammation in asymptomatic carriers of *Streptococcus equi*. *Equine Vet J* **32**(6):527–532.

Walshe N, Duggan V (2011) Equine strangles: a review. *Vet Ire J* **1**(8):459.

CASE 159

1 What condition should be suspected? Equine herpesvirus-1 (EHV-1) myeloencephalopathy. Although many horses with EHV-1 myeloencephalopathy will be febrile at the time of clinical disease, the absence of fever does not rule out this condition. The history of previous mild respiratory disease and fever in some of the other horses in the yard is suspicious of EHV-1 infection.

2 How can the suspicion be connfirmed? PCR on whole blood and nasal swabs are the preferred ante-mortem test, and should provide a rapid diagnosis. Analysis of cerebrospinal fluid may show xanthochromia (yellow discoloration), elevated TP concentration, and a low number of nucleated cells. A 4-fold rise in serum neutralizing antibody titer in samples drawn 10 days apart is supportive of the diagnosis, but does not occur in every case.

3 What is the underlying pathogenesis of this disease? The pathogenesis of EHV-1 myeloencephalopathy involves infection by EHV-1 (by inhalation or ingestion), infection of the nasopharyngeal and respiratory epithelium, viremia (associated with mononuclear cells), and infection of vascular endothelial cells in the CNS. Subsequent vasculitis and thrombosis of small arterioles in the brain and spinal cord results is hemorrhage and hypoxic damage to the adjacent neural tissue (especially the white matter). A 'neuropathic' strain of EHV-1, which has a single point mutation, is associated with a greater potential to cause neurologic disease, although neurologic disease can also occur with 'non-neuropathic' strains.

Discussion

Herpesviruses are enveloped double-stranded DNA viruses and are among the most successful pathogenic viruses; all animal species are affected by at least one herpesvirus. Viral DNA replicates in the nucleus of the host cell and histopathology often reveals intranuclear inclusion bodies. An important distinguishing feature of herpesvirus is its ability to become latent within the host (typically in neurons), meaning that the animal harbors the virus for its entire life following initial infection. Recrudescence and associated viral shedding may occur without clinical evidence of disease and at times that are most likely to optimize transmission. As a family, herpesviruses are further broken down into three subfamilies: alpha, beta, and gamma herpesviruses. The alpha subfamily is by far the most important from an equine medical standpoint. Alpha herpesviruses usually destroy the host cell during infection and replication, setting them apart from beta and gamma herpes. There are at least nine equine herpesviruses known, numbered chronologically based on date of isolation.

Of the equine herpesviruses, EHV-1, EHV-3, and EHV-4 are the most clinically significant; all are alpha herpesviruses. The initial stages of EHV-1 infection may include mild respiratory signs, but the ability of this virus to cause potentially life-threatening neurologic disease in any horse and/or abortion in late-term pregnant mares makes it by far the most concerning. As with all alpha herpesviruses, EHV-1 is epitheliotropic and the pathogenesis is as described above, often with a particular affinity for the lower spinal cord. It remains unknown whether the vasculitis is the direct result of viral infection or is immune-mediated. Latency establishes primarily in the trigeminal and sacral ganglia, and recrudescence may be brought about in times of stress (e.g. trailering, showing, racing).

Clinical signs may be asymmetrical and can range from mild to more severe (as in this case) in horses that develop neurologic disease. Bladder distension without incontinence is common and catheterization may be necessary. Treatment of EHV-1 myeloencephalopathy is largely symptomatic but may include dexamethasone and antibiotics (if bladder function is compromised). Dexamethasone should be used in animals with rapid progression of clinical signs or acute recumbency (0.1–0.2 mg/kg IV for 1–2 days). Valacyclovir (20–40 mg/kg PO q8h) may be helpful if used early in the course of the disease (before the development of neurologic signs). Currently there is no reliable way to predict which horses in an outbreak will develop neurologic disease, although some evidence suggests older horses (>20 years) are at greater risk. As noted above, a neuropathic strain has been identified based on a point mutation. There is a greater incidence of neurologic disease in horses infected with this strain, but not all horses infected with neuropathic EHV-1 will develop neurologic signs, and some horses infected with non-neuropathic EHV-1 will develop neurologic signs. Vaccines are available, with the modified live vaccine eliciting stronger immunity than the killed vaccine, but both provide a narrower spectrum of protection than natural infection and neither is labelled to prevent EHV-1 myeloencephalopathy. The modified live vaccine is not labelled for pregnant mares due to the risk of abortion. Further research is necessary in these areas to better predict at-risk populations and develop more effective vaccines.

References

Goodman LB, Wimer C, Dubovi EJ *et al.* (2012) Immunological correlates of vaccination and infection for equine herpesvirus 1. *Clin Vaccine Immunol* **19**(2):235–241.

Pronost S, Cook RF, Fortier G *et al.* (2010) Relationship between equine herpesvirus-1 myeloencephalopathy and viral genotype. *Equine Vet J* **42**(8): 672–674.

CASE 160

1 What is the likely diagnosis based on the gross appearance? Hemangiosarcoma.

2 What treatments should be recommended? Lamellar keratectomy or enucleation.

Discussion

With the exception of SCC, tumors of the eye and adnexa are rare in the horse. They include lymphosarcoma, amelanotic melanoma, melanoma, adenocarcinoma, hemangioma, and hemangiosarcoma. The case presented here is probably a hemangiosarcoma, which is a rare, generally very aggressive, tumor of vascular endothelial origin that usually carries a very poor prognosis regardless of therapy. Very few cases of ocular hemangiosarcoma have been reported in the literature. Based on the reports, there does not appear to be any breed or sex predilection, but all reported cases have been in horses at least 9 years of age. The etiology is unknown, but UV radiation is thought to play a role. Development of hemangiosarcoma in the same eye as SCC has been reported previously. The most common site of origin has been the limbal conjunctiva, but tumors are locally invasive, often involving the nictitans, and frequently metastasize via the lymphatics to mandibular, retropharyngeal, and superficial cervical lymph nodes. Metastasis to the globe, orbit, sinuses, and facial dermis has also been reported. Masses are raised, irregularly shaped, non-pigmented to red in color, and are associated with a serosanguineous discharge from the affected eye and ipsilateral nostril. The masses may or may not be friable or contain areas of necrosis.

Definitive diagnosis is accomplished by biopsy or (preferably) removal of the mass for histopathology and immunohistochemical staining to differentiate hemangiosarcoma from other possible neoplasms, particularly SCC, hemangioma, and lymphosarcoma. One important distinction to be made (and hence a good argument for histopathology and immunohistochemical staining) is between hemangiosarcoma and hemangioma; hemangioma looks very similar grossly, but is benign and carries a far better prognosis than does hemangiosarcoma. Treatment is complete surgical excision (if possible) or enucleation, with or without follow-up radiation. The prognosis is typically very poor regardless of therapy, with recurrence or metastasis often occurring within 6 months.

References

Gearhart PM, Steficek BA, Peterson-Jones SM (2007) Hemangiosarcoma and squamous cell carcinoma in the third eyelid of a horse. *Vet Ophthalmol* **10:** 121–126.

Pinn TL, Cushing T, Valentino LM *et al.* (2011) Corneal invasion by hemangiosarcoma in a horse. *Vet Ophthalmol* **14:**200–204.

CASE 161

1 What abnormality can be seen in the thoracic radiograph, and what is the provisional diagnosis? There is diffuse radiodensity caudoventrally indicative of ventral consolidation or pleural effusion. The most likely cause is aspiration pneumonia.

2 What other diagnostic imaging modality could be considered to help confirm the diagnosis? Thoracic ultrasound should be used to exclude the possibility of pleural effusion and confirm the presence of ventral consolidation of the lungs.

3 What can cause this condition? Aspiration of foreign material classically results in pneumonia with ventral consolidation. Meconium aspiration may occur preterm if the fetus experiences distress before or during birth. Although meconium is sterile, the large particles cause airway obstruction, regional lung atelectasis, and chemical pneumonitis. Aspiration of milk post-natally may occur if the foal is dysphagic or is being supplemented with milk. Dysphagia in the newborn foal usually results in milk regurgitation from the nose; common causes include perinatal asphyxia syndrome, weakness, cleft palate, subepiglottal or pharyngeal cysts, and megaesophagus. In this foal, the hyperfibrinogenemia is suspicious of an *in-utero* infection as fibrinogen takes approximately 48 hours to increase following the onset of inflammation.

Discussion

Pneumonia in neonatal foals can result from meconium aspiration, bacterial infection/sepsis, or viral infection, although parasitic migration may mimic bacterial pneumonia in older foals. Meconium is a concretion of glandular secretions from the developing fetus's GI tract, the amniotic fluid, mucus, and cellular debris. Acute pre- or intrapartum asphyxiation and severe fetal distress often triggers premature passage of meconium by the fetus, which may result in aspiration. Often, affected foals also have marked surfactant dysfunction. The presence of meconium incites an inflammatory reaction that ranges from mild to severe and may result in death. In humans, severe cases are generally associated with *in-utero* chronic asphyxia or infection.

Bacterial pneumonia is most often secondary to sepsis or aspiration of milk. Aspiration pneumonia tends to have a cranioventral distribution, while sepsis may be more generalized. Transtracheal aspirate samples submitted for culture and cytology and blood cultures help guide therapy, but broad-spectrum antibiotics are appropriate pending results, as multiple organisms may be isolated. Radiography and ultrasonography along with auscultation and percussion are helpful in determining the extent of disease, although auscultation may not correlate well with disease severity. For sick foals fed via nasogastric tube, or those being fed by bottle, care must be taken to ensure that the foal is in sternal recumbency or standing during feeding and for at least 5 minutes after feeding.

Viral pneumonias are typically the result of EHV-1, EHV-4, or equine viral arteritis (EVA), with EHV-1 being the most clinically important. Sporadic cases of adenovirus have been reported, typically in Arabian foals with severe combined immunodeficiency disease. Neonatal infection with EHV-1 or EVA has a very high mortality rate and ante-mortem diagnosis is challenging. Clinical signs may include icterus, leukopenia, neutropenia, and petechial hemorrhages, all signs that may also be associated with sepsis. Mildly affected foals may respond to acyclovir. Foals infected with EHV-1 or EVA often develop non-compliant lungs and potentially pulmonary edema, and most succumb to the disease.

Reference

Wilkins PA (2003) Lower respiratory problems of the neonate. *Vet Clin North Am Equine Pract* **19(1)**:19–33, v.

CASE 162

1 What is the most likely diagnosis? Limbal squamous cell carcinoma has a characteristic appearance but must be differentiated from other neoplasms, eosinophilic keratitis, and granulation tissue.

2 What treatment is indicated? Lamellar keratectomy/conjunctivectomy with adjunctive therapy (strontium-90 irradiation, cryoablation, radiofrequency hyperthermia, and CO_2 laser ablation) is recommended.

CASE 163

1 What abnormalities can be identified on the CT images, and how should they be interpreted? 163.1 shows a well-defined, irregularly shaped focal hyperattenuation in the right caudal maxillary sinus. This is located caudal to the 111 tooth periapex. This may be consistent with a focal osteoma, a cementoma, or possibly a dysplastic supernumerary tooth. This lesion does not appear to be clinically significant at present. 163.2 shows disruption of the contour of the right ventral conchal scroll and there is a defect in the scroll that has created a communication between the ventral conchal sinus and the nasal passage. In the absence of this defect being created iatrogenically, the disruption in the concha may be associated with trauma, infection (e.g. fungal rhinitis), conchal necrosis or, less likely, neoplasia. The soft tissue attenuation within the middle meatus may represent tissue, inspissated discharge, or a foreign body.

2 What further examinations should be undertaken? Endoscopic examination of the nasal passages.

Follow up/discussion

Endoscopy revealed an accumulation of necrotic material within the right middle meatus and a fistula between the ventral conchus and the ventral conchal sinus (163.3, 163.4). The necrotic material was removed from the nasal passage using esophageal grasping forceps; histopathology of tissue samples revealed evidence of fungal infection (i.e. fungal rhinitis).

Diagnostic imaging of the equine head, and in particular of the maxilla and sinuses, is challenging. Good quality diagnostic radiographs of the mandible can be obtained relatively easily, but the same cannot be said of radiographs of the maxilla and sinuses or calvarium and tympanic bullae. The superimposition of sinuses and structures of the nasal cavities makes interpretation difficult and relatively unreliable. This is especially true for soft tissue/mucosal disease. In some cases, fluid lines or a soft tissue opacity (mass) may be discernible in the sinuses, but disease is often advanced by the time these can be recognized. Endoscopy can be utilized to investigate the nasal passages, ethmoid turbinates, and some sinus cavities and their mucosal surfaces, but is limited by the diameter of the scope and location/size of the sinuses.

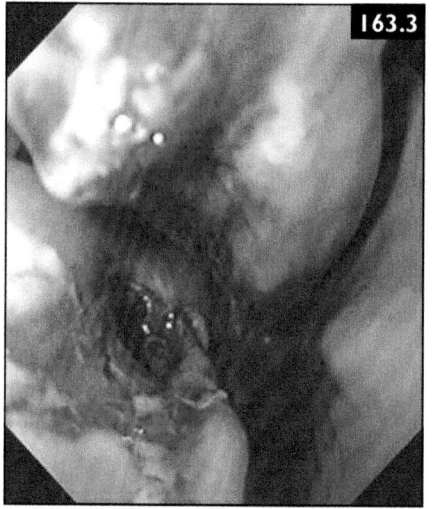
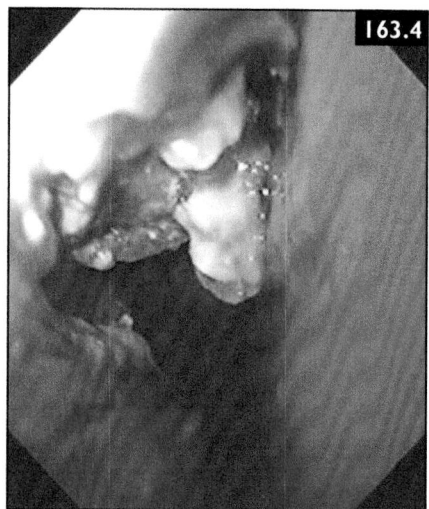

CT and MRI are available at many large referral and academic hospitals and allow more detailed imaging of the sinuses and nasal cavities. Unfortunately, these modalities are not as low risk as endoscopy and radiography as they usually require and carry the associated risks of GA. A recent study indicated that MRI was very good for identification/localization and surgical planning for soft tissue masses (i.e. cysts and neoplasia) and potentially for other non-dental sinus disease, but was relatively poor at correctly identifying the teeth involved with dental-related sinus disease or bony sequestra. They also noted the sensitivity of MRI in detecting minor changes that may not be of clinical significance. CT has also been utilized to more clearly image sinus and nasal cavities and determine the location and extent of sinonasal disease. It is commonly used for surgical planning and is superior to MRI for the investigation of dental-related sinus disease. In recent years, the ability to extrapolate 3-D images from CT has further aided diagnostics, but this technology is new and not yet available at all clinics using CT.

References

Brinkschulte M, Bienert-Zeit A, Lüpke M *et al.* (2013) Using semi-automated segmentation of computed tomography datasets for three-dimensional visualization and volume measurements of equine paranasal sinuses. *Vet Radiol Ultrasound* **54**(6):582–590.

Cehak A, vonBorstel M, Gehlen H *et al.* (2008) Necrosis of the nasal conchae in 12 horses. *Vet Rec* **163**(10):300–302.

Tessier C, Brühschwein A, Lang J *et al.* (2013) Magnetic resonance imaging features of sinonasal disorders in horses. *Vet Radiol Ultrasound* **54**(1):54–60.

CASE 164

1 What condition is suspected? Panniculitis. This is a chronic inflammatory condition of the subcutaneous fat resulting in hard subcutaneous nodules or plaques of inflamed fat.

2 How could the diagnosis be confirmed? Excisional biopsy.

3 What treatment options should be considered? There is no reliably effective treatment. Some cases have responded to corticosteroids. Small lesions may be surgically debrided. Debridement by medical maggots has also been reported.

Discussion

Panniculitis is an idiopathic inflammatory condition of the subcutaneous fat and is rare in horses; the disease appears to be more common in donkeys. It manifests as nodules that may become cystic, ulcerated, and/or develop draining tracts. There appear to be at least two forms: pansteatitis (a.k.a. yellow fat disease) and sterile panniculitis (as in this case). No breed, age, or sex predilection has been noted with sterile panniculitis and the pattern and number of nodules is very variable. Pain may or may not be associated with the nodules, they may wax and wane periodically and other clinical signs (pyrexia, depression, weight loss) may or may not be present. Diagnosis is by excisional biopsy of one or more nodules for histopathology and culture (which is usually negative because this is an aseptic inflammatory condition). Therapy with systemic corticosteroids has been documented, with variable success and recurrence rates after cessation of therapy. Surgical removal/debridement may be curative for small areas/few nodules, but in many cases it is difficult to determine margins and the nodules return. Successful treatment using medical sterile maggots has been reported.

References

Bell NJ, Thomas S (2001) Use of sterile maggots to treat panniculitis in an aged donkey. *Vet Rec* **149(25)**:768–70.

Menzies-Gow NJ, Patterson-Kane JC, McGowan CM (2002) Chronic nodular panniculitis in a three-year-old mare. *Vet Rec* **151(14)**:416.

Scott D, Miller W (2011) *Equine Dermatology*, 2nd edn. Elsevier, Maryland Heights, pp. 449–453.

CASE 165

1 What is the most likely diagnosis? Splenic neoplasia. The commonest neoplasm affecting the spleen is lymphoma. The absence of exfoliated neoplastic cells in the sample of peritoneal fluid does not rule out lymphoma. The ultrasonographic image of the spleen shows a heterogeneous echogenicity with multiloculated areas of anechoic/hypoechoic fluid; this appearance could also be compatible with a splenic hematoma.

2 What further diagnostic tests should be considered? Transcutaneous splenic biopsy (ultrasound-guided) and exploratory surgery (laparoscopy or celiotomy).

Follow up/discussion
Exploratory laparoscopy showed a multinodular mass of the spleen; biopsy confirmed lymphoma. On postmortem examination, a large hematoma was identified in the substance of the spleen as well as areas of tissue necrosis (**165.3**).

Splenic neoplasia is rare in horses, but lymphoma is the most common splenic neoplasm and is generally part of a multicentric presentation, although very rare primary splenic lymphoma cases have been reported. In cases involving the spleen, concurrent hepatic involvement is not uncommon, but this does not appear to be a feature of this case. The clinical signs of mild intermittent or persistent colic along with anorexia, lethargy, and weight loss are consistent with abdominal neoplasia. Anemia is not uncommon with abdominal lymphoma and occasionally thrombocytopenia may be evident, although not in this case. Abdominal ultrasound and abdominal palpation per rectum allow the best non-surgical evaluation of the equine abdomen and are very useful in obtaining accurate transcutaneous biopsies. Abdominocentesis may yield neoplastic cells in a small percentage of cases of lymphoma, but their absence does not rule out neoplasia. Exploratory laparotomy or laparoscopy may be warranted to obtain adequate biopsies. Prognosis is poor given that splenic lymphoma is likely part of a multicentric presentation and by the time horses begin to show clinical signs the disease is typically advanced.

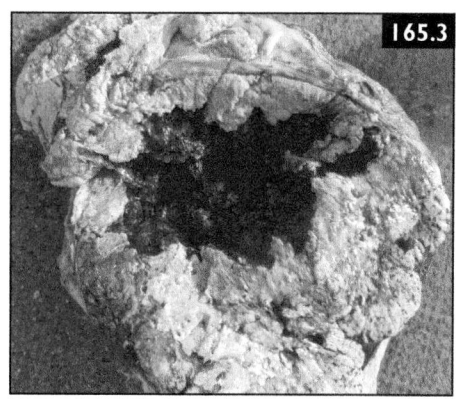

References
Roccabianca P, Paltrinieri S, Gallo E et al. (2002) Hepatosplenic T-cell lymphoma in a mare. *Vet Pathol* **39**(4):508–511.
Tanimoto T, Yamasaki S, Ohtsuki Y (1994) Primary splenic lymphoma in a horse. *J Vet Med Sci* **56**(4):767–769.

CASE 166
1 What are the differential diagnoses? Include lymphosarcoma, sarcoids, habronemiasis, SCC, papilloma, melanoma, and orbital fat prolapse. Lymphosarcoma was diagnosed based on histopathology.

CASE 167

1 What are the common causes of swellings in the proximal cervical region?
Potential causes include branchial apparatus anomalies, goitre, thyroid tumors, retropharyngeal masses (metastases from other tumors of the head or lymph node abscesses), and parotid salivary gland disease (tumor or abscess).

2 What should be suspected based on the physical examination and ultrasound findings? A thyroid tumor. The ultrasound image confirms a solid mass with slightly heterogeneous echogenicity. Common tumor types include adenoma, adenocarcinoma, and medullary carcinoma (C cell or parafollicular cell tumor). Adenomas are common in older horses but rarely cause clinical disease.

3 What further tests are recommended? Percutaneous biopsy, needle aspirate, or surgical excision should be performed. Hyperthyroidism has been reported in association with thyroid adenocarcinoma, but appears to be rare. Measurement of resting thyroid hormone concentrations or evaluation using a stimulation test (TSH or TRH stimulation test) may be considered. Occasionally, advanced imaging techniques, such as nuclear scintigraphy and CT scanning, may yield helpful information.

Discussion

Thyroid enlargement in horses is typically unilateral and the result of a neoplastic process (the most common are listed above). Presentations vary, but most often thyroid tumors occur in older horses (>10 years) with the majority of cases being 20+ years of age. It is not uncommon for owners to report a recent history of rapid enlargement (over a 2–6 month period). The clinical signs generally associated with pharyngeal masses include dysphagia (as in this case) and, more commonly, respiratory signs including upper respiratory noise (likely from pressure on the trachea) or increased exercise intolerance. Ultrasound is very useful in identifying enlargements in this area including thyroid tumors, but ultrasonographic appearance does not correlate with specific thyroid tumor types. A biopsy must be submitted if differentiation between tumor types is desired. If treatment is needed or requested by the owner, surgical unilateral thyroidectomy can be relatively easily performed, although a potential risk is potential damage to the recurrent laryngeal nerve and resulting laryngeal hemiplasia.

Hypothyroidism is exceptionally rare in horses and has not been reported in horses undergoing unilateral thyroidectomy. On rare occasions, clinical signs of hyperthyroidism have been reported to accompany a thyroid tumor, including hyperactivity, weight loss with increased appetite, and polydypsia. In light of these cases, it may be prudent to consider measurement of serum T3, T4, and free T4 levels and/or perform a thyroid stimulation test prior to thyroid gland removal if hyperthyroidism is suspected. In the majority of cases there is no clinically evident endocrinopathy.

References

Alberts MK, McCann JP, Woods PR (2000) Hemithyroidectomy in a horse with confirmed hyperthyroidism. *J Am Vet Med Assoc* **217**(7):1051–4, 1009.

Elce YA, Ross MW, Davidson EJ *et al.* (2003) Unilateral thyroidectomy in 6 horses. *Vet Surg* **32**(2):187–190.

Tan RH, Davies SE, Crisman MV *et al.* (2008) Propylthiouracil for treatment of hyperthyroidism in a horse. *J Vet Intern Med* **22**(5):1253–1258.

Ueki H, Kowatari Y, Oyamada T *et al.* (2004) Non-functional C-cell adenoma in aged horses. *J Comp Pathol* **131**(2–3):157–165.

CASE 168

1 What condition is suspected? Mammary neoplasia. The commonest forms of mammary neoplasia are mammary carcinoma and mammary adenocarcinoma.

2 How would the diagnosis be confirmed? Cytology of fine needle aspirates is often unrewarding because inflammatory cells are frequently present in mammary neoplasms. A core or excisional biopsy away from the ulcerated areas is recommended.

3 What is the prognosis? Most mammary tumors in mares are malignant. The prognosis is therefore guarded even after complete surgical ablation of the mammary gland and regional lymph nodes, because metastasis has often already occurred by the time the diagnosis is made.

Discussion

While mammary neoplasia is common in carnivores, it is very uncommon in horses, with very few cases reported in the last 30 years, although all were in horses over the age of 12. Reported cases have included solid carcinoma, invasive ductal carcinoma, tubular adenocarcinoma, and papillary ductal adenocarcinoma. Mammary enlargement with ulcerated skin appears to be a common clinical presentation. As stated above, the prognosis is guarded to poor, as mammary neoplasia is likely to have metastasized to other organs or lymph nodes by the time the diagnosis is made. Surgical removal of the mammary glands (mastectomy) with or without removal of local lymph nodes is likely to be only palliative unless the disease is detected very early in its course. Late-stage horses may exhibit weight loss, depression, and ventral edema.

References

Gamba CO, Araújo MR, Palhares MS *et al.* (2011) Invasive micropapillary carcinoma of the mammary glands in a mare. *Vet Q* **31**(4):207–210.

Hirayama K, Honda Y, Sako T *et al.* (2003) Invasive ductal carcinoma of the mammary gland in a mare. *Vet Pathol* **40**(1):86–91.

CASE 169

1 What abnormalities can be seen in the endoscopic image? Bilateral laryngeal paralysis.

2 What disease process is suspected? Liver failure.

3 How could the suspicion be confirmed? Serum biochemistry to identify elevated liver enzymes (including ALT, GGT, and ALP), ammonia, and bile acids. Liver ultrasound and biopsy could be considered.

Discussion

While laryngeal hemiplasia is relatively common in horses, bilateral laryngeal paralysis is rare. Cases have been reported in relation to toxicosis (lead, organophosphate, or pyrrolizidine alkaloid), hyperkalemic periodic paralysis, or GA, as well as without a known cause. Multiple reports indicate an association with hepatic disease and especially in conjunction with hepatic encephalopathy and hyperammonemia. Ponies appear to be overrepresented in these reports, followed by Cobs and other breeds of horses, although all of the cases reported in the literature are mature animals.

The pathophysiology connecting liver failure (and associated hyperammonemia and/or hepatic encephalopathy) and bilateral laryngeal paralysis is unclear, but it appears to be a functional neuropathy. This is based on a lack of structural changes or histopathologic lesions in the laryngeal muscles or recurrent laryngeal nerves of affected horses/ponies. Histopathology of the liver (pre- or postmortem) of affected horses shows evidence of chronic liver disease. Postmortem histopathology of the brain of affected horses reveals reactive astrocytosis and increased Alzheimer type II cells along with cerebral edema, supporting the idea of a functional neuropathy. Additionally, in some cases where therapy was able to improve signs of hepatic encephalopathy, the laryngeal paralysis also improved, supporting the idea of a potential role of hyperammonemia and hepatic encephalopathy. The potential for laryngeal improvement/recovery is likely linked to the cause of hepatic disease, as lead and organophosphate toxicity are known to cause irreversible paralysis. Interestingly, bilateral laryngeal paralysis has never been reported and does not appear to occur in cases of hyperammonemia from a non-hepatic etiology. As with most cases of hepatic failure, especially chronic hepatic failure where hepatic fibrosis is present, the prognosis is typically poor. There are a few reported cases of short-term improvement, but most horses are euthanized due to poor response to therapy and worsening clinical signs.

References

Bergero D, Nery J (2008) Hepatic diseases in horses. *J Anim Physiol Anim Nutr (Berl)* 92(3):345–355.

Dixon PM, McGorum BC, Raliton DI *et al.* (2001) Laryngeal paralysis: a study of 375 cases in a mixed-breed population of horses. *Equine Vet J* 33:452–458.

Hughes KJ, McGorum BC, Love S *et al.* (2009) Bilateral laryngeal paralysis associated with hepatic dysfunction and hepatic encephalopathy in six ponies and four horses. *Vet Rec* **164**(5):142–147.

CASE 170

1 What abnormality can be identified in this radiograph? Unilateral pneumothorax.
2 What treatment should be recommended? The pneumothorax in this case is likely due to air entering the pleural space during surgery when the diaphragmatic hernia was being repaired. Unless the horse is dyspneic or the tachycardia worsens, no specific treatment is necessary since this relatively small pneumothorax will probably resolve spontaneously.

Discussion

Pneumothorax is an uncommon to rare condition in horses whereby air or gas enters the pleural space, causing collapse of the lung and leading to clinical signs of tachypnea, dyspnea, cyanosis, restlessness, tympany, expansion of the rib cage, and, if severe, respiratory distress. The majority of cases are bilateral, owing to horses having an incomplete mediastinum; however, unilateral cases have been reported in association with pleuropneumonia. Other etiologies for pneumothorax include thoracic trauma, invasive medical procedures (including surgery), and recurrent airway obstruction. Pneumothorax is classified as open if the defect is in the thoracic wall or closed if the defect is in the lung parenchyma. They are further classified as primary or secondary, spontaneous, traumatic or iatrogenic, and by the extent of lung collapse. In this case the horse had a mild, iatrogenic pneumothorax secondary to surgical repair of the diaphragmatic hernia. Tension pneumothorax is a specific, life-threatening type of pneumothorax; air enters the pleural cavity but does not leave, resulting in a continuous and increasing accumulation of air, which prevents the lungs from expanding (e.g. primary lung lesion with rupture of the visceral pleura).

Diagnosis of pneumothorax is by history, clinical signs, thoracic auscultation, and radiography and/or ultrasonography. Auscultation of the dorsal lung fields is of limited value in many cases as the frequent presence of subcutaneous emphysema or abnormal lung sounds associated with pleuropneumonia may hinder the accuracy. Radiographs are also considered superior to ultrasound as a diagnostic tool, as they are more reliable and are unaffected by the presence of subcutaneous emphysema. Treatment is dependent on the underlying cause and severity. In mild cases, keeping the horse in confinement and monitoring may be all that is needed, but removal of the air dorsally via chest tube and suction may be necessary one or many times in more severe cases or in those with tension pneumothorax. Treatment of the underlying cause (i.e. pleuropneumonia or other injury) is imperative.

Diaphragmatic hernias are rare in horses and may be congenital (foals) or traumatic (most adult cases). Clinical signs are generally related not to the defect in the diaphragm, but to viscera becoming entrapped and strangulated by the defect. Signs are usually acute, but chronic diaphragmatic hernias can also occur. Signs include moderate to severe colic. Dyspnea or respiratory distress may be present depending on the volume of viscera displaced into the thoracic cavity. The presence of gas-filled viscera in the thoracic cavity on ultrasound or radiography is pathognomonic for a diaphragmatic hernia. Treatment is surgical (attempts to salvage incarcerated bowel and repair the diaphragmatic rent) but prognosis is poor with a reported survival rate of 23%. Pneumothorax is a common postoperative complication associated with diaphragmatic hernia repair.

References

Boy MG, Sweeney CR (2000) Pneumothorax in horses: 40 cases (1980–1997). *J Am Vet Med Assoc* **216(12):**1955–1959.

Mochal CA, Brinkman EL, Linford RL *et al.* (2009) What is your diagnosis? Pneumothorax and pneumomediastinum. *J Am Vet Med Assoc* **234(12):**1533–1534.

Röcken M, Mosel G, Barske K *et al.* (2013) Thoracoscopic diaphragmatic hernia repair in a warmblood mare. *Vet Surg* **42(5):**591–594.

Romero AE, Rodgerson DH (2010) Diaphragmatic herniation in the horse: 31 cases from 2001–2006. *Can Vet J* **51(11):**1247–1250.

CASE 171

1 What is the diagnosis? Grain overload.

2 What other complications are likely with this disease? Laminitis.

3 How should this case be managed? First, remove all feed. Pass a stomach tube to check for the presence of gastric reflux. If there is no gastric reflux, then water, magnesium sulfate, and activated charcoal may be administered via the nasogastric tube. Di-trioctahedrol smectite can also be given every 12–24 hours by nasogastric tube. Flunixin meglumine should be administered (1.1 mg/kg IV followed by 0.3 mg/kg IV q8h). Prophylactic therapies for laminitis should include proper foot support and continuous cryotherapy of the lower limbs. Diphenhydramine or doxylamine succinate have also been recommended for this disease. IV fluid therapy (polyionic fluids supplemented with KCl and calcium borogluconate) is indicated, and hyperimmune plasma (2–4 liters) should be administered if possible. Lidocaine given by CRI may help to provide analgesia, reduce neutrophil migration, and subsequently improve intestinal motility. Polymixin B can be administered to bind endotoxin in the circulation.

Discussion

Grain overload is a common problem that can have very serious consequences. The relative term 'grain overload' can be applied to any situation where the horse ingests a significant amount of concentrate feed compared with what he normally eats at a feeding. The amount may seem insignificant to an owner but, if it is a significant increase for the animal, a cascade of events takes place that can lead to metabolic acidosis, endotoxemia, colitis, and laminitis. The severity of clinical signs depends on many factors including the amount and type of concentrate, the promptness of treatment, and the individual horse's physiology.

The equine digestive system is designed to digest high-fiber, low-carbohydrate feeds. The relatively low amounts of carbohydrate should be digested in the small intestine and the bulk of the fibrous feed should be fermented in the cecum (and to some extent in the large colon). If the amount of concentrate a horse is fed is increased slowly over time, the capacity to digest greater amounts of carbohydrate in the small intestine increases. In the case of grain overload, the sudden increase in carbohydrate intake overwhelms the small intestine's ability to absorb it, leading to an overflow of excess carbohydrate to the cecum and large colon. This sudden increase in easily digestible material triggers rapid proliferation of the bacteria that produce lactic acid. The lactic acid production leads to increased permeability of the gut and rapid die-off of gram-negative bacteria, resulting in increased systemic absorption of both d- and l-lactic acid and endotoxin through the gut wall. The result is metabolic acidosis and endotoxemia. The resulting inflammation of the large bowel, along with an increased osmotic gradient, may result in diarrhea.

These processes combine to create a rapidly declining condition from which the horse may not be able to recover. This involves fluid losses from diarrhea and vasculitis leading to shock, activation of coagulation cascades potentially resulting in DIC, and the potential for severe laminitis. It is therefore imperative that these horses be recognized and treated as soon as possible following excess grain consumption. Therapy is typically as described above, although a recent study indicated that, in addition to standard therapies, a 1 g/kg dose of sodium bicarbonate dissolved in 3 liters of water, given orally through a nasogastric tube, has the potential to increase the pH of the cecum following high carbohydrate intake. This might buffer the initial lactic acid production, particularly if administered early in the course of disease and/or treatment.

References

Gomez DE, Arroyo LG, Stämpfli HR *et al.* (2013) Physicochemical interpretation of acid-base abnormalities in 54 adult horses with acute severe colitis and diarrhea. *J Vet Intern Med* **27**(3):548–553.

Taylor EA, Beard WL, Douthit T *et al.* (2014) Effect of orally administered sodium bicarbonate on caecal pH. *Equine Vet J* **46**(2):223–226.

CASE 172

1 What is the likely diagnosis? The cytologic findings of eosinophils and mast cells are supportive of a diagnosis of mast cell tumors (MCTs). Differential diagnoses would include other tumors (e.g. melanoma, nodular sarcoid, eosinophilic collagenolytic granulomas, cutaneous amyloidosis, cutaneous lymphoma).

2 What treatment should be recommended? Surgical excision with a wide margin. Other treatments that have been reported, but for which there is currently little scientific evidence, include intralesional injection of sterile water, intralesional injection of triamcinolone or methylprednisolone acetate, oral cimetidine, and intralesional cisplatin.

3 What is the prognosis following treatment? Generally good. Most MCTs are benign, but rare malignant forms have been recorded. The dangers associated with surgical handling of MCTs (as recorded in other species) appear to be minimal in the horse. There is a small risk of recurrence at the same or different sites.

Discussion

MCTs are rare, typically benign, non-painful, non-pruritic subcutaneous firm nodules accounting for approximately 3–7% of cutaneous or mucocutaneous masses in the horse. While there is no breed or age predilection, males are overrepresented. Nodules most often appear on the head, neck, trunk, or limbs (where they are often noted in proximity to synovial structures) and rarely are noted in association with other structures such as the eye and adnexa. The overlying skin is usually intact, but in some cases ulceration or a draining tract may be present. Chronic lesions may also contain variable areas of mineralization, which can be seen radiographically. Prognosis is very good to excellent as surgical removal is typically curative. Recurrence is rare, even if surgical removal is incomplete. A condition similar to urticaria pigmentosa in people has rarely been reported in newborn foals, where there is widespread appearance and spontaneous regression of MCTs of varying sizes (with larger masses having a tendency to ulcerate) in otherwise healthy foals.

Fine needle aspirate is often helpful in identifying MCTs, but definitive diagnosis should still be by histopathology. Characteristics include single to multiple coalescing nodules of mast cells with rare mitotic figures, large aggregates of eosinophils, areas of necrosis, and involvement of the subcutis and possibly underlying musculature with older lesions exhibiting variable amounts of fibrosis and dystrophic mineralization. The presence of large numbers of eosinophils or fibrosis and dystrophic mineralization may lead to misdiagnosis as either eosinophilic granuloma or calcinosis circumscripta, respectively.

References
Millward LM, Hamberg A, Mathews J *et al.* (2010) Multicentric mast cell tumors in a horse. *Vet Clin Pathol* 39(3):365–370.

Scott D, Miller W (2011) *Equine Dermatology*, 2nd edn. Elsevier, Maryland Heights, pp. 317–324.

CASE 173

1 What condition is suspected? Hemangiosarcoma. Spontaneous hemoabdomen and hemothorax in a middle-aged or older horse should immediately raise suspicion of disseminated hemangiosarcoma. Hemangiosarcoma can occur as a solitary mass, as a single mass with local invasion, or as disseminated masses. Since any organ can be affected, the presenting complaint and clinical signs often include the presence of a visible mass and/or reflect dysfunction of the organ involved; hemorrhagic effusions are common with intra-abdominal or intrathoracic hemangiosarcomas.

2 How could the diagnosis be confirmed? Percutaneous ultrasound-guided biopsy of the spleen could be attempted. Alternatively, pleuroscopic or laparoscopic examinations of the body cavities might identify lesions that could be biopsied.

Follow up/discussion

The owner elected to euthanize the horse rather that attempt to confirm the clinical suspicion of hemangiosarcoma. At postmortem examination, multiple tumor masses (confirmed as hemangiosarcoma) were found throughout the abdominal and thoracic viscera, including the lungs and spleen (**173.2, 173.3**).

 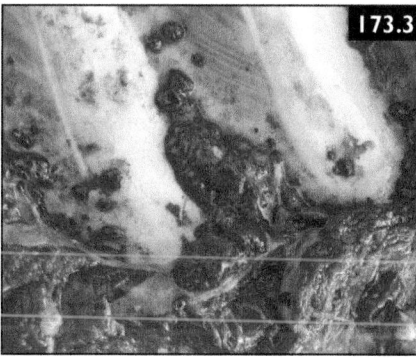

Hemangiosarcomas are rare, highly malignant neoplasms of vascular endothelial origin. Recognized forms in the horse include single mass, locally invasive, and disseminated. Care must be taken to differentiate a benign hematoma or hemangioma from malignant hemangiosarcoma, and there has been some suggestion that the locally invasive form may be an earlier presentation of the disseminated form. No sex or breed predilection has been identified, but most cases of disseminated hemangiosarcoma occur in middle-aged to older animals, with rare cases affecting juveniles (<3 years). Definitive diagnosis is by histopathology and immunohistochemical staining. As hemangiosarcoma can be very similar in histologic appearance to hemangioma or hematoma, multiple biopsies are prudent whenever possible. Hemangiosarcoma may originate in, or metastasize to, any organ, leading to a wide variety of clinical signs, but should be suspected particularly in cases involving hemoabdomen and/or hemothorax without a history of trauma. (Note: Fluid sampled from hemothorax or hemoabdomen related to hemangiosarcoma resembles hemorrhage and is very unlikely to contain neoplastic cells.) Worsening or non-resolving hemorrhage of skeletal muscle should also raise suspicion of neoplasia. Treatment is most often supportive and is generally unsuccessful, with the vast majority of horses continuing to deteriorate despite therapy. Although laparoscopy/-otomy or thoracoscopy/-otomy may be necessary to identify or access masses for biopsy, in most cases surgical removal of a primary mass is not likely to improve the prognosis due to the aggressive nature of the tumor and the likelihood that metastasis has already occurred.

References

Johns I, Stephen JO, Del Piero F *et al.* (2005) Haemangiosarcoma in 11 young horses. *J Vet Intern Med* **19**(4):564–570.

Southwood LL, Schott HC 2nd, Henry CJ *et al.* (2000) Disseminated haemangiosarcoma in the horse: 35 cases. *J Vet Intern Med* **14**(1):105–109.

CASE 174

1 What is amyloid, and why does it occur? Amyloidosis is a group of diseases characterized by the extracellular deposition of amyloid in tissues. Amyloid deposits are homogeneous hyaline protein infiltrates that cause distortion of normal anatomic architecture and can lead to functional impairment of organs. Classification of the various types of amyloidosis is based on the precursor protein that forms the fibrillar deposit. Most cases of amyloidosis in animals are idiopathic, but appear to be of the reactive systemic or secondary type, in which amyloid apoprotein (AA) is derived from serum amyloid A protein. Less commonly seen is immunocytic or primary amyloidosis, in which unstable monoclonal immunoglobulin light chains, produced by plasma cell dyscrasia, lead to the formation and deposition of light chain fibrils (termed amyloid light chain or AL type). Amyloidosis in horses is most commonly of the cutaneous form, although conjunctival and nasal amyloidosis have been reported. The amyloid in these localized forms of equine amyloidosis appears to be primarily of the AL type, whereas systemic amyloidosis, often infiltrating the liver and/or other organs, generally is of the AA type.

2 What further evaluations should be recommended in this case? A full clinical and laboratory evaluation should be undertaken in an attempt to identify the presence of an underlying primary inflammatory or neoplastic disease that may require treatment. In some cases the underlying disease process may not be identified.
3 What is the prognosis for this mare? Poor.

Discussion

Amyloidosis is not one but a group of diseases that result from the deposition of homogeneous amyloid fibrils in tissues, and includes localized amyloidosis, where fibrils are deposited in the specific organs where the precursor proteins are synthesized, and systemic amyloidosis, where precursor proteins circulate in the blood and polymerize prior to tissue deposition. Both AA and AL are forms of systemic amyloidosis and typically arise secondary to long-term inflammation. Both localized and systemic forms are recognized in humans, with AL being the more common systemic presentation compared with AA. In domestic animals, including the horse, AA is the more common systemic presentation. Diagnosis is by Congo Red staining viewed under polarized light or electron microscopic identification of the fibrils. Prognosis is poor but variable depending on the tissues/organs involved, the extent of disease and clinical signs, and the underlying cause.

While amyloidosis does not appear to be transmissible in humans (with the exception of prion disease), there is some evidence that it may be orally transmissible in animals.

References

Kim DY, Taylor HW, Eades SC *et al.* (2005) Systemic AL amyloidosis associated with multiple myeloma in a horse. *Vet Pathol* **42(1)**:81–84.

Merlini G, Bellotti V (2003) Mechanisms of disease: molecular mechanisms of amyloidosis. *N Engl J Med* **349(6)**:583–596.

Murakami T, Ishiguro N, Higuchi K (2014) Transmission of systemic AA amyloidosis in animals. *Vet Pathol* **51(2)**:363–371.

Nout YS, Hinchcliff KW, Bonagura JD *et al.* (2003) Cardiac amyloidosis in a horse. *J Vet Intern Med* **17(4)**:588–592.

CASE 175

1 What is the most likely diagnosis? Regional metastasis of SCC. The masses on the floor of both guttural pouches are likely to be enlarged retropharyngeal lymph nodes.
2 How could the diagnosis be confirmed? Biopsy of the ocular tumor and ultrasound guided biopsy of the subcutaneous masses.

3 What treatment options are available? If the biopsies confirm SCC (which they did in this case), it is unlikely that any treatment will change the prognosis. Radiotherapy could be considered, but in view of the extensive metastases, this is also unlikely to improve the horse's quality of life.

4 What is the relevance of the history of SCC of the third eyelid of the left eye? The current SCC is likely to be a delayed metastasis of the original tumor that was treated by excision of the third eyelid and topical chemotherapy (mitomycin). Although local recurrence at the site of the original tumor is more common, local metastasis to the regional lymph nodes (parotid, retropharyngeal, and submandibular) may also occur, and this may be delayed by months or years following the initial tumor development.

Discussion

It is well recognized that SCC may be locally invasive and has metastatic potential, but that metastasis is slow. Local recurrence is reportedly more likely than metastasis. SCC is the most common ocular/adnexal tumor in horses and may affect the cornea, conjunctiva, limbus, nictitans, or eyelid. Treatment is typically by surgical excision or laser ablation alone or in combination with chemo- or immunotherapy, irradiation, or cryo- or photodynamic therapy. Secondary local invasion to other structures (including the paranasal sinuses, nasolacrimal duct, temporomandibular joint, guttural pouches) is well described and metastasis to regional lymph nodes, salivary glands, or to a distant site has been reported to occur in up to 18% of cases. Clinical signs associated with metastasis are varied depending on the metastatic extent and location, but may include weight loss, depression/dullness, and anorexia. Uncommonly, metastasis may not be recognized for months or years after apparently successful treatment of ocular SCC. Whether it is the metastatic process itself or the clinical recognition of metastasis that is delayed is unknown, but given the slow tumor growth associated with SCC, delayed recognition may be more likely. No predisposing factors for metastasis (delayed or otherwise) have been identified and once metastasis has occurred prognosis is poor. In recent years, a link between *Equus caballus* papillomavirus type 2 (EcPV-2) and SCC has been investigated, but to date there does not appear to be any link between EcPV-2 and ocular/adnexal SCC, although the virus may play a role in the pathogenesis of SCC in other locations.

References

Knight CG, Dunowska M, Munday JS *et al.* (2013) Comparison of the levels of *Equus caballus* papillomavirus type 2 (EcPV-2) in equine squamous cell carcinomas and non-cancerous tissues using quantitative PCR. *Vet Microbiol* **166(1–2):**257–262.

Mair TS, Sherlock CE, Pearson GR (2012) Delayed metastasis of ocular squamous cell carcinoma following treatment in five horses. *Equine Vet Educ* Article first published online: 24 Aug 2012. DOI: 10.1111/j.2042-3292.2012.00435.x

CASE 176

1 What condition is suspected? Progressive ethmoidal hematoma.

2 What is the etiology of this disease? The etiology is unknown. There is no evidence that any form of neoplastic process is responsible for the repeated submucosal hemorrhages that cause these expanding lesions to develop on the surfaces of the ethmoidal turbinate labyrinth. The mucosal capsule splits intermittently to release a bloody discharge, but the overall trend is towards expansion.

3 What treatment options should be considered? Chemical ablation using intralesional 4% formaldehyde solution passed though a transendoscopic needle catheter is the treatment of choice for those progressive ethmoidal hematoma lesions that can be seen by endoscopy. Treatment is repeated at 2-week intervals until resolved and at least three treatments are usually needed. Lesions arising within the sinuses are problematic for chemical ablation because repeated access may be needed. Lesions >3 cm in diameter are unlikely to be successfully resolved by this technique. Surgical removal via a frontal facial flap may be necessary when intralesional chemical ablation is inappropriate or unsuccessful. Transendoscopic laser destruction is suitable for small lesions.

Discussion

Progressive ethmoidal hematomas are uncommon, but well known to most practitioners despite their unknown, but apparently non-neoplastic, pathogenesis. Thoroughbreds and geldings are reportedly overrepresented and hematomas are diagnosed more frequently, but not exclusively, in older horses. As in the above case, the primary clinical signs include uni- or bilateral epistaxis and/or mucopurulent nasal discharge. Respiratory obstruction may occur as the mass enlarges. While generally considered to be non-neoplastic, they are locally destructive to the ethmoidal turbinates, paranasal sinuses, and potentially the cribiform plate. While they often arise from the ethmoids (hence the name), they have also been reported with some frequency arising from the paranasal sinuses either uni- or bilaterally. Bilateral disease is common, so both sides should always be assessed.

Clinical diagnosis is typically by endoscopy with or without radiographs or CT. Endoscopy of the upper airway alone may miss paranasal sinus masses, and radiography or CT may be helpful in some cases. Sinoscopy may be performed, but investigation of the sinuses via radiography or CT is preferred prior to this more invasive procedure. Evaluation by CT, especially with the available 3-D modeling, may be very helpful in evaluating cases with sinus, bilateral, or very large masses that cannot be completely visualized by endoscopy. CT carries the risks associated with GA (unless facilities for standing CT are available) and there is significantly increased cost to the client.

Treatment is by intralesional formalin injection (as described above) or surgical resection. Intralesional injection has many advantages over surgical procedures including, but not limited to: the ease and cost effectiveness of the procedure; the ability to perform in the standing animal; and greatly reduced risk of hemorrhage and post-surgical complications. Severe neurologic complications, including death, have been reported following formalin injection and generally correspond with erosion of the cribiform plate by the mass, creating an abnormal communication to the ventral cranial vault. Regardless of method, recurrence rates are reported to be between 30 and 50% overall.

References

Conti MB, Marchesi MC, Rueca F *et al.* (2003) Diagnosis and treatment of progressive ethmoidal haematoma (PEH) in horses. *Vet Res Commun* **27**(Suppl 1):739–743.

Frees KE, Gaughan EM, Lillich JD *et al.* (2001) Severe complication after administration of formalin for treatment of progressive ethmoidal hematoma in a horse. *J Am Vet Med Assoc* **219**(7):950–952, 939.

Textor JA, Puchalski SM, Affolter VK *et al.* (2012) Results of computed tomography in horses with ethmoid hematoma: 16 cases (1993–2005). *J Am Vet Med Assoc* **240**(11):1338–1344.

Vreman S, Wiemer P, Keesler RI (2013) Bleeding in the subarachnoid space: a possible complication during laser therapy for equine progressive ethmoid haematoma. *Tijdschr Diergeneeskd* **138**(10):30–33.

CASE 177

1 What conditions commonly cause sudden-onset dysphagia, dyspnea, and stridor? Epiglottal, pharyngeal and laryngeal swelling due to anaphylaxis or trauma, arytenoid chrondropathy, guttural pouch (GP) empyema or tympany, GP mycosis, strangles, proximal esophageal obstruction, pharyngeal/laryngeal foreign body.

2 What conditions should be suspected in this case? Pharyngeal/laryngeal trauma or foreign body.

3 What further investigations should be undertaken? Endoscopic examination of the GPs. This revealed hemorrhagic fluid in the ventral aspect of the left GP (**177.3**) and a foreign body penetrating through the medial wall of the medial compartment of the left pouch (**177.4**). Laryngeal/pharyngeal radiography revealed

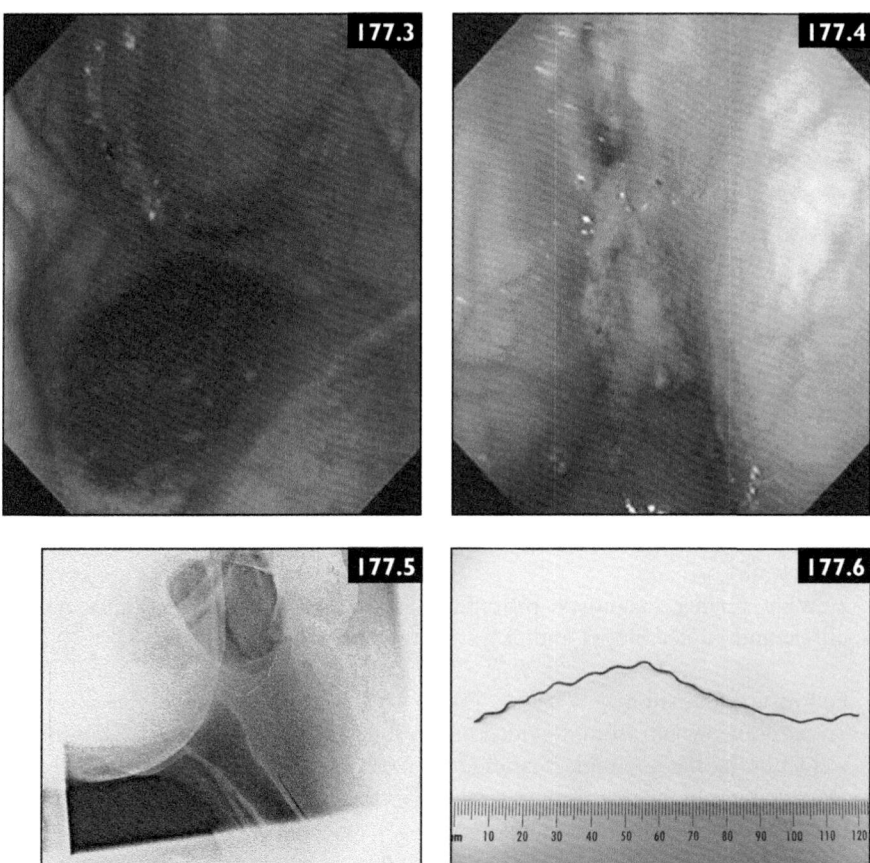

a linear radiodensity compatible with a wire foreign body (177.5). The wire was successfully removed using forceps introduced per nasum into the GP (177.6).

Discussion
GP foreign bodies are rare and typically caused by migration of a penetrating metallic foreign body from the naso- or oropharynx. The GP has a lateral and medial compartment separated by the stylohyoid bone. Several important structures are present in the GP including: CNs IX, X, XI, and XII; the cranial cervical ganglion; the cervical sympathetic trunk; and the internal and external carotid arteries (and branches: caudal auricular, superficial temporal, and maxillary arteries). Clinical signs are related to inflammation and/or damage to these structures and/or abscess formation/drainage and may include nasal discharge, epistaxis, Horner's syndrome, CN deficits (related to IX, X, XI,

or XII [i.e. dysphagia or tongue paralysis), dyspnea, pain in the parotid region on palpation, and restricted movement of the head due to pain. Empyema may develop if an abscess forms and drains into the pouch. Endoscopy may only reveal soft tissue swelling and irritation or a source (but not always the cause) of the epistaxis. A metallic foreign body will be easily identified on radiographs, but multiple angles are needed to better identify the exact location. Depending on the location, the foreign body may be accessible by endoscopy or may require surgical removal.

Reference
Bayly WM, Robertson JT (1982) Epistaxis caused by foreign body penetration of a guttural pouch. *J Am Vet Med Assoc* 180:1232–1234.

CASE 178
1 What abnormalities can be seen in the ultrasound image? There is a soft tissue mass with homogeneous echogenicity ventral to the liver and adjacent to the right body wall.
2 What further diagnostic procedures should be considered? Transcutaneous ultrasound-guided biopsy and/or a standing laparoscopy.

Follow up/discussion
Transcutaneous ultrasound-guided biopsy of the mass was attempted, but failed to yield a useful tissue sample. Standing laparoscopy revealed a 10 cm diameter fatty mass compatible with a lipoma attached to the right dorsal colon wall. Biopsy of the mass confirmed it to be a lipoma.

The lipoma might cause recurrent colic due to intermittent compression of the bowel lumen, but this does not explain the weight loss and the abnormal blood results (in particular, the hyperfibrinogenemia). Exploratory celiotomy identified a large neoplastic mass involving the distal ileum, ileocecal junction, and cecal base, which was probably the mass palpated on the abdominal palpation per rectum. Histopathologic examination revealed adenocarcinoma.

Intestinal neoplasia is rare in horses, but adenocarcinoma is the second most common type after lymphoma. Adenocarcinoma typically presents in older horses as discrete solitary or multiple masses rather than diffuse infiltration and thickening of the intestine, although they do deeply invade the intestinal wall and may protrude into, or occlude, the lumen. The small intestine is commonly affected, but the large intestine or cecum may also be affected. While lymphoma often causes thickening of the small intestine due to diffuse infiltration, adenocarcinoma has been noted to cause hypertrophy or dilation of segments of intestine orad to the tumor(s) due to partial occlusion by the mass. Metastasis to other organs or segments of the intestine is very common and masses may adhere to nearby structures.

Most cases of intestinal neoplasia have very similar clinical signs and diagnostic findings, making them difficult to impossible to differentiate without biopsy. Commonly, they present with some combination of often non-specific clinical signs including weight loss, anorexia, colic of varying severity and duration, fever, ventral edema, anemia (from chronic inflammation), hypoalbuminemia, and/or hyperglobulinemia (from chronic inflammation). Masses or dilated or thickened bowel may be palpated on abdominal palpation per rectum or observed by ultrasound, aiding in the diagnosis of intestinal neoplasia. With the exception of some lymphoma cases, abdominocentesis is generally unrewarding. Definitive diagnosis is by biopsy (either transcutaneous or by laparoscopy or laparotomy). Laparoscopy or laparotomy may be necessary to identify a cause for worsening cases or those that present with colic as the primary complaint but for whom masses were not identified on rectal or ultrasound examination. In general, intestinal neoplasia carries a very poor prognosis owing primarily to the fact that disease is generally advanced by the time clinical signs are evident and a diagnosis is made.

Reference
Taylor SD, Pusterla N, Vaughan B *et al.* (2006) Intestinal neoplasia in horses. *J Vet Intern Med* **20**(6):1429–1436.

CASE 179
1 What are the differential diagnoses? Include retrobulbar/orbital neoplasia, abscess/cellulitis, extension of a sinus cyst, myositis (extra ocular muscles or masticatory), hemorrhage, zygomatic salivary mucocele, orbital bone abnormalities, congenital malformation.

CASE 180
1 What are the possible causes of this epistaxis? Hemorrhage from the GP is most likely to be caused by GP mycosis. Epistaxis may be the only clinical sign of this condition or there may be additional signs of CN dysfunction (most commonly pharyngeal paralysis and/or laryngeal hemiplegia). There may be 1 or 2 minor hemorrhages prior to a severe bleed. Rupture of the longus capitis muscle may also cause GP hemorrhage.
2 What could be done to obtain a definitive diagnosis of the cause? Endoscopic examination of the affected GP. The mycotic lesion is often a yellowish–green or gray diphtheritic plaque, most commonly located dorsally in the medial or

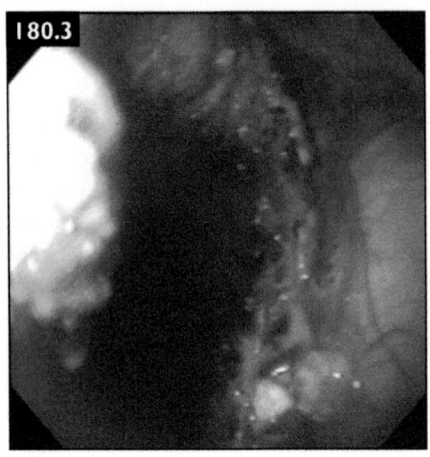

180.3

lateral compartment (**180.3**). A definitive endoscopic diagnosis of a mycotic plaque in the GP is not always straightforward for two main reasons. First, the stress of handling the horse may precipitate a fatal epistaxis. Second, endoscopic visibility within the affected pouch may be poor after a recent hemorrhage and accurate location of the lesion may not be possible. If the epistaxis has been recent, it is sufficient to identify the stream of blood flowing from the pharyngeal drainage ostium. In all cases of mycosis, the contralateral pouch should be checked for extension of the disease through the midline septum and for concurrent bilateral mycosis. A full endoscopic assessment of laryngeal and pharyngeal function is required for an accurate prognosis.

3 How should this case be manageed? If necessary, a blood transfusion to support the circulation should be carried out. Medical treatment of GP mycosis is unlikely to be successful. Surgical occlusion of the branches of the carotid artery is recommended and should be undertaken as soon as possible. Topical antimycotic medication as an adjunct to arterial occlusion is probably not necessary.

Discussion

GP mycosis is an uncommon, but well-known, potentially fatal disease in horses. There are no age, sex, or breed predilections and much of the pathogenesis is not well defined. Clinical presentation, however, is easily recognized by both veterinary professionals and horse owners. Mycosis may be unilateral or bilateral, with bilateral including either co-infection of both pouches or invasion of the fungus across the septum. It is not understood why infection usually occurs in only one pouch or why it occurs in a specific individual at all. Fungal plaques (described above) are typically located dorsally in the pouch and reside over an artery. Erosion through the wall of the affected artery causes hemorrhage. There are typically no hematologic or biochemical abnormalities, except a possible anemia if significant bleeding has occurred. The most common clinical sign is epistaxis or mucopurulent nasal discharge (either uni- or bilateral) and dysphagia, with other CN deficits being less common.

Surgical treatment (with or without medical therapy) offers a better prognosis than medical therapy and involves occlusion of the affected artery(ies) with either ligation or coil embolization (both proximal and distal to the fungal lesion).

In recent years, the use of transarterial nitinol vascular plugs has been described. All of these require GA, but coil embolization in a standing horse has also been described. Medical therapy generally involves topical administration of an antifungal agent, which is both difficult to achieve and generally considered to be far less effective. There is some thought that post-surgical resolution of the mycosis may be hastened by the addition of medical therapy. Culture of the fungal organism is usually not performed due to the risk of bleeding and the known likelihood that the organism is *Aspergillus* spp.

Prognosis is somewhat dependent on the stage of disease, with the best prognosis given to those horses presenting with nasal discharge only. However, owners may overlook nasal discharge, leading to the high percentage of horses with this disease that present with epistaxis. History or presentation with only epistaxis carries a better prognosis than horses with nasal discharge or epistaxis and another clinical sign, often related to CN deficits (typically dysphagia) or, even worse, horses with multiple CN deficits. Dysphagia will likely resolve over time if the mycosis resolves, but horses may develop secondary aspiration pneumonia depending on the severity. Persistent dorsal displacement of the soft palate and laryngeal hemiplasia (of the affected side) are the two most common findings after the dysphagia has resolved.

References
Benredouane K, Lepage O (2012) Trans-arterial coil embolization of the internal carotid artery in standing horses. *Vet Surg* **41**(3):404–409.

Borges AS, Watanabe MJ (2011) Guttural pouch diseases causing neurologic dysfunction in the horse. *Vet Clin North Am Equine Pract* **27**(3):545–572.

Delfs KC, Hawkins JF, Hogan DF (2009) Treatment of acute epistaxis secondary to guttural pouch mycosis with transarterial nitinol vascular occlusion plugs in three equids. *J Am Vet Med Assoc* **235**(2):189–193.

Dobesova O, Schwarz B, Velde K *et al.* (2012) Guttural pouch mycosis in horses: a retrospective study of 28 cases. *Vet Rec* **171**:561–564.

CASE 181

1 What parasite groups affecting adult horses would fenbendazole and pyrantel pamoate be expected to control? Both fenbendazole and pyrantel pamoate should be effective against large and small strongyles unless resistance is documented on the farm. Pyrantel pamoate may be effective against tapeworms at higher dose rates.

2 How is the high fecal worm egg count explained in a horse that received a routine dose of pyrantel pamoate 4 weeks ago? It suggests possible resistance of the strongyle parasites to this drug. Alternatively, it may indicate a shorter than expected egg reappearance time (which may also be an indicator of anthelmintic resistance). Following treatment with pyrantel pamoate, an egg reappearance time of 5–6 weeks is expected if the anthelmintic is effective against this population of parasites. Another potential reason for a high fecal worm egg count at this time would be that the mare was not dosed correctly (i.e. she did not receive the correct dose of anthelmintic drug). There is widespread resistance of small strongyles to fenbendazole in many countries and a growing rate of resistance to pyrantel pamoate in some parts of the world.

3 What further tests might be performed to evaluate this problem? A fecal egg count reduction test may be performed as a crude assessment of anthelmintic resistance. In order to perform a fecal egg count reduction test for pyrantel pamoate, the mare's body weight is measured and a baseline fecal egg count is performed immediately prior to dosing with pyrantel pamoate at the correct dose rate (6.6 mg/kg). A second fecal egg count is performed 14 days later, at which time >90% reduction in the egg count is expected. If the reduction in the egg count is <90%, resistance to pyrantel pamoate should be considered.

Discussion

Anthelmintic resistance in parasites is becoming increasingly more common, particularly in *Parascaris equorum* and the cyathostomins. Both practitioners and owners in the past have been reluctant to check for resistance or implement changes to their deworming program before the appearance of parasite-related disease, making it difficult to slow down the development of resistance within populations. In recent years, anthelmintic resistance has been a topic of much debate. One aspect of this debate centers around the need for a more standardized way of approaching the development of a deworming program. Furthermore, there is a need to standardize testing methods and cut-off values for egg count tests so that results can be more accurately compared.

There are varying schools of thought regarding slowing or preventing anthelmintic resistance, but virtually all recommend utilizing some combination of fecal egg counts, fecal egg reduction tests, and fecal egg reappearance times to develop more individualized programs. Most are also aimed at reducing the amount of anthelmintic used on a farm and using specific drugs for specific parasites instead of the combinations often used. The latter may shed excess drug into the environment, increasing the selection pressure for drug-resistant parasites. There are also some who advocate allowing some animals to go untreated to allow shedding of unexposed refugia parasites so that genetic diversity can help maintain susceptibility to anthelmintics.

Some research indicates there may be only a few horses in a given herd shedding a disproportionately large number of eggs compared with their herdmates. One proposal suggests that the high egg shedding horses be identified via the testing noted above. Then a specific, more frequent worming program should be developed for them while the low egg shedders should be placed on a less frequent program. The egg count tests allow a program to be tailored not only to the horse, but also to the parasites of interest. The reduction of drug use also means less drug shed into the environment where the larvae are developing, reducing the selective pressure for anthelmintic resistance.

References

Lester HE, Matthews JB (2014) Fecal worm egg count analysis for targeting anthelmintic treatment in horses: points to consider. *Equine Vet J* **46**:139–145.

Peregrine AS, Molento MB, Kaplan RM *et al.* (2014) Anthelmintic resistance in important parasites of horses: does it really matter? *Vet Parasitol* **201**:1–8.

von Samson-Himmelstjerna G (2012) Anthelmintic resistance in equine parasites: detection, potential clinical relevance and implications for control. *Vet Parasitol* **185**:2–8.

CASE 182

1 What abnormalities can be identified in the ultrasound image? There is an irregularity at the visceral surface of the spleen with hypoechoic/anechoic fluid and irregular tissue lying deep to this. This appearance is compatible with a hematoma of the visceral surface of the spleen.

2 How should this case be managed? Analgesia should be provided as necessary and circulatory support provided with IV crystalloids. There is a risk of rupture of the hematoma with subsequent hemorrhage/hemoabdomen, in which case hemostatic drugs such as tranexamic acid or ε-aminocaproic acid could be administered, along with blood transfusion if necessary.

Follow up/discussion
This horse demonstrated increasing amounts of pain that could not be controlled with analgesic drugs; an exploratory celiotomy was performed, which confirmed a hematoma at the visceral surface of the spleen. The hematoma ruptured during surgery, resulting in uncontrollable and fatal hemorrhage. Postmortem examination confirmed a ruptured splenic hematoma (**182.2**).

Splenic hematomas are very rare in horses and clinical signs may include depression, tachycardia, pale mucous membranes, colic, and/or abdominal distension.

Answers

Abdominal palpation per rectum may or may not be able to identify a splenic mass depending on its location, but localization within the abdomen due to pain response may be possible. Ultrasound is likely to be helpful in determining the location of a non-palpable mass as well as the potential etiology of the mass and possible presence of hemoabdomen, which may be confirmed or ruled out by abdominocentesis. In rare cases, exploratory laparoscopy or laparotomy may be necessary to identify the cause of the clinical signs. Most cases of splenic hematoma are idiopathic, and the prognosis may be good if the capsule does not rupture and the horse remains stable. Treatment therefore is supportive in nature. Signs of colic may result either expansion of the hematoma pressuring other abdominal organs or rupture of the hematoma and resulting hemoabdomen. Regardless of the cause, worsening or uncontrollable colic or other clinical signs is indicative of a very poor prognosis.

References
Ayala I, Rodríguez MJ, Martos N et al. (2004) Nonfatal splenic haematoma and pancytopenia in an ass. Aust Vet J 82(8):479–480.
Conwell RC, Hillyer MH, Mair TS et al. (2010) Haemoperitoneum in horses: a retrospective review of 54 cases. Vet Rec 167:514–518.

CASE 183

1 What is the cause of the dermatitis of the muzzle, and what is the underlying pathogenesis of this disease? Photosensitization. Secondary photosensitization occurs in liver failure due to the inability of the damaged liver to excrete phylloerythrin (a metabolite of chlorophyll). Phylloerythrin is a photodynamic agent that becomes reactive when activated by ultraviolet light. Photosensitization affects non-pigmented areas of skin, such as the muzzle in this pony.

2 What is the likely cause of the colic and esophageal obstruction? Gastric impaction occurs in some cases of liver failure, probably as a result of dysmotility. Gastric impaction and swelling of the liver are likely the cause of the signs of colic in this pony. Esophageal impaction may also be a result of dysmotility and/or the gastric impaction.

Follow up/discussion

Both gastric and esophageal impaction were confirmed in this pony following subsequent euthanasia and postmortem examination (**183.3**).

Hepatic disease, creating increases in hepatic enzyme concentrations, often arises secondary to GI disease and can progress to hepatic failure. Primary hepatic failure (defined as a loss of >70% of function) is rare and carries a very poor prognosis. Serum enzymes, conjugated bilirubin, and bile acids measurements are the best biochemical indicators of hepatic disease or failure (SDH, AST, GGT), with increases in GGT being the most reliable for hepatic disease. Clinical signs of primary hepatic disease or failure are variable, often non-specific, and may include anorexia, abdominal pain, weight loss, photosensitization, abnormal intestinal motility, icterus, dermatitis, bilateral laryngeal paralysis, and hepatic encephalopathy, among others. Treatment for hepatic disease or failure is supportive. This case

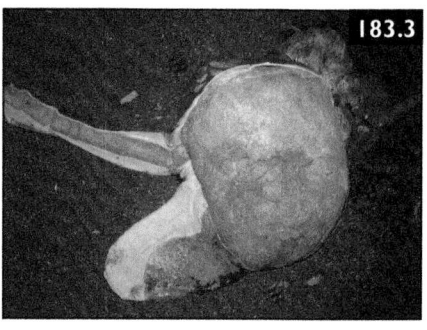

illustrates several of these clinical signs of hepatic failure, but discussion will focus on the two most prominent: photosensitization and gastric impaction.

Photosensitization is not uncommon in horses with liver disease and is likely to be the first clinical sign recognized, although it may be overlooked or dismissed by owners as sunburn. Phylloerythrin photosensitivity (secondary to liver failure) is most often associated with pyrrolizidine alkaloid toxicosis, although photosensitization has also been associated with alsike or red clover (likely a mycotoxin or the toxin saponin in fresh forage) as well as with hypericin from ingestion of St. John's wort (*Hypericum* spp.). While not related to photosensitivity, other dermal presentations may include severe pruritis or generalized seborrhea. The pathogenesis is not understood.

Gastric impaction is not an uncommon sequela to hepatic failure in horses, but does tend to be associated with a very poor prognosis. While the cause of gastric impaction in hepatic failure is not definitively known, theories include mechanical obstruction by a rapidly enlarging liver in acute failure or neuromuscular dysfunction leading to delayed gastric emptying and general GI dysfunction (especially if hepatic encephalopathy is present). While the pathogenesis is unknown, theories are that it is a similar process to the development of bilateral laryngeal paralysis and generalized skeletal muscle weakness in horses with hepatic encephalopathy. In less severely affected horses, the gastric impaction may be subclinical.

References

Bergero D, Nery J. (2008) Hepatic diseases in horses. *J Anim Physiol Anim Nutr (Berl)* **92**(3):345–355.

McGorum BC, Murphy D, Love S *et al.* (1999) Clinicopathological features of equine primary hepatic disease: a review of 50 cases *Vet Rec* **145**:134–139.

CASE 184

1 What abnormalities can be identified in the abdominal ultrasonogram? There is an abnormal soft tissue structure with heterogeneous echogenicity lying against the right ventral body wall. There appears to be fluid (anechoic area) within it, surrounded by a thick wall/capsule that also contains hyperechoic foci, possibly representing gas pockets.

2 Based on the history and clinical findings, what disease process should be considered? Tumor or abdominal abscess.

3 What further diagnostic tests should be recommended? Repeated attempts to obtain peritoneal fluid for cytology should be considered. Exploratory surgery (laparoscopy or celiotomy) to obtain a definitive diagnosis.

Follow up/discussion

An exploratory celiotomy revealed a 15 cm diameter mass involving the wall of the jejunum (**184.2**) with distended intestine proximal to it. On cutting into the mass, it was apparent that it contained fluid and necrotic material, and appeared to be a diverticulum of the thickened jejunal wall (**184.3**). A diagnosis of lymphoma was made.

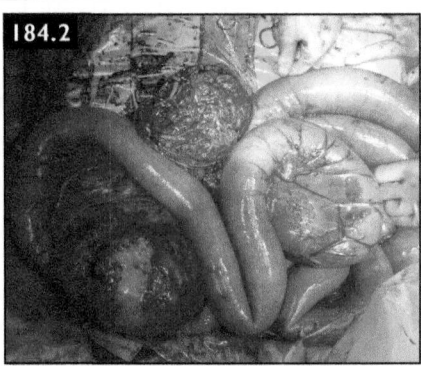 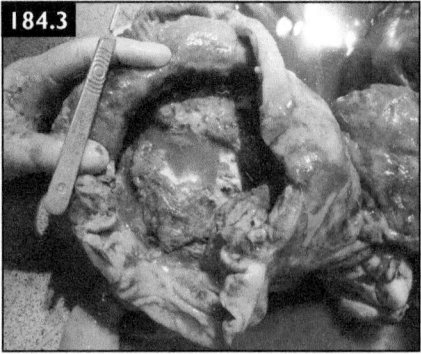

Small intestinal diverticuli are uncommon, congenital, usually incidental findings involving an outpouching of all layers of the intestinal wall along the anti-mesenteric (Meckel's diverticula) or mesenteric border of the small intestine. This is in contrast to pseudodiverticuli, which are acquired outpouchings of mucosa and submucosa through the muscularis layer. Pseudodiverticuli have been reported in association with GI lymphoma.

Lymphoma, while rare in the horse, is the most commonly diagnosed malignancy, with GI lymphoma comprising approximately 19% of all lymphomas. GI lymphoma typically affects the small intestine (particularly the ileum) and may manifest either as a generalized thickening of the bowel wall or as distinct masses. Mucosal ulcers with raised margins have also been reported. This case is interesting as it appeared to be a mass on palpation, on ultrasound, and at the time of surgery, but on cut surface was more consistent with the diffuse transmural wall thickening of a diverticulum.

The most common presenting complaints for horses with intestinal neoplasia are mild to moderate and include weight loss, colic (typically recurring or chronic), and fever. Abdominal palpation per rectum and transabdominal ultrasound are most helpful in identifying thickened or distended segments of intestine or abdominal masses. Abdominocentesis may be helpful in identifying lymphoma in cases of suspected abdominal neoplasia, but most lymphoma cases will not yield neoplastic cells in peritoneal fluid samples, and most other malignancies will also yield a negative sample. Anemia (associated with chronic inflammation), hyperglobulinemia, and/or hypoalbuminemia may be present. In some cases, exploratory laparotomy or laparoscopy with biopsies or resection may be the only way to reach a diagnosis. Prognosis is poor for most cases of abdominal neoplasia unless there is an easily resectable single mass, but in most cases the disease is advanced by the time clinical signs are evident and a diagnosis made; this case was euthanized due to extensive disease.

References

Mair TS, Pearson GR, Scase TJ (2011) Multiple small intestinal pseudodiverticula associated with lymphoma in three horses. *Equine Vet J* **39(Suppl)**:128–132.

Matsuda K, Shimada T, Kawamura Y *et al.* (2013) Jejunal intussusception associated with lymphoma in a horse. *J Vet Med Sci* **75(9)**:1253–1256.

Taylor SD, Pusterla N, Vaughan B *et al.* (2006) Intestinal neoplasia in horses. *J Vet Intern Med* **20(6)**:1429–1436.

Wefel S, Mendez-Angulo JL, Ernst NS (2011) Small intestinal strangulation caused by a mesodiverticular band and diverticulum on the mesenteric border of the small intestine in a horse. *Can Vet J* **52(8)**:884–887.

CASE 185

1 What abnormalities can be identified? The horse has a large unilateral palatal defect and aspiration of food material into the lower airways.

2 What advice should be offered to the client? The prognosis for future athletic performance is very poor. Defects of the hard and soft palate may be congenital or iatrogenic. Although the overwhelming majority of animals with palatal defects are presented as foals with a history of dysphagia and nasal return of milk, more subtle signs of long-standing clefts include a cough, not invariably related to feeding, and respiratory noise associated with palatal instability. Epiglottal entrapment is sometimes seen as a concurrent finding with long-standing congenital defects.

Discussion

Palatal defects are rare and typically recognized in foals because of post-prandial milk dripping from the nose, with or without pneumonia. Fusion of the left and right sides of the palate is usually complete by day 47 of gestation, but approximately

0.1–0.8% of all foals will have a palatal defect. Palatal defects account for approximately 4% of the congenital defects reported in foals. Typically, the defect involves a varying length of the soft palate, but in more severe cases the hard palate is also involved. Surgical repair is difficult and has a relatively high rate of at least partial failure, most commonly due to dehiscence at the caudal border of the palate. Many affected foals have concurrent complications (e.g. pneumonia), which contributes to a poorer prognosis.

Most of the available literature is focused on individual case studies and most of those are focused on surgical techniques for repair. The likelihood of athletic performance at a high level is very low, but the limited information available for horses that survive to at least 2 years of age suggests that they may perform reasonably well as low-level pleasure horses and companions if the defect is small and clinical signs are minor. With small defects, the rate of pneumonia appears to be very low despite the common presence of feed material in the trachea on presentation. In addition, these mildly affected horses do not exhibit poor body condition or growth, although they are frequently noted to have an abnormal respiratory noise during exercise and a persistent cough. Surgical correction has been attempted on older animals, but with less success than with foals. Some animals with larger defects are reported to have improvement but not resolution of clinical signs. In general, surgical correction is not recommended in adult horses for smaller defects or mild clinical signs, as it is unlikely to produce a benefit. Management of these horses should include keeping all feed and water low or at ground level to reduce aspiration.

References
Barakzai SZ, Fraser BLS, Dixon PM (2014) Congenital defects of the soft palate in 15 mature horses. *Equine Vet J* **46**:185–188.
Murray SJ, Elce YA, Woodie JB *et al.* (2013) Evaluation of survival rate and athletic ability after non-surgical or surgical treatment of cleft palate in horses: 55 cases (1986–2008). *J Am Vet Med Assoc* **243**(3):406–410.

CASE 186

1 What abnormalities can be seen in the radiograph and the endoscopic images? There is gross thickening/swelling of the epiglottis. The endoscopy images also show ulceration of the tip of the epiglottis, with white, necrotic-looking submucosal tissue.

2 What are the differential diagnoses? Neoplasia or severe inflammation (epiglottitis).

3 How should this case be further investigated? Biopsy of the epiglottis should help to differentiate between neoplasia and inflammation. In this case, pinch biopsies confirmed chronic, ulcerating, necrotizing epiglottitis.

Discussion

Epiglottitis is an uncommon, poorly understood condition resulting in reddening and thickening of the epiglottis and aryepiglottic membranes, with or without ulceration, and even possible exposure of the tip of the epiglottic cartilage. Typically, epiglottitis is treated with local and/or systemic medical therapy and carries a relatively good prognosis, although epiglottic deformity (usually mild) may occur as a secondary complication. Septic chondritis of the epiglottic cartilage, as in the case above, is less common and equally poorly understood, but carries a worse prognosis as deformation and shortening of the epiglottis is far more likely and more severe. Common clinical signs and history include abnormal respiratory noise (especially while exercising), cough, dorsal displacement of the soft palate, and in some cases dysphagia. Physical examination is likely to be normal in most horses with abnormalities only being noted on upper airway endoscopy. While the pathogenesis is unknown, septic epiglottitis likely results from bacterial infection of the epiglottic cartilage secondary to epiglottitis or mucosal ulceration/abscessation. Treatment is by debridement of the necrotic areas and long-term local and systemic medical therapy (antibiotics and systemic and topical anti-inflammatory drugs). Cartilage shortening and deformity is likely, increasing the probability of other upper respiratory abnormalities. This translates into a poor prognosis for high-level performance horses (such as racehorses), but depending on severity, pleasure horses and companion animals may have a better prognosis due to lighter exercise demands.

Reference

Infernuso T, Watts AE, Ducharme NG (2006) Septic epiglottic chondritis with abscessation in 2 young Thoroughbred racehorses. *Can Vet J* **47**(10):1007–1010.

CASE 187

1 What is suspected to have caused the skin condition? This horse appears to have had a reaction to nettles that were present in the fence line where he got caught up. The urticarial reaction on only one side of the body suggests that he was lying with this side down when he was stuck in the fence.

2 What treatment(s) should be recommended? No treatment is necessary, since signs generally resolve spontaneously in a few hours. Symptomatic treatment with anti-inflammatory drugs may be beneficial in the short term.

Discussion

Urticaria, or hives, are a reaction mediated by the degranulation of IgE-primed mast cells (i.e. type I hypersensitivity reaction). Type I hypersensitivities cover a broad range of reactions to exogenous (e.g. insects, plants, topical medications) or endogenous (inhaled or ingested) antigens. Lesions present as raised, round, wheals or plaques of varying diameter most commonly along the back, flank, legs, and neck, although other areas may be affected. Lesions are often flat-topped and may contain an area of central depression. Urticaria develops rapidly and generally resolves in the same rapid fashion without medical intervention, although in more severe cases, or if there is evidence of anaphylaxis, a glucocorticoid injection may be helpful. If the causative antigen is known, removal of the agent from the horse's environment if possible is recommended, as additional contact will incite repeat reactions.

As with IgG, IgE is produced after exposure to a specific antigen – a process known as sensitization or priming. Unlike IgG, IgE exits the circulation and binds nearly irreversibly to the FcεRI receptors on mast cells in tissues to create an antigen-specific receptor for the mast cell. While no reaction occurs during sensitization/priming, once the IgE is bound to the mast cells, subsequent exposures result in antigen binding to mast cell-bound IgE, triggering mast cell degranulation and release of histamine, tryptase, chymotryptase, TNFα, and other inflammatory mediators. Histamine binds to the H1 receptors of vascular endothelium, triggering inflammation and increased permeability and resulting in the rapid visible swelling and signs of inflammation that constitute the lesions. Additionally, cytokine, chemokine, prostaglandin, and leukotriene production by the mast cell is triggered and these act locally to further incite inflammation and recruit other cells, primarily eosinophils. The degranulation and histamine release are responsible for the rapid onset of the reaction, but the inflammatory mediators produced after degranulation are responsible for the duration. Eosinophils recruited to the area are activated, then go through a similar but slower process of degranulation and production of inflammatory cytokines, as do mast cells.

The cause of the neurologic signs in horses with nettle rash is uncertain, but they are often acute and dramatic, with affected horses showing ataxia, distress, and muscle weakness.

References

Bathe AP (1994) An unusual manifestation of nettle rash in three horses. *Vet Rec* **134**:11–12.

Parham P (2005) *The Immune System*, 2nd edn. Garland Science, New York, pp. 311–330.

CASE 188

1 What are these lesions? Ectopic melanosis.

2 What advice should be given to the potential purchaser about the relevance of these lesions? Ectopic melanosis is a common feature of the mucosa of the guttural pouches of gray horses. The usual site for this is in the lateral compartment over the internal maxillary vessels. Primary melanomas can arise at this site, but there is currently no evidence that horses with ectopic melanosis are at significantly increased risk of developing melanoma.

Discussion

Ectopic melanosis in the guttural pouch of a gray horse is common and not of clinical concern. The area is of greater concern if it exhibits elevation or bulging, and could potentially represent an extension of parotid melanoma. Typically, these more mass-like areas are present high in the lateral compartment of the guttural pouch.

Reference

McGorum B, Dixon P, Robinson E *et al.* (2007) (eds) *Equine Respiratory Medicine and Surgery*. Elsevier, New York.

CASE 189

1 What abnormality is suspected? Post-anesthetic myopathy.
2 What is the cause of this syndrome? Affected horses are usually large and well-muscled and have been exposed to prolonged or deep anesthesia on a hard surface. Muscle compression and prolonged immobility result in muscle ischemia and hypoperfusion. A history of gaseous anesthesia, mechanical ventilation, and mean arterial pressures of <65 mmHg for an extended period of time predispose to the condition. A localized form ('compartmental syndrome') is associated with damage within myofascial compartments.
3 What can be done to reduce the risk of this disease? Proper padding and positioning under GA; the lower forelimb should be pulled forward and the hindlimbs kept apart. Anesthesia should be maintained on the lightest plane possible and mean arterial pressures maintained at >70 mmHg. Dantrolene sodium may be administered 1–2 hours prior to surgery and may aid in reduction of the myopathy risk.

Discussion

Post-anesthetic myopathy is much less common today than in earlier decades, with incidence decreasing as our understanding of contributory factors increases. Hypotension, prolonged anesthesia with insufficient padding, and the horse's overall body mass are the most influential factors. All of these contribute to increased intracompartmental muscle pressure and decreased venous return, which in turn leads to poor muscle perfusion. Affected muscles include the masseter, triceps, gluteals, and the muscles overlying the ribs. Appropriate padding and positioning of the limbs is essential to reduce incidence. Swollen, painful muscles or groups of muscles, typically those in the dependant position during the procedure, are the hallmark of this condition. It is important to remember that post-anesthesic myopathy may occur with any procedure requiring anesthesia, including imaging and dental procedures. Therapy is supportive and generally includes IV fluids (to increase perfusion and for diureses) and NSAIDs. If the triceps are involved and the horse has difficulty supporting weight, splinting may be helpful. Prognosis is variable depending on severity of the condition and promptness of treatment. Affected horses may remain recumbent for many hours or days. Some horses develop varying degrees of fibrosis, which may limit future athletic ability.

In rare cases, post-anesthetic hemorrhagic or thrombotic myelomalacia may occur in horses, usually <2 years old, that are positioned in dorsal recumbency with a relatively short anesthetic period. This condition is characterized by an inability to stand or move the hindlimbs with normal or increased tail and anal tone, which may initially appear similar to post-anesthetic myopathy; however, the horses deteriorate neurologically despite medical intervention.

When the triceps muscles are involved, particularly if the horse is having difficulty bearing weight or extending the limb and is exhibiting a 'dropped elbow', olecranon fracture and radial neuropathy must also be considered. Careful examination and radiographs may be necessary to differentiate between the three conditions. The facial nerve may also be affected by neuropathy, with clinical signs of unilateral partial paralysis of portions of the face. Supportive therapy, including administration of anti-inflammatories (potentially initially including steroids) and splinting of the limb if necessary, typically results in resolution of the neuropathy within a few days, assuming nerve damage is not severe.

References

Franci P, Leece EA, Brearley JC (2006) Post anaesthetic myopathy/neuropathy in horses undergoing magnetic resonance imaging compared to horses undergoing surgery. *Equine Vet J* 38(6):497–501.

Wagner AE (2008) Complications in equine anesthesia. *Vet Clin North Am Equine Pract* 24(3):735–752.

CASE 190

1 What differential diagnoses should be considered? Folliculitis (*Staphylococcus* spp. infection), dermatophilosis, photosensitization, vasculitis, actinic dermatitis, pemphigus foliaceus.

2 What further investigations should be recommended? Bacterial cultures, skin biopsy.

3 How should this condition be treated? Clip the affected areas, apply topical chlorhexidine or another antibacterial wash, protect from sunlight, take off pasture (to avoid sunlight), and administer systemic corticosteroids daily for up to 2 weeks.

Discussion

Leukocytoclastic vasculitis is a histopathologic classification of neutrophilic vasculitis and not a specific disease in itself. It is associated with hypersensitivities, connective tissue disorders, or purpura hemorrhagica, or may be an idiopathic disorder. Characteristic features include perivascular neutrophil infiltration with neutrophil fragmentation and nuclear debris in and around affected blood vessels.

This histopathologic pattern is commonly seen in a specific subset of pastern dermatitis that, while being poorly understood, is believed to be immune-mediated. The condition primarily affects unpigmented skin of the lower hindlimbs, although lesions on pigmented skin have been reported. The pasterns are often affected, but in the case above the cannon bones were primarily affected. It appears to be a photosensitivity reaction, but occurs in horses with no prior exposure to photosensitizing compounds or compromised liver function.

Lesions primarily appear during the summer months, supporting the photosensitivity hypothesis. Chronic cases are visually indistinguishable from other chronic pastern dermatitis. In contrast to other presentations of pastern dermatitis, these lesions are not pruritic but are instead painful and may result in more extensive vasculitis and edema of the limb, leading to lameness. Diagnosis and treatment is the same as for other causes of pastern dermatitis, but relies more heavily on biopsy and histopathologic identification of the leukocytoclastic vasculitis pattern. Management is also similar (secondary bacterial infections are common) but in addition, these horses should be kept out of the sun (long term) and systemic corticosteroids should be instituted.

References

Risberg AI, Webb CB, Cooley AJ *et al.* (2005) Leucocytoclastic vasculitis associated with *Staphylococcus intermedius* in the pastern of a horse. *Vet Rec* **156**(23): 740–743.

Scott D, Miller W (2011) *Equine Dermatology*, 2nd edn. Elsevier, Maryland Heights, p. 87.

Yu AA (2013) Equine pastern dermatitis. *Vet Clin North Am Equine Pract* **29**(3): 577–588.

CASE 191

1 What are the differential diagnoses for unilateral facial swelling? Common causes of swelling of the sinonasal region (and maxillary, nasal, and frontal bones) include inflammation of the facial bone sutures (suture periostitis), sinus cysts, sinus neoplasia, and trauma. Bacterial sinusitis does not typically cause externally visible swelling.

2 What is the most likely cause of the facial swelling in this case? The diffuse unilateral swelling, epiphora, and narrowed nasal meati suggest distension of the left maxillary sinuses, most likely due to a maxillary sinus cyst. The radiographic finding of diffusely increased radiopacity in the maxillary sinuses is compatible with this diagnosis. Neoplasia is also possible (commonest forms of sinus neoplasia are squamous cell carcinoma and adenocarcinoma).

3 What further diagnostic tests should be considered? Oblique and dorsal-ventral radiographic views of the sinuses; further diagnostic imaging (CT or MRI); oral examination (most squamous cell carcinomas originate from the hard palate); trephination/needle centesis or sinoscopy of the maxillary sinus.

Discussion

Sinus cysts account for approximately 13% of paranasal sinus disease in horses and are expansile, fluid-filled structures of unknown etiology that typically develop in the frontal or caudal maxillary sinus. Prognosis is very good if treated. Histologically, they have a thick fibrous capsule lined with ciliated, pseudostratified, columnar epithelium with goblet cells. Cytology of the yellowish fluid produced by the cyst should be sterile and culture should be negative. Non-degenerate neutrophils, mild erythrophagocytosis, and a proteinaceous background (mucus) are common cytologic findings. As the cyst expands, it may fill the sinus and protrude through the communication into another sinus or through the communication with the nasal conchae. Distortion of the nasal conchae, facial bone, or sinomaxillary opening may occur. Bone destruction, presumably from pressure necrosis, is possible in some cases, potentially resulting in exophthalmos.

Clinical signs include epiphora, nasal discharge (ipsilateral to epiphora), obstruction of nasal airflow, abnormal upper respiratory noise, and submandibular lymphadenopathy. Secondary rhinitis may be present, leading to a mucoid or mucopurulent nasal discharge. Radiographs reveal a soft tissue or fluid opacity in one or more sinuses and, depending on the cyst size, rounded or oval borders may be visible. CT has been used in some cases and offers better detail regarding the size and extent of cysts, as well as any bony destruction present. Rhinoscopy reveals the cyst protruding through the nasomaxillary opening in roughly 25% of cases, otherwise it may be unremarkable or only suggestive of sinus disease, but not conclusive for sinus cyst. When possible, sinoscopy via trephination is likely to yield the best results.

Treatment is surgical in nature and the approach depends in part on the size of the cyst. Smaller cysts may be removed via trephination or endoscopic sinus surgery, but a sinus flap may be necessary for larger or more expansive cysts. The cyst lining should be removed as completely as possible, but in many cases this is not possible owing to the limited access to the entire equine sinus. Interestingly, cyst recurrence appears to be rare despite incomplete removal of the cyst lining, although occasionally a persistent unilateral nasal discharge will remain.

References

Annear MJ, Gemensky-Metzler AJ, Elce YA *et al.* (2008) Exophthalmus secondary to a sinonasal cyst in a horse. *J Am Vet Med Assoc* **233**(2):285–288.
Silva LC, Zoppa AL, Fernandes WR *et al.* (2009) Bilateral sinus cysts in a filly treated by endoscopic sinus surgery. *Can Vet J* **50**(4):417–420.

CASE 192

1 What conditions are suspected? Acute laryngeal trauma or infection, laryngeal foreign body.

2 What other investigations should be considered? Diagnostic imaging (radiography, CT, MRI, ultrasonography), oral examination, palpation under GA.

3 How should this case be managed?
Treatment involves removal of the foreign body and elimination of infection. The approach to removal of the foreign body is dependent on identifying its location. If a foreign body can be identified by endoscopic examination, it may be possible to remove it in the standing patient using forceps passed *per nasum*. Alternatively, removal may be achieved under GA. Perioperative antibiotics and NSAIDs should be administered.

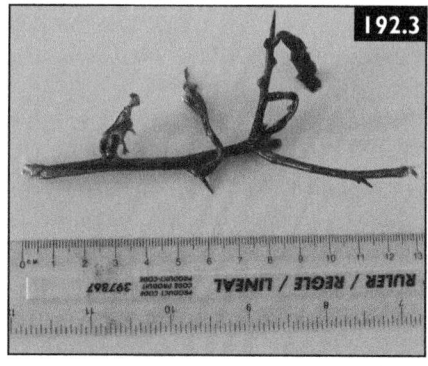

192.3

Discussion

Horses ingest foreign bodies much less frequently than other large animals with their sensitive prehensile lips. Foreign bodies may be very rarely inadvertently inhaled. Foreign bodies are categorized as either obstructive or penetrating and by composition. Metal, wood, bone, hair, or plant material have all been reported as foreign bodies in horses. The above case was a penetrating wood (thorn) foreign body (**192.3**). Clinical signs depend on the location of the foreign body and the chronicity, as eventual abscess formation is likely. In the oral or sinonasal cavities and pharyngeal area these may include dysphagia, respiratory obstruction, nasal discharge, laryngeal hemiplegia, swelling of the tongue, hypersalivation, pain, other localized tissue swelling, and Horner's syndrome. Diagnosis is often best accomplished with a combination of endoscopy, radiography, and ultrasound to localize the foreign body and any abscessation. Foreign bodies tend to lodge in the soft tissues; this, combined with inflammation of surrounding tissues, may make identification by endoscopy alone difficult. While some materials (i.e. metal or bone) may be readily identified on radiographs, foreign bodies composed of other materials (i.e. wood or other plant material, as in this case) may be less radiodense and missed. Often, multiple angles may be necessary to discern the location of a foreign body by radiography. This, along with searching for less radiodense materials, is where ultrasound may be of benefit.

Depending on the location, the foreign body may be removed endoscopically or may require surgical removal and drainage of any abscesses. In either case, antibiotics should be instituted with the route of administration (oral, IM, or IV) determined by the location of the foreign body, presence of abscesses, and clinical signs. In some cases it may be beneficial to institute antibiotic therapy for a period of time prior to attempting removal. Bacteria likely involved in abscess formation are oral/pharyngeal commensals, usually aerobic and facultative anaerobic bacteria including β-hemolytic *Streptococcus,* spp. Pasteurellaceae, Enterobacteriaceae, *Pseudomonas* spp., *Staphylococcus* spp., *Bordetella brochiseptica*, and, potentially, *Bacterioides* spp. *or Fusobacterium* spp.

References

Bell RJ, Dart AJ, Smith CL. (2007) Treatment of a metallic foreign body in the cranial cervical region of a horse. *Aust Vet J* 85(12):517–519.

Gutierrez-Nibeyro SD, Keoughan CG. (2007) What is your diagnosis? A metallic foreign body in the dorsal aspect of the pharyngeal recess. *J Am Vet Med Assoc* 230(3):347–348.

KiperML, WrigleyR, Traub-DargatzJ *et al.* (1992) Metallic foreign bodies in the mouth or pharynx of horses: seven cases (1983–1989). *J Am Vet Med Assoc* 200(1):91–93.

Pusterla N, Latson KM, Wilson WD *et al.* (2006) Metallic foreign bodies in the tongues of 16 horses. *Vet Rec* 159(15):485–488.

CASE 193

1 What condition us suspected? Pemphigus foliaceus.

2 How could the diagnosis be confirmed? Skin biopsy is diagnostic.

3 What treatment should be recommended, and what is the prognosis? Long-term immunosuppressive therapy with corticosteroids: high doses of prednisolone (2.2–4.4 mg/kg PO q24h administered in the mornings for 1 week, followed by 1 mg/kg PO q24h for a further 2 weeks, followed by a reducing dose to achieve an effective minimum dose). Dexamethasone may also be useful, especially early in the course of treatment. If corticosteroids alone are ineffective, the addition of gold injections (e.g. sodium aurothiomylate 1 mg/kg once a week for 4–6 weeks, followed by monthly injections) may be beneficial. Azathioprine and pentoxifylline (typically as an adjunctive to steroids) may also be used. The prognosis is poor with full recovery occurring only rarely; younger horses may respond better.

Discussion
Pemphigus foliaceus, while rare, is the most commonly diagnosed autoimmune dermatosis in the horse. As with all autoimmune diseases, the body produces antibodies against a protein expressed in the affected tissue (i.e. the skin). This is in contrast to immune-mediated disease, in which antibodies are made to a foreign antigen (i.e. drugs). In humans and some dogs the immunogenic protein is known, but in horses it has yet to be elucidated. Pemphigus foliaceus shows no breed, sex, or age predilection, although younger horses may show better responses to therapy and longer periods of control or 'remission'. Lesions often begin on the face or limbs, but may rapidly become generalized. Occasionally, the preputial or mammary area will be affected first and in some animals lesions may be restricted to the coronary band. The lesions are pustules, but as these are fragile and transient, what owners and veterinarians typically recognize are the crusts, scaling, and resulting alopecia that occur when the pustule is broken. Clinical signs of edema, fever, pruritis, or pain may be noted and occasionally other systemic signs, including weight loss or depression, may be present. The condition may wax and wane or exhibit a seasonal worsening during warm, humid times of the year.

Diagnosis is by complete history, physical examination, and skin biopsy for histopathology. Cytology may be used to support the diagnosis and can be taken from either intact pustules or crusted lesions; biopsies, on the other hand, should be taken of intact pustules if at all possible, with crusted lesions being the second choice. In both cases there should be no surgical preparation as this will likely break the pustules or remove the crusts, potentially rendering the sample non-diagnostic. Differentials include dermatophytosis, dermatophilosis, systemic granulomatous disease, and primary keratinization disorders. Treatment is as above and affected horses should be kept indoors if the condition is noted to be worse in sunny, humid weather. Many horses will 'relapse' after apparently successful treatment and will potentially become more refractory to therapy each time. Due to the use of glucocorticoids, these horses are at risk of developing laminitis.

References
Camus MS, Austel MG, Woolums AR *et al.* (2010) Pathology in practice. Pemphigus foliaceous. *J Am Vet Med Assoc* **237**(9):1041–1043.

Rosenkrantz W (2013) Immune-mediated dermatoses. *Vet Clin North Am Equine Pract* **29**(3):607–613.

CASE 194

1 What further diagnostic techniques should be considered? Otoscopic, microbiological, and cytologic examination of the external ear canals. Otoscopic examination is difficult in the conscious horse, but limited evaluation may be possible. In this mare, endoscopic examination in the standing, sedated animal was found to be impossible due to apparent narrowing of both external ear canals. Endoscopic examination of the guttural pouches should be considered because of the association in some horses between otitis externa/media and temporohyoid osteoarthropathy; swelling and modeling of the proximal extremity of the stylohyoid bone may be observed in such cases. Lateral-lateral radiographs of the parotid region may also help show osseous modeling of the tympanic bullae, but dorsoventral views (difficult to obtain in standing horses) are likely to be more useful.

2 What abnormalities can be identified on the CT scans? There is soft tissue attenuation and swelling of both external ear canals with narrowing of the lumens. There is patchy hyperattenuation in the soft tissues adjacent to the external ear canal, compatible with mineralization/osseous metaplasia of the soft tissues. There is soft tissue attenuation in the ventral aspect of the right tympanic cavity. These findings are compatible with bilateral chronic otitis externa and right-sided otitis media. There is mild remodeling of the tympanohyoid articulations (likely an age-related change).

Discussion

In this mare, endoscopic examination showed mild/equivocal swelling of the right stylohyoid bone (**194.3**) compared with the left stylohyoid bone (**194.4**). A lateral-lateral radiograph showed no significant abnormalities (**194.5**).

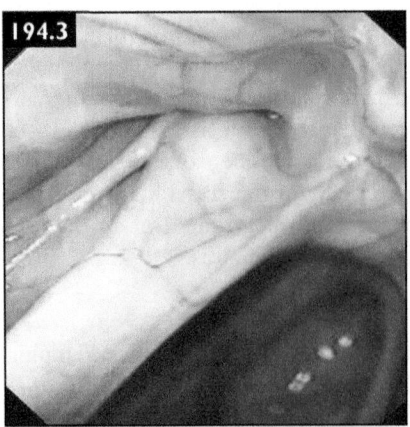
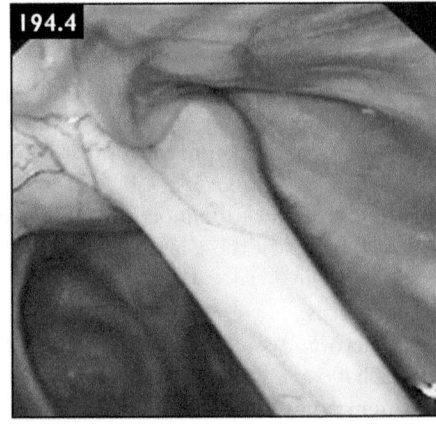

Very little has been reported regarding the visual appearance, cytology, or the presence/lack of normal flora of the normal equine external ear canal. Otic examinations are rarely performed, likely due to the need for chemical restraint as well as the difficulty presented by the long length of the external canal. A technique for otic examination in standing, sedated horses using regional anesthesia has been recently described. Anatomically/histologically the canal can be

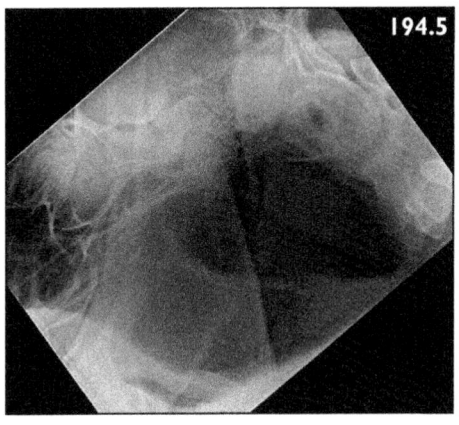

divided into a (distal) cartilaginous portion with thick, pigmented epithelium containing hair follicles and sebaceous and ceruminous glands and a (proximal) bony portion with thinner, non-pigmented, non-secretory epithelium. Unlike in other species, the horse does not have a vertical and horizontal portion of the external canal, but rather the entire canal traverses in a dorsolateral to ventromedial gradual slope. A narrowing of the canal exists at the junction of the proximal and distal portions of the canal, making visual inspection beyond this junction difficult. It may be possible to pass a pediatric video endoscope through the canal enough to visualize the tympanic membrane in some horses. CT gives an excellent level of detail of the external and middle ear but, as it often requires GA, it is likely to be reserved for more advanced cases or once other diagnostic options have been attempted without yielding a diagnosis.

Attempts to establish cytologic and fungal/bacteriologic baselines for the proximal portion of the equine external ear canal indicate that it is potentially a sterile environment, as very few cultures were positive and the few positive cultures that were found were likely contaminants from the distal portion of the canal. Likewise, cytology was negative except for a few instances where a few epithelial cells containing melanin were noted, also probably contamination from the distal canal. The likely sterile environment of the proximal canal may help explain the rarity of otitis externa in the horse.

Reference

Sargent SJ, Frank LA, Buchanan BR *et al.* (2006) Otoscopic, cytological, and microbiological examination of the equine external ear canal. *Vet Dermatol* **17**:175–181.

Sommerauer S, Snyder A, Breuer J *et al.* (2013) A technique for examining the external ear canal in standing sedated horses *J Equine Vet Sci* **33**:1124–1130.

CASE 195

1 What abnormalities can be seen in the radiograph and the ultrasonogram?
There are ill-defined radiopacities in the craniodorsal lung fields and a generalized interstitial pattern. The ultrasonogram shows loss of the normal hyperechoic line of the visceral pleural surface with a hypoechoic lesion extending into the lung parenchyma bounded by irregular hyperechoic foci. These findings are indicative of pulmonary abscessation.
2 What is the most likely diagnosis? *Rhodococcus equi* pneumonia.
3 How can the diagnosis be confirmed? Culture +/− PCR of a tracheobronchial aspirate.

Discussion

R. equi is a gram-positive, facultative intracellular, pathogenic actinomycete. This soil-dwelling encapsulated coccobacillus is strictly aerobic and virulence is conferred via a plasmid that allows replication and survival within macrophages. Inhalation is the major route of infection and experimental incubation periods range from 9 days to 4 weeks, depending on concentration of exposure. *R. equi* is phagocytized by macrophages, but the virulence plasmid allows for modification of the phagocytic vacuole and replication within the alveolar macrophage. Eventually, the host cell is killed and becomes necrotic. Disease usually occurs in foals between 3 and 16 weeks of life. Given the typical age of foals presenting with pneumonia and the experimental incubation period, it is likely that foals become infected very early in life. The prevalence appears to be extremely variable between farms.

The most common presentation for *R. equi* in foals is pyogranulomatous bronchopneumonia with abscess formation, which may be subclinical, subacute (rapid onset of respiratory distress), or chronic, and clinical signs are dependent on the severity and chronicity of lung lesions. Clinical signs most commonly include pyrexia, lethargy, and cough early in the course of disease, with anorexia, tachypnea, dyspnea, increased abdominal effort in respiration, failure to grow, or weight loss appearing in more severely affected foals. A myriad of extrapulmonary disorders (EPDs) have been associated with *R. equi* and up to 74% of foals may have at least one EPD, some of which may worsen the prognosis. The most common EPDs include pyogranulomatous typhlocolitis or enterotyphlocolitis (resulting in diarrhea in up to 50% of foals), polysynovitis (multiple joint effusion without lameness in up to 33% of foals), ocular lesions (uveitis, keratouveitis, panophthalmitis), and osteomyelitis and/or septic synovitis (effusion with lameness).

Definitive diagnosis is by culture/sensitivity from a tracheobronchial aspirate, with or without PCR, from foals with clinical, cytologic, or radiographic/ultrasonographic evidence of disease. A positive PCR alone is discouraged as a solitary diagnostic test because it neither allows for the identification of other bacteria that may be present nor determines sensitivity patterns. Current recommended therapy combines a macrolide (azithromycin or clarithromycin) with rifampin alongside supportive care (including keeping foals in a cool, well-ventilated area). The duration of therapy varies depending on the severity of disease and clinical signs, but may range from 3–12 weeks. **Note:** Up to 36% of foals will develop an often self-limiting diarrhea as a result of treatment and mares may develop severe colitis if even small amounts of macrolide are ingested (e.g. from licking foals lips or contamination of hay during administration). On some farms resistance to this combination has been noted. These isolates have been found to be susceptible to other antimicrobials, highlighting the importance of culture and sensitivity testing. On farms where *R. equi* is endemic, screening by various methods (e.g. thoracic ultrasound or CBCs) may be prudent so that early therapeutic intervention can take place. However, care must be taken as many subclinical foals will clear the infection without the use of antibiotics. Isolation of sick foals does not appear to be necessary, and prognosis for future performance is very good for foals that respond to therapy. Prophylactic use of *R. equi* hyperimmune plasma may be considered on endemic farms.

References

Giguère S, Cohen ND, Chaffin MK *et al.* (2011) *Rhodococcus equi*: clinical manifestations, virulence, and immunity. *J Vet Intern Med* **25**(6):1221–1230.

Giguère S, Cohen ND, Chaffin MK *et al.* (2011) Diagnosis, treatment, control, and prevention of infections caused by *Rhodococcus equi* in foals. *J Vet Intern Med* **25**(6):1209–1220.

Venner M, Credner N, Lämmer M *et al.* (2013) Comparison of tulathromycin, azithromycin and azithromycin-rifampin for the treatment of mild pneumonia associated with *Rhodococcus equi*. *Vet Rec* **173**(16):397.

CASE 196

1 What is the diagnosis? These signs are typical of tetanus.

2 How should this horse be treated? The pony should be kept in a quiet environment with good footing. The lighting should be kept subdued and cotton wool should be packed into the ears. Muscle relaxant and tranquillizer drugs should be administered: acepromazine, phenobarbital, diazepam, or methocarbamol can be used. Any wound should be cleaned and debrided. Crystalline penicillin should be administered; broad-spectrum antibiotics should also be provided as a preventive therapy for aspiration pneumonia. Tetanus antitoxin (100–200 units/kg IV or IM) should be administered. Intrathecal administration of tetanus antitoxin has been suggested, but there is no good evidence of efficacy in horses.

3 What is the prognosis? Fair if the pony remains standing but becomes poor if he becomes recumbent. Clinical signs can persist for weeks or months.

Answers

Discussion

Tetanus is a neuromuscular disease that results from the colonization of a deep wound (typically a puncture, although surgical wounds and umbilici have also been reported as sources of infection) with the gram-positive, spore-forming bacteria *Clostridium tetani*. The bacteria typically reside in the soil and spores may persist for years in the environment. The anaerobic conditions established at the wound site allow the bacteria to proliferate and produce the tetanolysin toxin, which incites tissue necrosis to enhance bacterial growing conditions, and the tetanus neurotoxin, which blocks inhibitory neurotransmitter release at the level of the spinal cord. The result is spastic paralysis of skeletal muscle. The severity of clinical signs is dependent on the sensitivity of the species to the toxin and concentration of toxin present. Horses are the most sensitive of all the domestic animals to the neurotoxin. The incubation period is variable, but typically is approximately 7–10 days. Common clinical signs may include a stiffening of the head and neck, stiff gait, prolapse of the nictitans, elevation of the tail head, and potentially an inability to open the jaw (hence the colloquial term 'lock-jaw'). Horses become hypersensitive to visual and auditory stimuli and very little is needed to incite tetanic muscle spasms. In severe cases, horses may become recumbent, significantly worsening their prognosis.

Treatment is generally supportive in nature, aimed at reducing external stimuli and keeping the horse comfortable. Tetanus antitoxin is available and may help stabilize the horse but will not reverse or improve any clinical signs, as the neurotoxin once bound cannot be removed. (**Note:** Antitoxin may be prohibitively expensive for many owners.) Growth of new nerve terminals is required for recovery, a process that may take weeks to months. Antitoxin may best be reserved for unvaccinated horses due to the concern of potentially developing serum hepatitis (Theiler's disease). Prognosis for tetanus is variable and depends greatly on vaccination history, toxin concentration, and promptness of treatment, but studies often report high mortality rates. Routine vaccination with tetanus toxoid has severely reduced the number of cases in the last couple of decades, but rare cases still occur.

Serum hepatitis (Theiler's disease) is a condition of hepatic failure in adult horses that has been linked to a history of administration of tetanus antitoxin or other equine biologics (e.g. hyperimmune serum or plasma) several weeks prior to the onset of clinical signs. Horses present with hepatic encephalopathy, icterus, variable signs of colic, increased hepatocellular enzymes, but no pyrexia. Care for affected horses is supportive and the prognosis is guarded. Recently, a possible connection between a Flavivirus and Theiler's disease was suggested, but more research is needed before a connection can be confirmed.

References

Chandriani S, Skewes-Cox P, Zhong W *et al.* (2013) Identification of a previously undescribed divergent virus from the Flaviviridae family in an outbreak of equine serum hepatitis. *Proc Natl Acad Sci USA* **110**(15):E1407–1415.

Mykkänen AK, Hyytiäinen HK, McGowan CM (2011) Generalised tetanus in a 2-week-old foal: use of physiotherapy to aid recovery. *Aust Vet J* **89(11):** 447–451.

van Galen G, Delguste C, Sandersen C *et al.* (2008) Tetanus in the equine species: a retrospective study of 31 cases. *Tijdschr Diergeneeskd* **133(12):**512–517.

CASE 197

1 What abnormalities can be identified in the CT image? There is an oromaxillary sinus fistula at the site of the extracted 109 tooth. There is heterogeneous soft tissue attenuation and patchy areas of gas attenuation within the right rostral maxillary and ventral conchal sinuses. There is loss of alveolar bone resulting in direct communication between the alveolar cavity and the right rostral maxillary sinus. The right rostral maxillary and ventral conchal sinuses are likely filled with food material and caseous pus.

Discussion

Oromaxillary fistulae form either as a complication of repulsion of cheek teeth or, less commonly, secondary to a cheek tooth diastema, fracture, or wear (in older animals). This complication is much less common following intraoral extraction of teeth since the alveolar bone remains undamaged by the extraction procedure. Up to 33% of repulsed cheek teeth will result in an oromaxillary fistula and horses present with a history of tooth removal via repulsion and clinical signs consistent with sinusitis, occasionally with feed material in the nasal discharge. Following repulsion, the oral portion of the alveolus is filled with a packing material (e.g. polymethylmethacrylate) to prevent feed material and bacteria from entering the sinus. Over time the alveolus fills with granulation tissue and the packing can be removed. Fistulae typically develop as a sequela to one of the following: over-packing the alveolus (resulting in plug extension into the sinus, preventing healing); inadequate sealing of the alveolus; early loss of the alveolar plug; incomplete removal of dental material; sequestra development; or, as was likely in this case, residual infection. Treatment involves flushing the affected sinus, alveolar curettage/removal of sequestra, and sealing the oral aspect of the alveolus. Surgical transposition of the temporal or levator labii superioris muscle has been reported as a method to seal the fistula in refractory cases.

References

Brink P (2006) Levator labii superioris muscle transposition to treat oromaxillary sinus fistula in three horses. *Vet Surg* **35(7):**596–600.

Hawkes CS, Easley J, Barakzai SZ *et al.* (2008) Treatment of oromaxillary fistulae in nine standing horses (2002–2006). *Equine Vet J* **40(6):**546–551.

CASE 198

1 What are the diagnosis and treatment recommendations? Fluorescein staining of the cornea was negative, therefore a presumptive diagnosis of mid-stromal immune-mediated keratitis is made. Treatment options include topical steroids, NSAIDs, or cyclosporine. Keratectomy may be required for cases non-responsive to medical therapy.

CASE 199

1 What abnormality can be detected in the brain on the CT scans? There is a focal area of mottled hyperattenuation within the left lateral ventricle.

2 What is the likely diagnosis? The CT appearance is consistent with dystrophic mineralization of a mass within the left lateral ventricle. This is most likely to be a cholesterinic granuloma (a.k.a. cholesterol granuloma). These are benign growths of the choroid plexus occurring in the lateral and 4th ventricles. They are common in older horses. In many cases the lesions are asymptomatic (as in this horse), but in some cases the lesions can cause neurologic signs due to an associated obstructive hydrocephalus or direct brain compression. A CT-guided biopsy allows differentiation from less common tumor types.

3 What advice should be offered to the owners of this horse regarding this finding? In many cases, cholesterinic granulomas remain asymptomatic for the horse's entire life. No treatment is possible or necessary in these animals.

Discussion

Cholesterinic granulomas in the choroid plexus of horses are uncommon (accounting for approximately 1.3% of all neoplasms), benign, and most often incidental findings in older animals (up to 20%) on CT or at necropsy. Differentials include choroid plexus papilloma and papillary ependymoma. The most common location is in the 4th ventricle, but they are also commonly found bilaterally in the lateral ventricles where they are more likely to cause clinical signs. Pathogenesis is unknown, but the leading theory is that they develop as a granulomatous inflammatory response to the release of cholesterol (which acts as a foreign body) from RBCs during periods of vascular congestion and secondary hemorrhage. While 4th ventricle masses are likely to remain clinically silent, masses in the lateral ventricles have been associated with intermittent neurologic clinical signs including blindness, ataxia, anisocoria, lethargy, circling, paresis, tremors, seizures, and compulsive behavior. These signs are often relieved with anti-inflammatory or corticosteroid medications, but most horses have additional episodes following asymptomatic periods of unpredictable length. Clinical signs are thought to be due to impingement of cerebrospinal fluid drainage, compression of nervous tissue, intermittent release of cholesterol from the granuloma (creating a chemical meningitis), or a combination of these factors.

Cholesterinic granulomas have a typical CT appearance including heterogeneous attenuation, peripheral mineralization, homogeneous contrast enhancement, and location (usually bilaterally) within the rostral aspect of the lateral ventricle. Histopathology usually shows a well-demarcated nodule with poorly organized cholesterol clefts and 'foamy' macrophages. Diagnosis is by CT with possible biopsy (for cases showing clinical signs), and therapy is generally aimed at alleviating clinical signs when they appear. Prognosis is very good for horses in which this is an incidental finding; for horses that show clinical signs, a conversation regarding safety both of the horse and anyone working with the horse needs to occur.

References
Tofflemire KL, Whitley RD, Wong DM *et al.* (2013) Episodic blindness and ataxia in a horse with cholesterinic granulomas. *Vet Ophthalmol* **16**(2):149–152.
Vanschandevijl K, Gielen I, Nollet H *et al.* (2008) Computed tomography-guided brain biopsy for in vivo diagnosis of a cholesterinic granuloma in a horse. *J Am Vet Med Assoc* **233**(6):950–954.

CASE 200

1 What condition is suspected? Mycotic rhinitis. Mycotic opportunistic infection is more common after surgery or secondary to other suppurative conditions, such as dental periapical abscessation. However, occasionally horses can develop these infections on the sinonasal tissues without obvious underlying disease. The etiology is not known but the infection consists of a destructive rhinitis/sinusitis occasionally producing sinonasal fistulae. Horses with mycotic infections in the nasal region usually show a low-grade unilateral purulent discharge, which may be malodorous. There may be epistaxis on the rare occasions when erosion of a significant blood vessel has occurred, and occasionally bilateral infection occurs.
2 How can this suspicion be confirmed? Radiography may be normal or there may be small gas/fluid lines; such a finding in a horse with a notable nasal discharge is highly suggestive of a mycotic infection. The presence of mycotic plaques can be confirmed by endoscopy either per nasum or directly into the caudal maxillary sinus. Cytology and culture of swabs of the lesion will identify fungal infection.
3 How can this disease be treated? Topical medication with an antifungal agent, such as enilconazole, is usually effective for small lesions. A Foley balloon catheter can be placed into the caudal maxillary sinus; the sinus cavity acts as a reservoir for the medication, which is infused twice daily. Resolution may require prolonged treatment (4–6 weeks). Extensive mycoses are better treated by surgical curettage or laser ablation followed by topical medication to prevent recurrence.

Answers

Discussion

Mycotic or fungal rhinitis is a rare disease of unknown pathogenesis in the horse. Clinical signs can be variable but are similar to other upper respiratory diseases: dyspnea, uni- or bilateral mucoid or mucopurulent nasal discharge that may or may not have a foul odour, epistaxis, and head shaking. *Pseudallesheria boydii, Conidiobolus coronatus, Cryptococcus neoromans, Penicillum* spp., and *Aspergillus* spp. have all been cultured from fungal lesions in the upper airways of horses, with *Aspergillus* spp. being by far the most common isolate. All of these organisms commonly reside in the soil and it is unknown why some individuals are colonized while the vast majority of horses are not, although possibilities include an immune-compromised state or mucosal damage (via trauma or viral infection).

Identification of the causative fungus may be difficult, owing in part to the fastidious nature of many fungi and the opportunistic nature of *Aspergillus* spp. Fungal culture and identification can take several days and may not be successful, making it a less efficient diagnostic tool. Care must be taken in interpretation, as many cultures may be negative or contaminated with *Aspergillus* spp. or bacteria. However, it may be prudent to submit a sample for culture prior to initiation of therapy for potential sensitivity testing should it be needed. Confirmation of fungal hyphae by histopathology or by cytology from a direct smear may be more rewarding diagnostically.

Treatment depends on the location, type, and severity of the lesion. *Aspergillus* spp. tend to present as plaque lesions, while the other organisms are more likely to present with granulomatous lesions. Debridement and/or surgical removal when possible is very helpful, but is not likely curative. Surgical removal or debridement of any fungal lesion is likely to enhance the efficacy of medical therapy, particularly topical therapies. Antifungal therapy, either topical or systemic, is likely to be necessary. Both the choice of drug and route of administration should be carefully considered. Keep in mind that, while topical medications may penetrate a plaque lesion, they are unlikely to penetrate a granulomatous lesion enough to be of real benefit.

References

Cehak A, vonBorstel M, Gehlen H *et al.* (2008) Necrosis of the nasal conchae in 12 horses. *Vet Rec* **163(10)**:300–302.

Hunter B, Nation PN (2011) Mycotic encephalitis, sinus osteomyelitis, and guttural pouch mycosis in a 3-year-old Arabian colt. *Can Vet J* **52(12)**:1339–1341.

Kendall A, Bröjer J, Karlstam E *et al.* (2008) Enilconazole treatment of horses with superficial *Aspergillus* spp. rhinitis. *J Vet Intern Med* **22(5)**:1239–1242.

Perkins JD, Windley Z, Dixon PM *et al.* (2009) Sinoscopic treatment of rostral maxillary and ventral conchal sinusitis in 60 horses. *Vet Surg* **38(5)**:613–619.

CASE 201

1 What is this lesion? Thrush (oral candidiasis).

2 Why has this foal developed this disease? The thrush has likely developed because of a combination of immunodeficiency, associated with failure of passive transfer of immunity, and poor oral hygiene due to lack of sucking.

Discussion

Candida albicans is a polymorphic fungus that may be present as an oval budding yeast, form pseudohyphe or germ tubes, or develop into branching hyphae, depending on the growth conditions. Pathology is associated with a change from the yeast to the hyphal form. *C. albicans* and other *Candida* spp. are commensals of the mucous membranes of most mammals, but are opportunistic pathogens causing a variety of fungal diseases, some of which are associated with antibiotic use, which reduces the competition with bacteria for colonization. The fungus produces a neurominidase that thins mucus and allows contact with mucosal cells.

Oral lesions are common in foals presenting as in the case above. Treatment with topical antifungal rinses should be sufficient, especially if the foal is showing improvement; systemic antifungal therapy is rarely required.

References

Pirrone A, Castagnett iC, Mariella J *et al.* (2012) Yeast flora in oropharyngeal and rectal mucous membranes of healthy and critically ill neonatal foals. *J Equine Vet Sci* **32**(2):93–98.

Reilly LK, Palmer JE (1994) Systemic candidiasis in four foals. *J Am Vet Med Assoc* **205**:464–466.

Appendix: Reference Intervals – Adult Horses

	Conventional units	SI units
Albumin	2.9–4.1 g/dl	29–41 g/l
AST	102–350 U/l	102–350 U/l
Bile acids	1–8.5 µmol/l	1–8.5 µmol/l
Chloride	95–103 mEq/l	95–103 mmol/l
Cholesterol	<90 mg/dl	<2.45 mmol/l
Creatine kinase	110–250 U/l	110–250 U/l
Creatinine	0.8–1.8 mg/dl	85–165 µmol/l
Direct bilirubin	0.1–0.3 mg/dl	1.7–5.2 µmol/l
Fibrinogen	100–300 mg/dl	1.0–3.0 g/l
GGT	10–40 IU/l	10–40 IU/l
GLDH	1–10 IU/l	1–10 IU/l
Immunoglobulin G	1,372–3,032 mg/dl	13.7–30.3 g/l
Immunoglobulin M	63–143 mg/dl	0.63–1.43 g/l
Ionized calcium	5.6–6.9 mg/dl	1.4–1.7 mmol/l
Lactate-L	<1.78mmol/l	<1.78 mmol/l
Lymphocytes	$2.0–3.2 \times 10^3/\mu l$	$2.0–3.2 \times 10^9/l$
PCV	35–47%	0.35–0.47%
Platelets	$100–250 \times 10^3/\mu l$	$100–250 \times 10^9/l$
Potassium	3–5 mEq/l	3–5 mmol/l
Segmented neutrophils	$3.5–5.8 \times 10^3/\mu l$	$3.5–5.8 \times 10^9/l$
Serum amyloid A	0–20 ug/ml	0–20 mg/l
Sodium	134–142 mEq/l	134–142 mmol/l
Total bilirubin	0.5–2.5 mg/dl	8.5–42 µmol/l
Total protein	5.3–7.3 g/dl	53–73 g/l
Triglycerides	17–63 mg/dl	0.2–0.64 µmol/l
Troponin I	<0.04 ng/ml	<0.04 µg/l
Urea (BUN)	11–28 mg/dl	4.0–10.0 µmol/l
WBCs	$6.0–10.0 \times 10^3/\mu l$	$6.0–10.0 \times 10^9/l$

Normal values may vary slightly among laboratories and the methodology used for testing.

Normal TPR reference intervals for an adult horse. T, 99.0–101.5°F (37.2–38.6°C); HR, 25–45 beats per minute; RR, 10–20 breaths per minute.

Normal TPR reference intervals for a foal 24 hours of age. T, 99.0–102.0°F (37.2–38.9°C); HR, 80–100 beats per minute; RR, 20–30 breaths per minute.

Normal TPR reference intervals for a growing foal. T, 99.0–102.0°F (37.2–38.9°C); HR, 60–80 beats per minute; RR, 20–30 breaths per minute.

Index

Note: References are to case numbers, not page numbers.

Printed and bound by CPI Group (UK) Ltd, Croydon, CR0 4YY

23/10/2024

01777696-0006